The Earth's Magnetic Field

This is Volume 32 in
INTERNATIONAL GEOPHYSICS SERIES
A series of monographs and textbooks
Edited by WILLIAM L. DONN

A complete list of the books in this series appears at the end of this volume.

The Earth's Magnetic Field

Its History, Origin and Planetary Perspective

RONALD T. MERRILL

Geophysics Program
University of Washington
Seattle, USA

MICHAEL W. McELHINNY

Formerly, Research School of Earth Sciences
Australian National University
Canberra, Australia
and now with
Bureau of Mineral Resources
Canberra, Australia

1983

ACADEMIC PRESS

A Subsidiary of Harcourt Brace Jovanovich, Publishers

London New York
Paris San Diego San Francisco São Paulo Sydney Tokyo Toronto

ACADEMIC PRESS INC. (LONDON) LTD
24/28 Oval Road, London NW1 7DX

United States Edition published by
ACADEMIC PRESS INC.
111 Fifth Avenue, New York, New York, 10003

British Library Cataloguing in Publication Data

Merrill, R. T.
The earth's magnetic field.—(International geophysics series)
1. Magnetism, Terrestrial
I. Title II. McElhinny, M. W.
III. Series
538'.72 QC816

ISBN 0-12-491240-0
ISBN 0-12-491242-7 Pbk

LCCCN 83-70948

Filmset by Eta Services (Typesetters) Ltd, Beccles, Suffolk
and printed in Great Britain by
Galliard (Printers) Ltd, Great Yarmouth

Preface

Some of the earliest forms of scientific investigation, as we know them today, were studies of the earth's magnetic field. The *Epistola de Magnete* of 1259 by Petrus Perigrinus is now regarded as the first scientific treatise ever written—and was about geomagnetism. Yet it was not until historical records of the geomagnetic field could be extended back in time, through the use of archaeomagnetism and palaeomagnetism, that great progress was made in our understanding of the history and origin of the earth's magnetic field. One of the most significant discoveries of palaeomagnetism has been that the geomagnetic field has changed its polarity many times in the past. This discovery played a critical role in the development of the theory of plate tectonics and is a constraint on all models for the origin of the main field of the earth. However, the existence of two polarity states is not a powerful constraint on most magnetic field models. Indeed, the existence of two states of opposite polarity was compatible with theories for the origin of the earth's magnetic field that existed 30 years ago. Less startling observational results acquired subsequently appear likely to place far more powerful constraints on theories for the origin of the main field.

Over the past quarter of a century, scientific investigations of the geomagnetic field have been made on three broad fronts. There are the traditional aspects of geomagnetism relating to the present field and its secular variation. Also, palaeomagnetic and archaeomagnetic studies have been undertaken in many parts of the world. Often the geomagnetic aspects that arose from these studies were only secondary, the main thrust having been related to continental drift and plate tectonics. The third group of investigators have been theoreticians carrying out research into the origin of the main field and its secular variation. Unfortunately these three categories of investigators have tended to remain isolated from one another. For example, palaeomagnetists are generally experimentalists with relatively poor theoretical background. Their understanding of dynamo theory has tended to be rather cursory. At the same time theoreticians have often found it difficult to comprehend the results of palaeomagnetic studies and to appreciate just how they impinge on dynamo theories.

This book is an attempt to bring all three types of investigations in geomagnetism together and to relate them to one another. Theoreticians need to have palaeomagnetic results presented to them in a manner which is applicable to dynamo theory. In that way they may become more aware of some of the constraints which are imposed upon their theories by observations. Similarly, palaeomagnetists need to have dynamo theory explained to them in a way they can understand and so be able to present their data in a more useful form. We hope this book will serve that purpose. Thus after a historical introduction in Chapter 1, the more traditional aspects of geomagnetism relating to the present field and historical observations are presented in Chapter 2. The various methods and techniques and theoretical background of palaeomagnetism are given in Chapter 3. Chapters 4, 5 and 6 present the results of palaeomagnetic and archaeomagnetic studies in three topics. Chapter 4 relates to studies of the geomagnetic field roughly back to about 50,000 years ago. Chapter 5 is about reversals of the geomagnetic field and Chapter 6 presents studies of the field for times older than 50,000 years and on the geological time scale of millions or hundreds of millions of years. Chapters 7, 8 and 9 provide insight into dynamo theory. Chapter 7 is essentially a non-mathematical attempt to explain the physical basis of dynamo theories to palaeomagnetists. This is followed in Chapter 8 by a more advanced theoretical treatment for those with greater mathematical skills. Chapter 9 explains theoretical aspects of secular variation and the origin of reversals of the geomagnetic field. Chapter 10 is our attempt to bring the two sides together, to relate theory to experiment and vice versa.

The earth is but one of several planetary objects in the solar system that possess or possessed magnetic fields of internal origin. Thus the final two chapters consider the magnetic fields of the moon, sun, planets and meteorites, in an attempt to determine the necessary and sufficient conditions for magnetic field generation in large solar system bodies.

No book is ever written without help from colleagues. Here we should especially like to thank Phil McFadden for stimulating discussions and continued interest in much of the statistical theory that has been developed with him over the past 18 months. David Barraclough kindly provided the 1980 magnetic charts presented in Chapter 2. We are grateful for his assistance in providing these up-to-date charts; so many texts reproduce charts which are years out of date. Phil McFadden, Ross Taylor, Ted Lilley and Cathy Constable gave us critical reviews of selected chapters. Our thanks go to Gail Stewart who patiently typed the manuscript through many iterations and to Joan Cowley who drafted all the figures. Finally, we thank members of our families for understanding why extra long work days were sometimes required to finish this project.

It has taken us five years to complete this book because we found we were developing the subject as we went along. We found this to be immensely stimulating and we hope our readers will share our enthusiasm for this new integrated approach to one of the world's oldest sciences.

Seattle and Canberra
September 1983

R. T. Merrill
M. W. McElhinny

Contents

Chapter Ten **Palaeomagnetism and Dynamo Theory**

Chapter Eleven **Lunar Magnetism**

Chapter Twelve **Magnetic Fields of the Sun, Planets and Meteorites**

Chapter One

History of Geomagnetism and Palaeomagnetism

1.1 The Magnetic Compass

The attractive properties of lodestone (magnetite) were known to both the Chinese and Greeks in ancient times. The earliest observations on magnets are supposed to have been made by the Greek philosopher Thales in the sixth century B.C. The Chinese literature between the third century B.C. and sixth century A.D. is full of references to the attractive power of the magnet, but there is nothing as far back in time as Thales (Needham, 1962). The properties were explained by Thales in animistic terms. In China the most usual name was *tzhu shih*, the "loving stone" which can be compared with the French word for magnet, *aimant*.

The discovery of the magnetic compass was an event of immense importance in science. It represents the very first dial and pointer instrument to be invented and is the forerunner of all those self-registering instruments which play so important a part in modern scientific investigation. The sundial was much older but in this case it was only a shadow that moved and not the instrument itself. The windvane too was older but in all ancient forms there was no provision for precise readings on a circular graduated scale.

The earliest known form of magnetic compass was invented by the Chinese by at least the first century A.D. and probably as early as the second century B.C. (Needham, 1962). The compass comprised a lodestone spoon rotating on

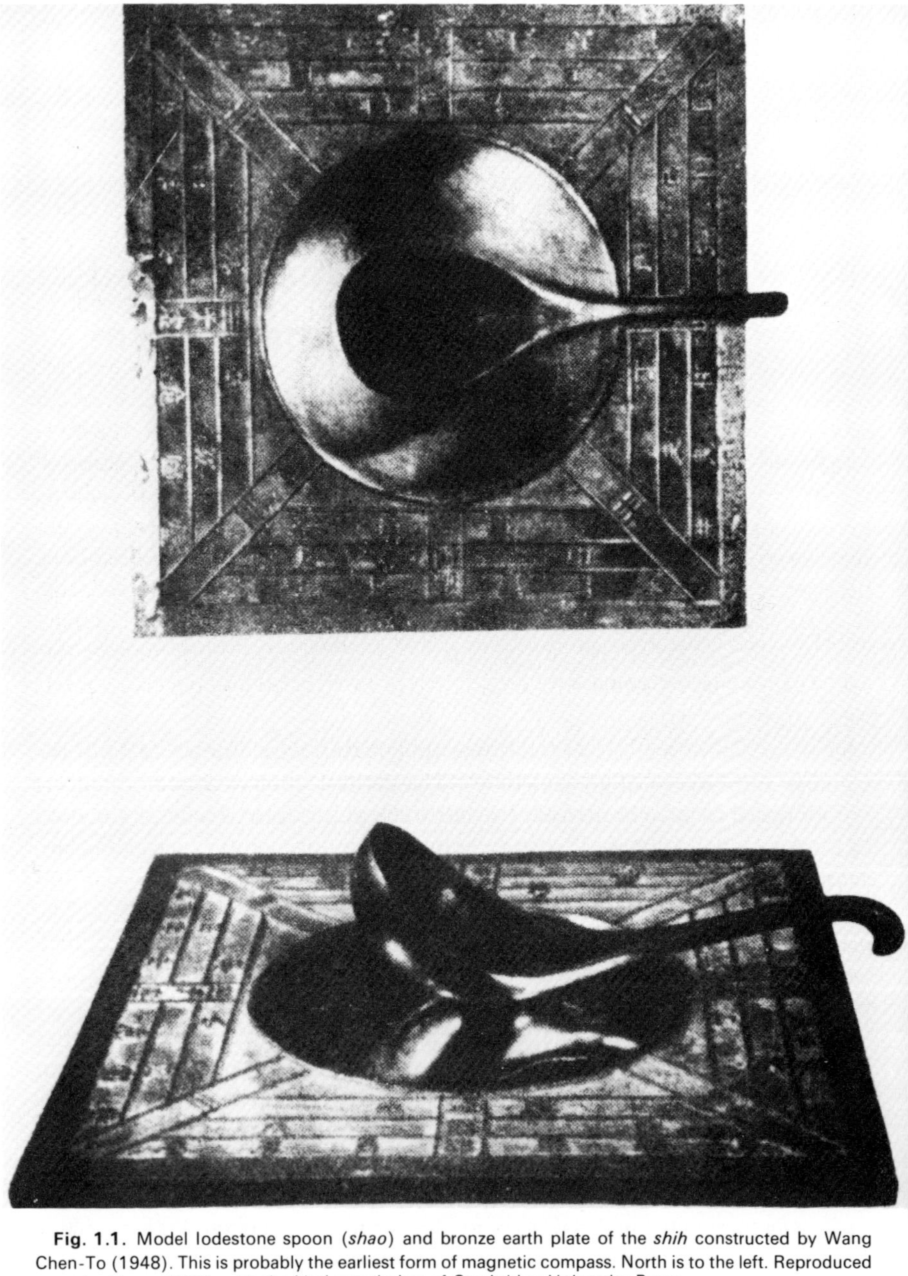

Fig. 1.1. Model lodestone spoon (*shao*) and bronze earth plate of the *shih* constructed by Wang Chen-To (1948). This is probably the earliest form of magnetic compass. North is to the left. Reproduced from Needham (1962) with the kind permission of Cambridge University Press.

a smooth board, a model of which has been constructed by Wang Chen-To (1948) (Fig. 1.1). Since the natural magnetization of such a spoon would establish itself in the direction of the main axis, this simple compass would automatically point north–south irrespective of the magnetization of the piece of rock from which it was cut. The spoon was probably given a thermoremanent magnetization by heating and cooling through the Curie Point whilst set in the north–south direction. The greater magnetization produced this way had the effect of reducing the drag between spoon and board. The disadvantage of this frictional drag was overcome during succeeding centuries by the invention of floating, dry-pivoted and silk or other fibre suspended versions.

The compass arrived in Europe during the twelfth century A.D., where the first reference to it is made by an English monk of St. Albans, Alexander Neckham, in 1190. By what means it reached Europe is unknown, the most commonly accepted view being that it came by the sea route in the hands of the Arabs (Benjamin, 1895). A land route is still however a possibility because the earliest Arab references to the compass are all somewhat later than the European ones and no Indian reference of any significance has been discovered (Needham, 1962). The Chinese regarded the south-pointing property of the magnet as fundamental whereas in Europe mariner's compass needles were regarded as north-pointers. This difference is of considerable significance because of the influence it had in Europe on theories of magnetic directivity.

The early thirteenth-century philosophy of magnetic directivity proposed that the compass needle pointed towards the pole star. Since, unlike other stars, the pole star was fixed it was concluded that the lodestone with which the needle was rubbed obtained its "virtue" from this star. In the same century the idea of polar lodestone mountains was propounded. The pole star gave its virtue to the lodestone mountains which in turn imparted its virtue to the needle which then directed itself towards the pole star. For the first time theories of magnetic directivity had descended from the heavens to the earth (Smith, 1968).

The idea of the universality of the north–south directivity of the compass needle was first questioned by Roger Bacon in 1266. A few years later Petrus Peregrinus (Peter the Wayfarer), an Italian scholar from Picardy, questioned the idea of polar lodestone mountains. Lodestone deposits occurred in many parts of the world so why should the polar ones have preference? However, Peregrinus went further than logical argument and performed a remarkable series of experiments with spherical pieces of lodestone. These are reported in his *Epistola de Magnete* written in 1269. Although written in 1269 and widely circulated in Europe during the succeeding centuries, the *Epistola* was not published in printed form under Peregrinus' name until 1558 (Smith, 1970a).

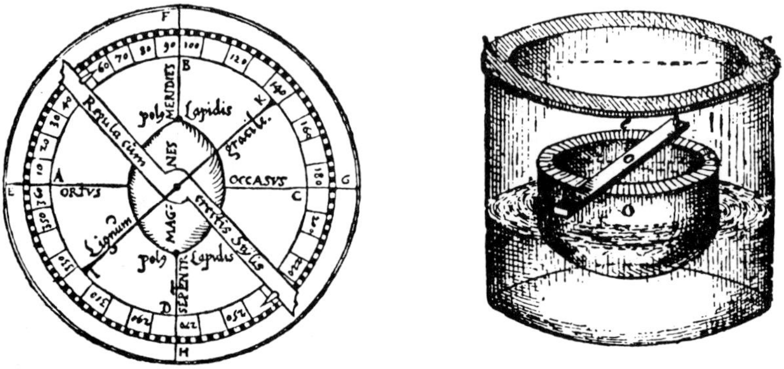

Fig. 1.2(a). The floating compass described in Peregrinus' *Epistola* and the reconstruction by Bertelli in 1868. An oval magnet (*magnes*) is placed in a bowl which floats in water in a larger transparent vessel. The top of the inner vessel is graduated so that north corresponds to east (*ortus*), 90° to south (*meridies*), 180° to west (*occasus*) and 270° to north (*septentrio*). These directions are with respect to the axis of the magnet each pole (*polus lapidis*) of which Peregrinus had previously determined. A ruler with erect pins (*regula cum erectis stylis*) rests on the edges of the graduated bowl, and is used to determine the azimuth of the sun or moon. The rule is turned until the shadows of the pins fall longitudinally down it. The magnetic azimuth of the sun (or moon) is then the angle between the ruler and the magnetic axis. The thin wooden strip (*lignum gracile*) is used to determine the fiducial line across the larger vessel. *Note.* The style and precise details shown by Peregrinus' illustrations vary slightly from manuscript to manuscript. These are from the first printed edition of 1558 and are taken from Smith (1970a). (From Smith–Atlas (News suppl. to *Earth Science Reviews*) **6**, 1, A15, 1970.)

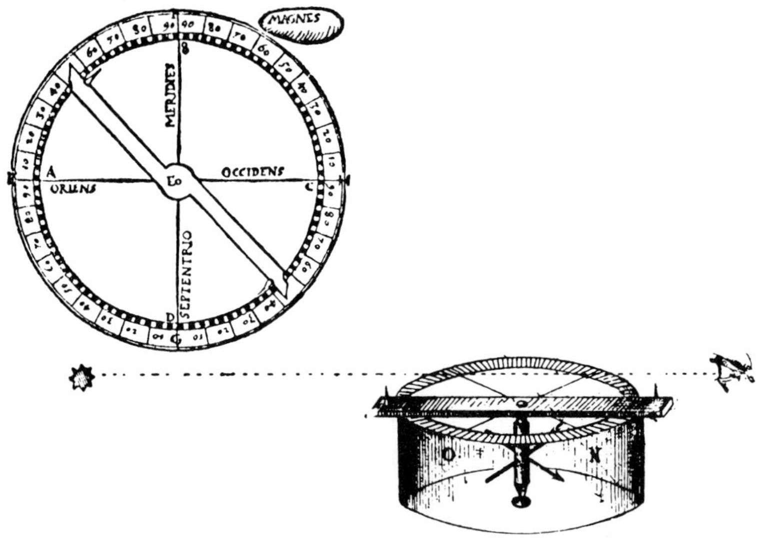

Fig. 1.2(b). Peregrinus' pivot compass and the reconstruction by Bertelli in 1868. The iron magnet is placed through the pivot which rotates in a transparent vessel (cf. the fixed pivot of the modern compass). The ruler with erect pins serves the same purpose as in the floating compass. The vessel is graduated as before. *Oriens* = east; *occidens* = west. The lodestone shown here was used to magnetize the iron needle. From Smith (1970a).

He defined the concept of polarity for the first time in Europe. He discovered magnetic meridians and showed several ways of determining the positions of the poles of a lodestone sphere, each of which demonstrated an important magnetic property. He thus discovered the dipolar nature of the magnet, that the magnetic force is strongest and vertical at the poles, and became the first to formulate the law that like poles repel and unlike poles attract. The *Epistola* bears a remarkable resemblance to a modern scientific paper. Peregrinus used his experimental data from which to draw conclusions, unlike his contemporaries who sought to reconcile facts with pre-existing speculation. His work is probably the first scientific treatise ever written. The second part of the *Epistola* contains descriptions of two types of magnetic compass, one floating and one pivoted. Peregrinus' version was the first pivoted compass to have been described in Europe (Fig. 1.2).

1.2 Declination, Inclination and Secular Variation

The first observation of magnetic declination appears to have been made in China by the Buddhist astronomer I-Hsing about A.D. 720 (Needham, 1962). However, knowledge of declination did not travel to Europe with the compass. Although the Chinese records reveal at least nine measurements of declination from A.D. 720–1280 (Fig. 1.3), the first measurement in Europe

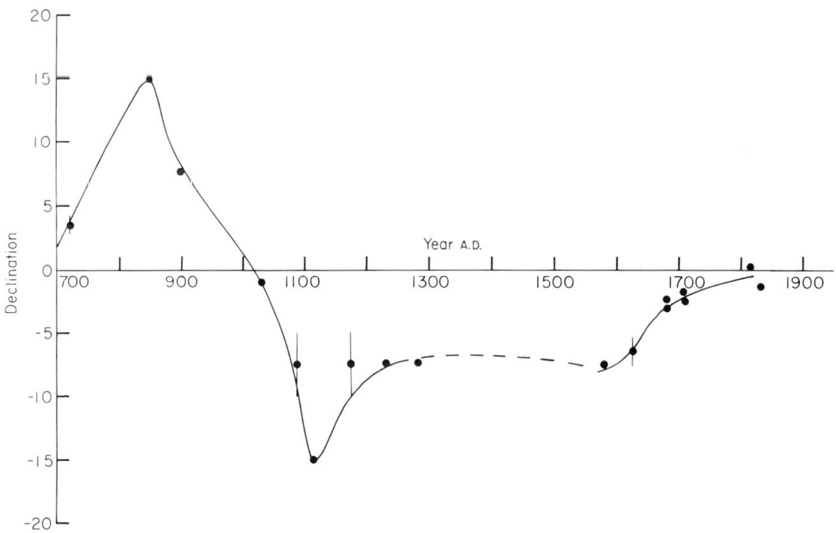

Fig. 1.3. Variation of magnetic declination with time in China between A.D. 720 and 1829. Error bars represent the range within which the declinations lie as quoted by Chinese texts. After Smith and Needham (1967).

appears to be that of Georg Hartmann, a Vicar of Nurenberg, made in Rome in about 1510. It has often been stated that Christopher Columbus first discovered declination in the European region during his first voyage to the West Indies in 1492. This claim now appears to be untrue (Chapman and Bartels, 1940), because there is strong evidence that declination was at least known in Europe by the early 1400s even if precise measurements were not recorded for another century.

Magnetic inclination was discovered by Georg Hartmann in 1544, as mentioned in a letter of that year, but which was only found in the Konigsberg archives in 1831. As a result his discovery had no influence on studies of the earth's magnetism. It was rediscovered by Robert Norman, an English hydrographer, in 1576. There appears to be no record of any discovery of inclination by the Chinese.

Henry Gellibrand, Professor of Astronomy at Gresham College, first discovered that magnetic declination changed with time in 1634. The observations in London on which Gellibrand based his conclusion were as follows:

Date	*Observer*	*Declination*
16 Oct. 1580	William Borough	11.3° E
13 June 1622	Edmund Gunter	6.0° E
16 June 1634	Henry Gellibrand	4.1° E

William Borough was Comptroller of the Navy and Edmund Gunter was Gellibrand's predecessor as Professor of Astronomy. Although Gunter noted the difference between his and Borough's measurements he failed to appreciate the significance, attributing it to possible inaccuracy of the earliest measurement.

Meanwhile Gerhard Mercator, mathematician and geographer, in 1546 first realized from observations of magnetic declination that the point to which the needle seeks could not lie in the heavens, leading him to fix the magnetic pole firmly on the earth. Norman and Borough subsequently consolidated the view that magnetic directivity was associated with the earth and even shifted the focus from the pole to a region closer to the centre of the earth. The culmination of this period of magnetic investigations was the publication in 1600 by William Gilbert, physician to Queen Elizabeth I, of his results of experimental studies in magnetism in a treatise entitled *De Magnete*.

In an impressive series of experiments he investigated the variation in inclination over the surface of a piece of lodestone cut into the shape of a sphere. He thus made the first practical dip-circle. Although Gilbert must have leaned heavily on the work of Petrus Peregrinus (Smith, 1970a), he had a much wider basis of general observation to draw on than Peregrinus. The most important gap in Peregrinus' knowledge was ignorance of magnetic inclination, and it was this extra information that enabled Gilbert to realize "*magnus magnes ipse est globus terrestris*" (the earth globe itself is a great magnet). Had Peregrinus known that his magnets behaved like compass needles he might well have been able to generalize from his spherical piece of lodestone to the earth as Gilbert did. Indeed Mitchell (1939) has suggested that without Norman's discovery of inclination, the *De Magnete* would never have been written. Gilbert's work was essentially the culmination of many centuries of thought and experimentation in geomagnetism. His conclusions put a stop to the wild speculations that were then current concerning magnetism and the magnetic needle. Apart from the roundness of the earth, magnetism was the first property to be attributed to the body of the earth as a whole. Newton's gravitation came 87 years later with the publication of his *Principia*.

1.3 Magnetic Charts and the Search for the Poles

The magnetic compass was clearly useful as an aid to navigation with the result that methods for determining declination at sea were developed by Portuguese navigators. The most remarkable of these navigators was João de Castro who commanded one of the 11 ships that sailed to the East Indies in 1538. Using an instrument with an arrangement like a sun-dial with magnetic needle, he determined the magnetic azimuth of the sun at equal altitudes before and after noon. The half difference of these azimuths measured clockwise and anticlockwise respectively was the magnetic declination. João de Castro carried on his observations also during his voyage along the west coast of India and in the Red Sea, obtaining 43 determinations of declination between the years 1538 and 1541. These observations represent the first and most significant attempts to chart the variation in declination world-wide. The method was soon universally introduced on ships.

The first geomagnetic chart was drawn up by Edmund Halley following two voyages between 1698 and 1700 in the North and South Atlantic Oceans. These were the first sea voyages made for purely scientific purposes and led to the first declination chart of the earth published in 1702. The first chart of inclination for the whole earth was published by Johann Carl Wilcke in Stockholm in 1768.

Alexander von Humboldt, the German explorer, first discovered the dependence of the intensity of the geomagnetic field with latitude. During his voyages through the Americas from 1799 to 1803 he swung a dip-needle in the plane of the magnetic meridian and measured the number of oscillations over ten minutes. In Peru, on the magnetic equator, the number of swings was 211 and it decreased to the north and south indicating a regular increase of magnetic intensity from the equator to the poles. Later von Humboldt used the observation of the time of oscillation of a compass needle in the horizontal plane to determine relative measures of horizontal intensity. This simpler method was generally adopted on scientific journeys. The first isodynamic-charts for **H** and **F** were published by Christopher Hansteen, Professor of Applied Mathematics at the Norwegian University in 1825 and 1826.

The first representation of the geomagnetic field in mathematical form was made by the German mathematician Carl Friederich Gauss in his treatise *Allgemeine Theorie des Erdmagnetismus* of 1838. He calculated the coefficients in the spherical harmonic expression for the potential of the geomagnetic field from values of X, Y, Z derived from isomagnetic charts at 84 points, spaced at 30° of longitude along seven circles of latitude. For this purpose he used three isomagnetic charts (Barlow's for declination, 1833; Horner's for inclination, 1836; and Sabine's for total intensity, 1837).

Gauss was able to forecast the positions of the geomagnetic poles, the points where the best fitting dipole axis cuts the surface of the earth. However, the positions of the North and South Magnetic Poles (or dip-poles where the inclination is vertical) are not so readily calculated and depend on extrapolation from existing maps. Their location was thus of considerable interest to early explorers of the polar regions. James Clark Ross discovered the position of the North Magnetic Pole on 31 May 1831 at 70°05′ N, 96°46′ W. He later attempted to reach the southern counterpart unsuccessfully and at least 78 years were to pass before an attempt at location occurred. During Shackleton's 1907–1909 expedition to the South Pole a party of Australians, including two geologists, Professors T. W. Edgeworth David and Douglas Mawson, set out from Cape Royds and thought they had located the position of the South Magnetic Pole on 16 January 1909. However, it now appears that because of the imprecise measurements, of necessity made in a hurry, their location was incorrect. The most probable location for 1909 was subsequently calculated to have been at 71°36′ S, 152°0′ E about 130 km north–west of David and Mawson's extreme station (Webb and Chree, 1925).

1.4 Fossil Magnetism and the Magnetic Field in the Past

The first observation that certain rocks were magnetized parallel to the earth's magnetic field were made by Delesse in 1849 and Melloni in 1853.

During the late eighteenth century it had already been observed that some rocks possessed extremely strong remanent magnetization because of their effect on the compass needle. These effects were attributed to lightning strikes by von Humboldt in 1797. Folgheraiter (1899) extended the work of Delesse and Melloni, but also studied the magnetization of bricks and pottery. He argued that if the position of the brick or pot in the firing kiln was known, then the remanent magnetization it acquired on cooling should provide a record of the direction of the earth's magnetic field.

Following on the work of Folgheraiter, David (1904) and Brunhes (1906) investigated material baked by lava flows rather than those baked by man. They compared the direction of magnetization of the flows with that of the underlying baked clay. They reported the first discovery of natural remanent magnetization roughly opposed to that of the present field. The fact that the baked clays were also reversely magnetized led to the first speculation that the earth's magnetic field had reversed itself in the past. Mercanton (1926) pointed out that if the earth's magnetic field had reversed itself in the past then reversely magnetized rocks should be found in all parts of the world. He obtained samples from Spitsbergen, Greenland, Iceland, the Faroe Islands, Mull, Jan Mayen Land and Australia and found that some were magnetized in the same sense as the present field and others were roughly reversed from it. At the same time Matuyama (1929) observed similar effects in lavas covering the last one or two million years from Japan and Manchuria. He noticed however that the reversed lavas were always older than those in the same sense as the present field (normal). This was the first suggestion of a time sequence associated with reversely magnetized rocks. This time sequence was then noted in detail from alternating polarities observed in the Massif Central, France by Roche (1951, 1953, 1956, 1958), who placed the time of the last reversal as the middle of the Lower Pleistocene. Similar observations were made by Hospers (1953–54) in lava sequences in Iceland, and by Khramov (1958) in sedimentary sequences in western Turkmenia.

Studies of secular variation in the past using fossil magnetization were first attempted by Chevallier (1925) using lava flows from Mount Etna (Fig. 1.4). The use of archaeomagnetic techniques on baked hearths and pottery was detailed by Thellier and Thellier (1951, 1952) in France and by Cook and Belshe (1958) in England. The first attempt at determining recent secular variation from sediments was made by Johnson *et al.* (1948) on some New England varied sediments covering a time span of about 5000 years over 10,000 years ago. A similar attempt using Swedish varied clays was made by Griffiths (1953). Studies of the intensity variations in the past using archaeological material were pioneered by Thellier (1937a, b). A summary of results and details of the Thellier technique was later published by Thellier and Thellier (1959a, b).

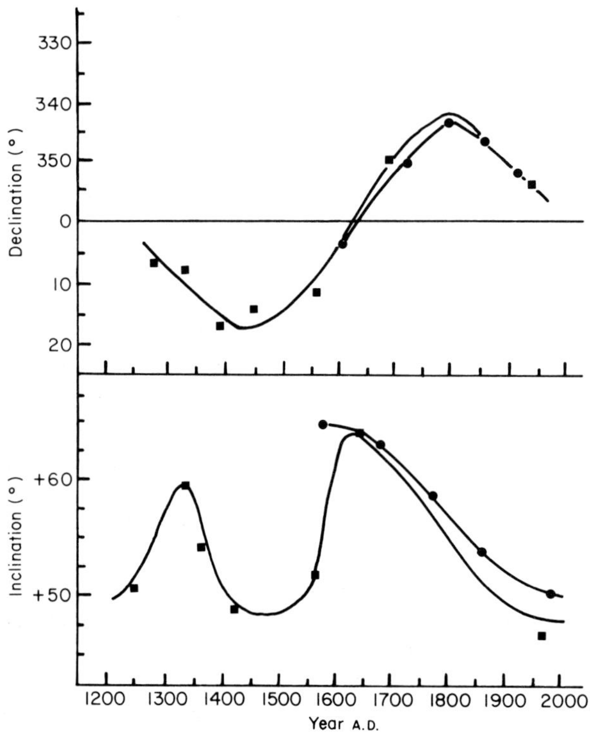

Fig. 1.4. Secular variation in Sicily determined by Chevallier (1925) from the magnetization of historic lavas of Mount Etna (shown as squares). The circles represent direct observations of the magnetic field.

The potential use of fossil magnetism in rocks was well established by 1930, the most astute suggestion having been made by Mercanton (1926). He proposed that, because of the approximate correlation of the geomagnetic and rotational axes at the present, it might be possible to test the hypotheses of polar wandering and continental drift. Coming at the time of Wegener's (1924) continental drift hypothesis, it might then have served as a simple way of settling the dispute. However, palaeomagnetism was a new technique, regarded by most as unpromising, and in any case Wegener's hypothesis was in general disrepute in both Europe and North America. Twenty years was to lapse until palaeomagnetism was again being studied to test the polar wandering and continental drift hypotheses. Graham (1949) in North America paved the way with a classic paper describing some of the field methods still in use today. Work carried out by groups in the United Kingdom led by Blackett and Runcorn during the early 1950s indicated polar differences of about 25° between Europe and North America. This suggested continental drift had occurred and that these continents had previously been very much closer (Runcorn, 1956; Irving, 1956). More dramatic differences were to emerge in data from southern Africa and Australia (Graham and Hales, 1957; Irving, 1957).

1.5 Transient Magnetic Variations—The External Magnetic Field

Careful observations of a compass needle with a microscope in 1722 led Graham, at London, to discover transient magnetic variations which are far more rapid than the slow secular variation. On some days he noted slow and regular changes of declination and on other days irregular changes sometimes larger and more rapid. He therefore made the important distinction between days that are magnetically quiet and those that are disturbed. In 1759 Canton, also at London, noted that the quiet daily variation was greater in summer than in winter. Later Gauss and his successors from 1834 onwards discovered the daily variation in the other geomagnetic elements. Many speculations were aroused but gradually the morphology of the quiet-day solar variation (S_q) became clear. S_q was found to vary in intensity by 50 % or more in correlation with rises and falls in the annual mean sunspot number. This led Stewart in 1882 to infer that S_q must have its origin in electric currents in the upper atmosphere. Schuster in 1889 applied spherical harmonic analysis to S_q and confirmed Stewart's inference that S_q arises from an external source, but found a smaller part of internal origin. He ascribed this correctly to electric currents induced in the earth by the external source.

The proposition that there is an electrically conductive upper atmospheric layer was the first suggestion of the existence of the ionosphere. Final proof of this had to await the first soundings with radio waves by Breit and Tuve (1925, 1926) and Appleton and Barnett (1925). In 1922 the Department of Terrestrial Magnetism of the Carnegie Institution of Washington set up a geomagnetic observatory at Huancayo in Peru, very close to the magnetic equator. This revealed a hitherto unsuspected enhancement of the horizontal component intensity of S_q by a factor of about two. Subsequently this enhancement was found to occur all along the magnetic equator within a narrow band of about 2° on either side. The narrow equatorial current system implied from these observations is called the *equatorial electrojet*.

In addition to the quiet solar daily variation S_q, there is the disturbance daily variation S_D. Both S_D, and the additional fields that arise during magnetic storms, have been intensely studied over the past century, especially the relation of magnetic storms to auroras. Many theories of magnetic storms and auroras have been proposed, and just as many have fallen by the wayside. In 1896 Birkeland showed that when a stream of electrons is projected towards a magnetized sphere they are deflected towards the poles by the magnetic field. Such a stream ejected from the sun would require a very high density of charged particles to produce auroras and magnetic storms. Such high densities of charged particles of one sign would however be dispersed by their mutual electrostatic repulsion long before the stream reached the earth. Lindemann (1919) therefore suggested that the solar corpuscular stream might be electrostatically neutral but ionized. Chapman and Ferraro (1931, 1932, 1933) developed a theory based on this assumption and deduced some of the consequences of the impact of such a stream upon the geomagnetic field. They found that a hollow or magnetic cavity would be formed in the stream and that the geomagnetic field would be confined within this region, labelled by Gold (1959) as the *magnetosphere*.

In the 1930s the existence of a plasma around the earth, extending to several tens of thousands of kilometres, was not known, nor was it known that there exists a steady plasma flow out of the sun. Biermann (1951) inferred the existence of such an outflow from comet observations, and Parker (1958) concluded that the source might be a hydrodynamic expansion of the ionized gas in the solar corona. This *solar wind* would of course be much enhanced during periods of solar activity leading to the disturbances observed during the initial phase of magnetic storms as originally proposed by Chapman and Ferraro. Finally, exploration of the upper atmosphere with rockets and satellites led to the discovery of the existence of geomagnetically trapped radiation by Van Allen *et al.* (1958). Within the year following this discovery the gross features of the trapped radiation in the so-called *Van Allen belts* had been determined.

1.6 Origin of the Earth's Magnetic Field

William Gilbert in his *De Magnete* first identified the earth's magnetic field as having its origin inside the earth. He believed that its origin lay in permanent magnetization (lodestone) at the centre of the earth. Later in the first half of the seventeenth century René Descartes developed a large following for his ideas on the origin of the earth's magnetic field (Mattis, 1965). Descartes believed that the earth's magnetism was associated with "threaded parts" that were channelled in one-way ducts through the earth whose main entrances and exits were the North and South Poles. There were two types of threaded parts, one entered the North Pole and exited the South Pole and the other was the inverse of this. In both cases the threaded parts had what in modern terms would be called "closed field lines". The threaded parts travelled through air to connect with the parts that travelled in the earth. If the threaded parts chanced upon some lodestone during the course of their voyage, they would abandon their trip and would pass through the lodestone. They would wind their way in and out of the lodestone giving rise to complex magnetic vortices. This is apparently the first explanation for the non-dipole field.

In more modern times Einstein, shortly after writing his special relativity paper in 1905, described the problem of the origin of the earth's magnetic field as being one of the five most important unsolved problems in physics. Blackett (1947) proposed that large astronomical bodies might have dipole moments that are directly proportional to their angular momentum. This was due to some new as yet unexplained law of physics. To test this hypothesis Blackett (1952) attempted to measure the weak magnetic field predicted from a pure gold sphere as it rotated with the earth. In spite of using a sensitive astatic magnetometer specially developed for this purpose, he obtained negative results and he suggested his hypothesis was therefore in error. The sensitive astatic magnetometer he produced was subsequently used to measure the weak magnetizations of sedimentary rocks and was an important instrument in the early days of palaeomagnetism (Chapter 3). Rotation of a body in which there was charge separation would produce a magnetic field external to the body and this could provide a possible physical explanation to the Blackett hypothesis. However, Inglis (1955) discounted the possibility that large magnetic fields in the earth might originate from charge separation due to temperature and/or pressure gradients. So although this process is probably unimportant in the earth, it may still play an important role in some other large bodies.

Sir Joseph Larmor (1919a, b) was the first to suggest that large astronomical bodies, such as the sun, might have magnetic fields that arise from a self-exciting dynamo process. However, enthusiasm for this suggestion was dampened by the theoretical finding of Cowling (1934) that a magnetic field symmetric about an axis cannot be maintained by a symmetric motion. This was the first of the anti-dynamo theorems and was a prelude to extensive theoretical work to produce a more general anti-dynamo theorem. This continued until some important theoretical breakthroughs by Childress (1970) and G. O. Roberts (1970) demonstrated that no such general anti-dynamo theorem exists. Further discussion of this point is given in Chapter 8.

The first important mathematical contributions to dynamo theory were made by Elsasser (1946a, b, 1947) and Bullard (1949a, b). They were the first to discuss dynamos in a more modern sense. They used magnetohydrodynamic dynamo theory in a liquid core to produce models for a self-sustaining dynamo, rather than the homopolar dynamo approach that relied on wires, rigid discs etc. However, the contributions of Alfven (1940) to magnetohydrodynamic dynamo theory are critical and played an important role in the theories for the earth's dynamo that followed. In particular, his frozen-in-field concept (Chapter 7) and the idea of magnetohydrodynamic waves are central to most dynamo theories today.

Chapter Two

The Present Geomagnetic Field: Analysis and Description from Historical Observations

2.1 Magnetic Elements and Charts

Direct measurements of the earth's magnetic field are made continuously at magnetic observatories, and are obtained from various oceanographic, land, aircraft and satellite surveys. A common way of describing the magnetic field is simply to plot different *magnetic elements*, such as total intensity or inclination. Various elements used in such plots are defined in Fig. 2.1. *Isomagnetic charts*, which are contour maps of equal values of a particular magnetic element, are most commonly used to display the data. These charts for declination are termed *isogonic charts*, for inclination *isoclinic charts*, and for equal intensity (or for intensity of a particular field component, say, the horizontal field) *isodynamic charts*. *Isoporic charts* refer to isomagnetic charts of secular variation. Examples of these charts, which give a pictorial representation of the earth's field and its secular variation, are shown in Fig. 2.2.

Following the work of Gilbert in the seventeenth century, geomagnetists have recognized that the magnetic field at the surface of the earth can be well approximated by a magnetic dipole placed at the earth's centre and tilted $11\frac{1}{2}°$ with respect to the axis of rotation. Such a dipole accounts for roughly

90% of the earth's magnetic field at the surface. The magnetic field inclination is downwards throughout most of the northern hemisphere and upwards throughout most of the southern hemisphere. A line that passes through the centre of the earth along the dipole axis intersects the earth's surface at two points, referred to as the *geomagnetic poles*. They differ from the *magnetic poles*, which are the two points on the earth's surface where the field is vertical. The geomagnetic poles and magnetic poles would coincide if the field at the earth's surface were perfectly described by a geocentric dipole. However, this is not the case because about 10% of the field at the earth's surface remains after the best fitting geocentric dipole field is removed. This remaining part of the field is referred to as the *non-dipole field*. Both the dipole and non-dipole parts of the earth's magnetic field change with time.

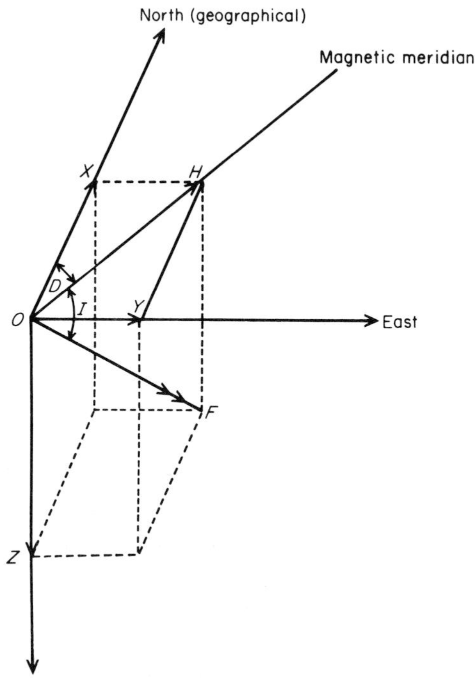

Fig. 2.1. The main elements of the geomagnetic field. The deviation, D, of a compass needle from true north is the *declination* (reckoned positive eastwards). The compass lies in the magnetic meridian containing the total field F which makes, with the horizontal, an angle I, termed the *inclination* (or dip). The inclination is reckoned positive downwards (as in the northern hemisphere) and negative upwards (as in the southern hemisphere). The horizontal and vertical components of F are denoted by $H = F \cos I$ and $Z = F \sin I$ respectively. Z is reckoned positive downwards as for I. The horizontal component can be resolved into two components X (northwards) $= H \cos D$, and Y (eastwards) $= H \sin D$. Then $\tan D = Y/X$ and $\tan I = Z/H$.

It is important to quantify the method of describing the earth's magnetic field for a variety of reasons that will become apparent later on in this book. Although other ways of representing the earth's magnetic field quantitatively are discussed in subsequent sections in this chapter, the spherical harmonic analysis description is most commonly used and emphasized in this book. Various combinations of global data, usually heavily weighted by data from continuously recording observatories, are used to obtain the coefficients of a truncated spherical harmonic analysis description of the earth's magnetic field. A considerable amount of information concerning the earth's magnetic field and its variation in time is contained in these coefficients. Often this is not well appreciated and just as often incorrect "physical" interpretations are drawn from the use of spherical harmonic representations of the field. Therefore, a considerable amount of material is devoted in this chapter to the understanding of spherical harmonic analysis, and its use in describing the earth's magnetic field. The present chapter also describes the actual picture of the field during the last 150 years or so.

Although the symbols for the magnetic elements given in Fig. 2.1 are widely referred to in describing the magnetic field, somewhat different symbols are used in other branches of geomagnetism. In particular $\boldsymbol{H}$ is generally used to describe the magnetic field, a notation which will be used throughout this text.

2.2 Spherical Harmonic Analysis Description of the Earth's Magnetic Field

2.2.1 Review of Spherical Harmonic Analysis

Any complete set of orthogonal functions can be used to describe the earth's magnetic field. *Surface harmonics* (to be defined mathematically shortly) are the obvious choice to use in the case when the quantities of interest are given on the surface of a sphere. Surface harmonics are orthogonal functions whose variables are θ, the colatitude (90° minus the latitude), and ϕ, the longitude (Fig. 2.3). Any reasonably behaved physical quantity (for example, one which has only a finite number of discontinuities) can be described by surface harmonics. When a radial component is included, then spherical harmonics are more appropriate to use; spherical harmonics essentially reduce to surface harmonics on any given sphere.

Although one could handle the problem of describing the earth's magnetic field through a mathematical discourse on orthogonal functions, considerably more physical insight is obtained by approaching the problem through Maxwell's equations in electricity and magnetism. As will be shown

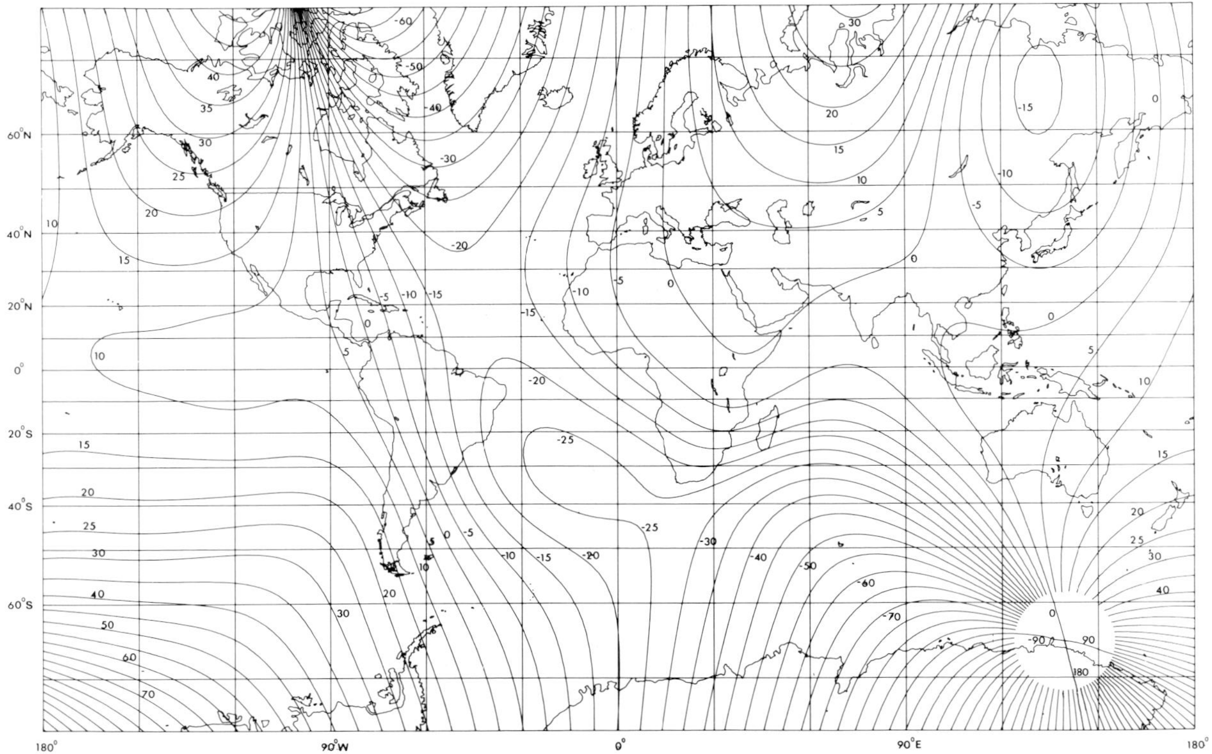

Fig. 2.2(a). Isogonic chart for 1980 showing the variation of declination in degrees over the earth's surface.

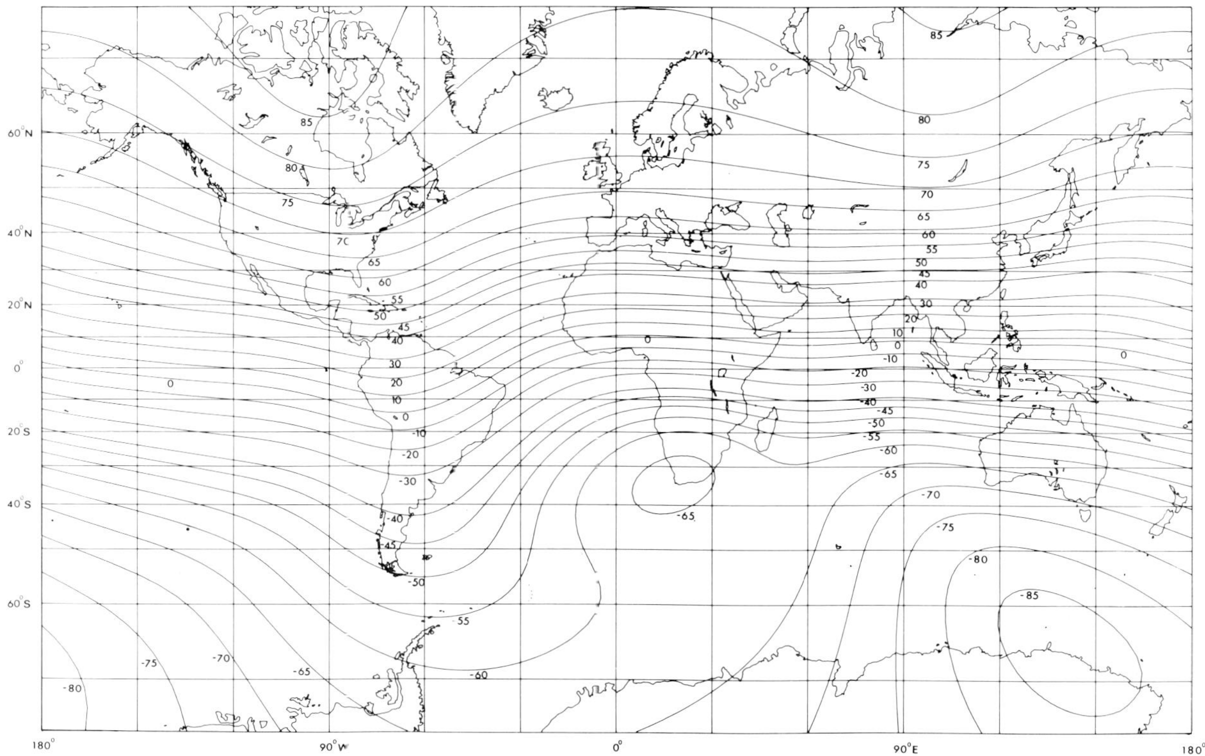

Fig. 2.2(b). Isoclinic chart for 1980 showing the variation of inclination in degrees over the earth's surface.

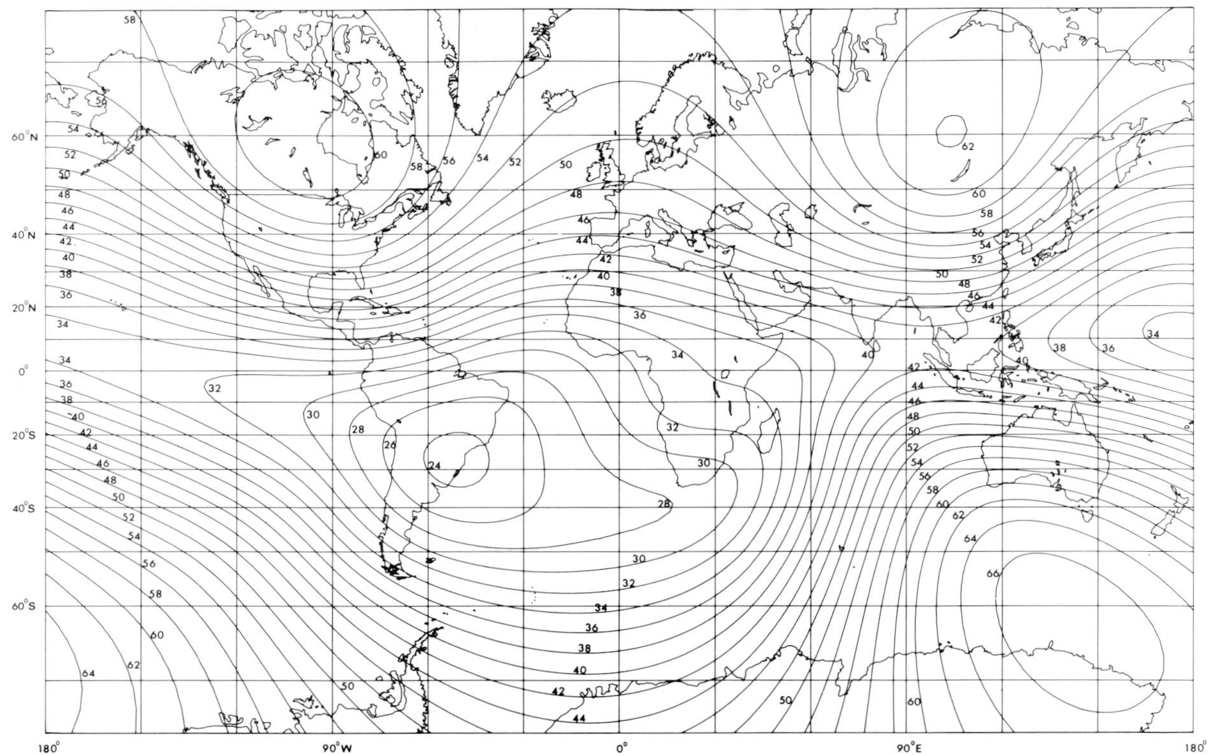

Fig. 2.2(c). Isodynamic chart for 1980 showing the variation of total intensity over the earth's surface. Contours are labelled in units of 1000 nT.

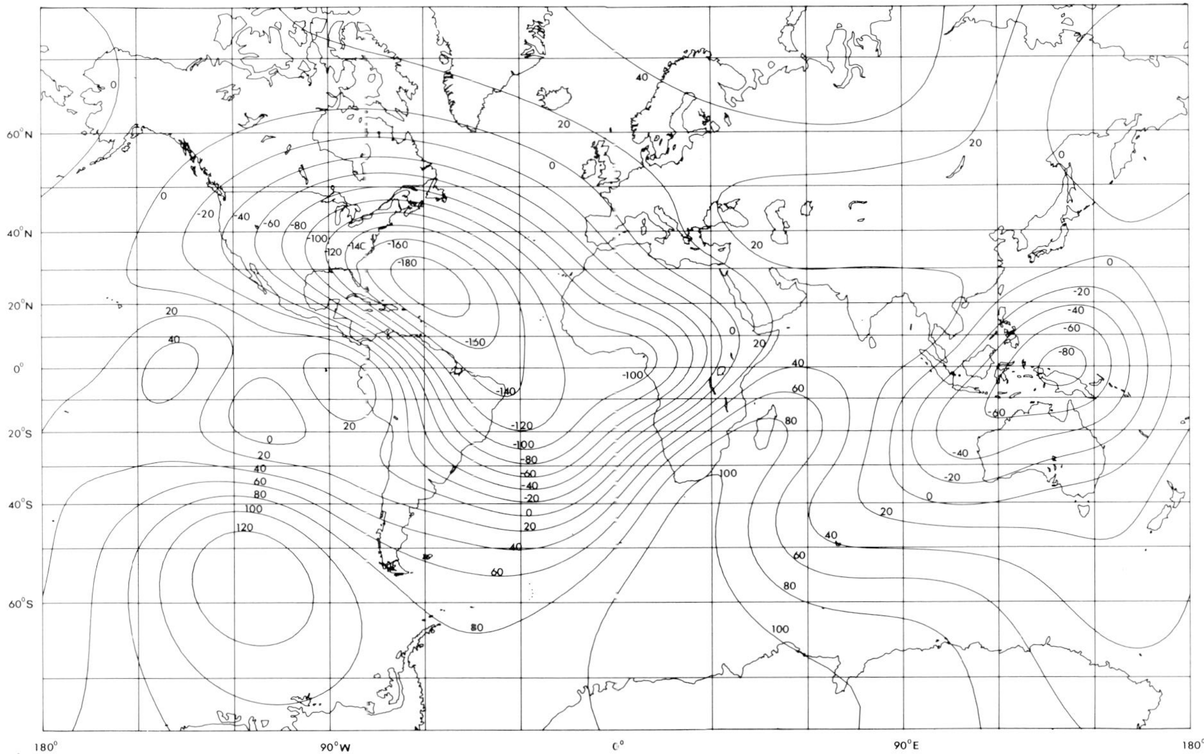

Fig. 2.2(d). Isoporic chart for 1980 showing contours of rate of secular change in intensity of the vertical component. Values are in nT yr^{-1}. (Maps for Fig. 2.2 provided by D. R. Barraclough. Courtesy of the Geomagnetism Unit, Institute of Geological Sciences, Edinburgh, Scotland.)

under certain simplifying assumptions that appear to be satisfied at the earth's surface, a scalar magnetic potential can be defined that satisfies Laplace's equation. One of the first advantages of using spherical harmonics comes when Laplace's equation is solved in spherical co-ordinates by the separation of variables. Spherical harmonics naturally fall out of the solution and, equally conveniently, the solution can be simply separated into two parts. The first part is associated with the external magnetic field sources, while the second part is associated with the internal sources. Thus, it is convenient to review briefly the solution of Laplace's equation in spherical co-ordinates.

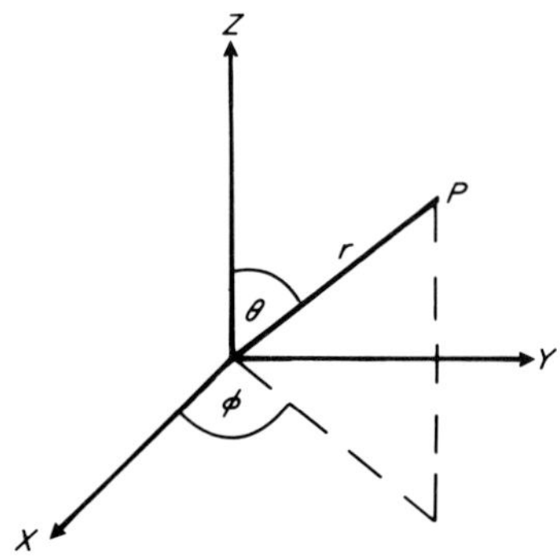

Fig. 2.3. Illustrating spherical polar co-ordinates.

The two Maxwell's equations relating to magnetic field are

$$\nabla \times \boldsymbol{H} = \boldsymbol{J} + \frac{\partial \boldsymbol{D}}{\partial t} \tag{2.1}$$

and

$$\nabla \cdot \boldsymbol{B} = 0 \tag{2.2}$$

where $\boldsymbol{H}$ is the magnetic field, $\boldsymbol{B}$ is the magnetic induction, $\boldsymbol{J}$ is the electric current density and $\partial \boldsymbol{D}/\partial t$ is the electric displacement current density. In the case of the earth it is assumed for the moment that the right side of (2.1) can be set to zero. Then $\boldsymbol{H}$ is a conservative vector field and there exists a scalar potential ψ such that:

$$\boldsymbol{H} = -\nabla\psi \tag{2.3}$$

Because $\boldsymbol{B} = \mu_0 \boldsymbol{H}$ above the earth's surface, where μ_0 is the permeability of free space ($4\pi \times 10^{-7}$ H m^{-1}), the divergence of $\boldsymbol{H}$ must vanish in (2.2). Combining this result with (2.3) yields Laplace's equation:

$$\nabla^2\psi = 0 \tag{2.4}$$

In spherical coordinates (r, θ, ϕ), shown in Fig. 2.3, Laplace's equation can be written in the form:

$$\frac{1}{r}\frac{\partial^2}{\partial r^2}(r\psi) + \frac{1}{r^2 \sin\theta}\frac{\partial}{\partial\theta}\left(\sin\theta\frac{\partial\psi}{\partial\theta}\right) + \frac{1}{r^2\sin^2\theta}\frac{\partial^2\psi}{\partial\theta^2} = 0 \tag{2.5}$$

This equation can be solved by separation of variables and by substituting

$$\psi = \frac{U(r)}{r}P(\theta)Q(\phi) \tag{2.6}$$

where $U(r)$, $P(\theta)$ and $Q(\phi)$ are functions of r, θ, ϕ, respectively, which are to be determined by the constraint that this potential satisfies (2.5).

The details of this substitution and subsequent analysis are given in many textbooks and will not be repeated here. For the special case in which there is no azimuthal asymmetry, the solution is:

$$\psi(r, \theta) = \sum_{l=0}^{\infty}\left(A_l r^l + \frac{B_l}{r^{l+1}}\right)P_l(\cos\theta) \tag{2.7}$$

where A_l and B_l are constants and P_l are the Legendre polynomials given by the recursion formula:

$$P_l(\cos\theta) = P_l(\chi) = \frac{1}{2^l l!}\frac{d^l}{d\chi^l}(\chi^2 - 1)^l \tag{2.8}$$

Figure 2.4 gives the first four Legendre polynomials and briefly describes some of their properties. The terms $A_l r^l$ in (2.7) refer to sources that are external to the point of reference, while the terms B_l/r^{l+1} refer to sources that are internal. As r gets smaller the internal source terms become larger, as

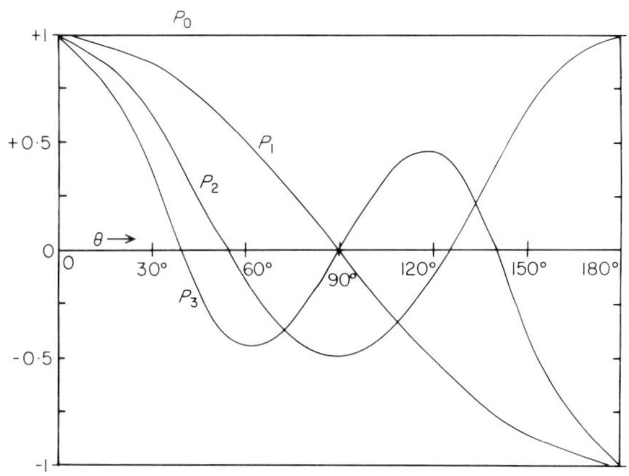

Fig. 2.4. The Legendre polynomials $P_l(\cos\theta)$ plotted as functions of θ up to $l = 3$.

expected, while the external source terms become smaller. These properties will be exploited later with respect to the origin of the earth's magnetic field.

A general solution to Laplace's equation is:

$$\psi(r, \theta, \phi) = \sum_{l=0}^{\infty} \sum_{m=-l}^{+l} [A_{lm} r^{l} + B_{lm} r^{-(l+1)}] Y_l^m(\theta, \phi) \tag{2.9}$$

where A_{lm} and B_{lm} are constants and $Y_l^m(\theta, \phi)$ are referred to as *surface harmonics*. Sometimes $Y_l^m(\theta, \phi)$ are referred to as spherical harmonics. However, the term spherical harmonics will be used here to include the r parts of (2.9) as well as the Y_l^m parts. Y_l^m is given by:

$$Y_l^m(\theta, \phi) = \frac{(2l + 1)(l - m)!}{4\pi(l + m)!} P_{lm}(\cos \theta) e^{im\phi} \tag{2.10}$$

where $P_{lm}(\cos \theta) = P_{lm}(\chi)$ are called the *associated Legendre polynomials* and are given by the recursion formula:

$$P_{lm}(\chi) = \frac{(-1)^m}{2^l l!} (1 - \chi^2)^{m/2} \frac{d^{l+m}}{d\chi^{l+m}} (\chi^2 - 1)^l$$

It can be shown that the Y_l^m defined by (2.10) are orthonormal, with the orthogonalization condition being:

$$\int_0^{2\pi} d\phi \int_0^{\pi} \sin \theta \, Y_{l'}^{m'*}(\theta, \phi) Y_l^m(\theta, \phi) \, d\theta = \delta_{ll'} \delta_{mm'} \tag{2.11}$$

where * refers to the complex conjugate, and $\delta_{ll'}$ and $\delta_{mm'}$ are Kronecker delta functions ($\delta_{ll'} = 0$ for $l \neq l'$ and $+1$ for $l = l'$).

In geomagnetism one uses *partially normalized Schmidt functions*, P_l^m, related to the associated Legendre polynominals, P_{lm} by:

$$P_l^m = P_{lm} \qquad \text{for} \quad m = 0$$

$$P_l^m \equiv \left| \frac{2(l - m)!}{(l + m)!} \right|^{1/2} P_{lm} \qquad \text{for} \quad m > 0$$

The surface harmonics $P_l^m(\cos \theta) \sin m\phi$ (or $\cos m\phi$) vanish along $(l - m)$ circles of latitude (corresponding to the zeros of P_l^m) and along $2m$ meridians at equal intervals of π/m (corresponding to zeros of $\sin m\phi$ or $\cos m\phi$). These zero lines divide up the surface of a sphere into regions; in each region the sign of the surface harmonic is constant, but changes on crossing a zero line into an adjacent region (Fig. 2.5). When $m = 0$, the surface harmonics are described by Legendre polynomials (see Fig. 2.4 also) and are referred to as *zonal harmonics*. When $l > m$ the surface harmonics are called *tesseral* or *non-zonal harmonics*. Finally, the term *sectorial harmonics* is sometimes used for the case when $l = m$ (Chapman and Bartels, 1940).

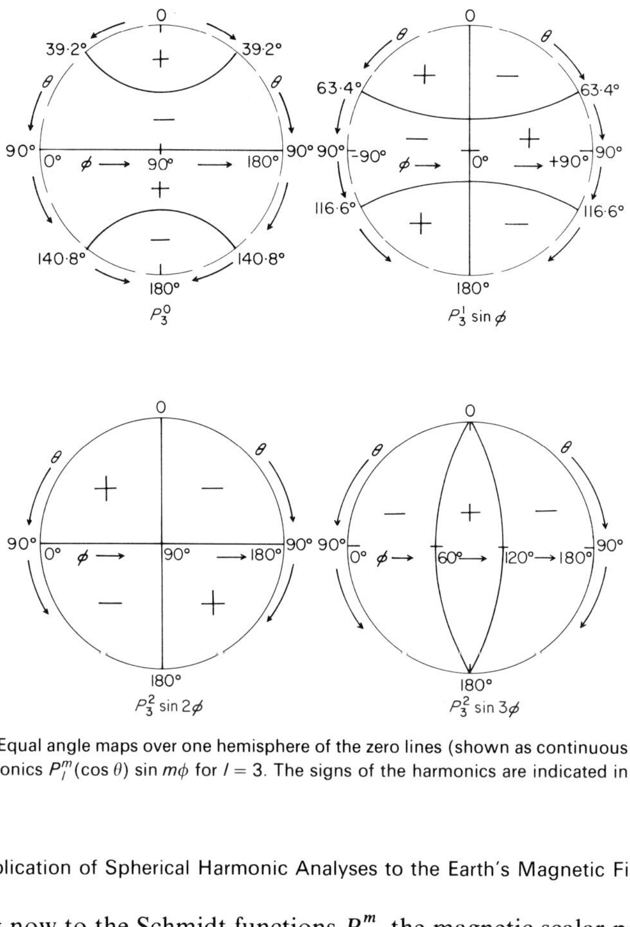

Fig. 2.5. Equal angle maps over one hemisphere of the zero lines (shown as continuous lines) for the surface harmonics $P_l^m(\cos\theta)\sin m\phi$ for $l = 3$. The signs of the harmonics are indicated in each region.

2.2.2 Application of Spherical Harmonic Analyses to the Earth's Magnetic Field

Reverting now to the Schmidt functions P_l^m, the magnetic scalar potential ψ is written (Chapman and Bartels, 1940, 1962):

$$\psi = \frac{a}{\mu_0}\sum_{l=0}^{\infty}\sum_{m=0}^{l} P_l^m(\cos\theta)\left\{\left[C_l^m\left(\frac{r}{a}\right)^l + (1 - C_l^m)\left(\frac{a}{r}\right)^{l+1}\right]g_l^m \cos m\phi\right.$$
$$\left. + \left[S_l^m\left(\frac{r}{a}\right)^l + (1 - S_l^m)\left(\frac{a}{r}\right)^{l+1}\right]h_l^m \sin m\phi\right\} \qquad (2.12)$$

where a is the earth's mean radius (6371 km; the radius of a sphere with the same volume as the earth) and C_l^m and S_l^m are positive numbers between 0 and 1 which indicate the fraction of the potential associated with sources of external origin ($r > a$). The coefficients $(1 - C_l^m)$ and $(1 - S_l^m)$ indicate the fraction of the potential associated with sources of internal origin ($r < a$). The coefficients g_l^m and h_l^m are to be evaluated and in SI units these are traditionally given in units of magnetic induction (commonly quoted in nanotesla, see Table 2.1). Hence in (2.12) the factor μ_0 is included to correct the dimensions on the right-hand side. In this way the coefficients in SI units have the same numerical values as in the cgs emu system of units where they were previously normally quoted in gamma. Schmidt (1939) also developed general formulae for a spheroidal body; however for the earth, with polar radius some 21 km smaller than average equatorial radius, the coefficients are only slightly modified by Schmidt's refinement (see Vestine, 1967).

Of course one does not measure the potential at the earth's surface, but the magnetic field, obtained from the potential by taking the gradient of ψ (2.3). If no sources exist outside the earth, then the radial component of the field, $-(\partial\psi/\partial r)$, must vanish at infinity and no positive powers of r can occur in ψ. Thus the internal source terms involve the terms multiplied by $(a/r)^{l+1}$ in (2.12). At the other extreme, if all the sources were outside the earth, then $-(\partial\psi/\partial r)$ must be finite throughout the interior and no negative powers of r can appear in ψ. The external source terms therefore involve those terms in (2.12) that are multiplied by $(r/a)^l$.

The components of the magnetic field

$$X = -\frac{1}{r}\frac{\partial\psi}{\partial\theta}, \quad Y = -\frac{1}{r\sin\theta}\frac{\partial\psi}{\partial\phi}, \quad Z = -\frac{\partial\psi}{\partial r}$$

can be obtained in terms of the Schmidt polynomials by substituting (2.12) in the above (recall the definitions of X, Y, and Z given in Fig. 2.1). By using measurements of the three components of the earth's magnetic field and by truncating the series at some value l, the relative importance of the internal (S_l^m and C_l^m) versus external ($1 - S_l^m$, $1 - C_l^m$) sources can be determined. The coefficients determined this way are not invariant in time, as changes in ionospheric currents, associated changes in induced electric currents in the crust and upper mantle, and currents in the earth's core lead to changes in the geomagnetic field at the earth's surface. However, if mean values of the field as determined over several years are used, the above method of analysis shows that C_l^m and S_l^m do not significantly differ from zero (see Vestine, 1967). In this case (2.12) reduces to

$$\psi = \frac{a}{\mu_0}\sum_{l=0}^{\infty}\sum_{m=0}^{l}\left(\frac{a}{r}\right)^{l+1} P_l^m(\cos\theta)(g_l^m\cos m\phi + h_l^m\sin m\phi) \tag{2.13}$$

The coefficients g_l^m and h_l^m are called the Gauss coefficients (Chapman and Bartels, 1940, 1962) appropriate to the Schmidt polynominals, P_l^m. Terms with $l = 0$ do not appear in the absence of magnetic monopole sources. Discussion of the external part of the earth's magnetic field will be taken up again in §2.5.

TABLE 2.1
1980 epoch World Chart model coefficients

l	m	Main field (nT) g	h	Secular change nT yr^{-1} $\dot{g}$	$\dot{h}$
1	0	−30,001	0	21.1	0.0
1	1	−1950	5634	12.1	−12.7
2	0	−2038	0	−21.6	0.0
2	1	3035	−2134	4.6	−12.0
2	2	1652	−179	4.5	−24.1
3	0	1293	0	−1.8	0.0
3	1	−2156	−38	−6.8	−0.1
3	2	1244	261	−2.4	2.0
3	3	851	−235	2.7	−2.9
4	0	919	0	−0.7	0.0
4	1	777	189	−2.5	1.5
4	2	411	−265	−5.1	0.1
4	3	−428	69	−3.5	3.9
4	4	224	−289	−2.2	−1.3
5	0	−216	0	−1.9	0.0
5	1	354	74	−1.3	3.7
5	2	261	147	−1.1	0.6
5	3	−66	−149	−4.0	−1.0
5	4	−173	−71	−1.2	1.9
5	5	−52	101	−0.1	1.3
6	0	51	0	1.3	0.0
6	1	57	−15	0.3	1.3
6	2	47	98	3.7	0.0
6	3	−194	75	2.6	−0.2
6	4	6	−44	1.1	−0.4
6	5	17	2	1.1	0.7
6	6	−104	27	0.4	2.1
7	0	65	0	0.0	0.0
7	1	−55	−71	−0.8	−2.5
7	2	7	−24	0.9	0.0
7	3	17	9	−0.4	1.0
7	4	−17	8	0.6	−0.5
7	5	−1	12	0.6	−0.9
7	6	16	−17	0.2	0.2
7	7	9	−14	0.9	−0.2
8	0	13	0	0.2	0.0
8	1	8	12	−0.1	−0.3
8	2	−4	−21	0.3	−0.6
8	3	−5	11	0.0	0.2
8	4	−12	−20	−0.2	−0.1
8	5	0	10	−0.7	0.6
8	6	−1	7	−0.2	−1.3
8	7	10	−13	−0.1	−0.2
8	8	3	−13	−0.5	0.2

2.2.3 Determination of the Gauss Coefficients

In principle, the number of measurements on the earth's surface required to obtain the Gauss coefficients up to $l = N$ is straightforward. Note that g_0^0 is zero because there is no monopole contribution to the potential. Then for $l = 1$ there are two g_l^m terms, g_1^0 and g_1^1; for $l = 2$ there are 3 terms g_2^0, g_2^1 and g_2^2 and for $l = N$ there are $N + 1$ terms. Therefore, the total number of terms g_l^m to degree N is:

$$\left[\frac{2 + (N + 1)}{2}\right]N = \frac{N^2 + 3N}{2} \tag{2.14}$$

Following a similar procedure for the h_l^m terms (note that all h_l^0 terms vanish), one finds that the total number of terms h_l^m to degree N is:

$$\left[\frac{1 + N}{2}\right]N \tag{2.15}$$

By summing (2.14) and (2.15) the total number of measurements required to determine the potential ψ given by (2.13) up to degree N is:

$$(N + 1)^2 - 1$$

Gauss (1838) calculated coefficients up to $l = 4$, while subsequently the series has often been computed to $l = 6$ to 10 (Vestine, 1967). Since 1965 there has been a tendency to calculate more terms, even up to $l = 23$, although the interpretation of terms greater than $l = 8$ is complex (Kolesova and Kropachev, 1973; Alldredge and Stearns, 1974; Cain, 1975; Harrison and Carle, 1982; Schure and Parker, 1982). Note that for $l = 6$ a minimum of 48 independent measurements of the earth's field are required. These measurements come from a variety of sources, including continuously recording magnetic observatories, repeat stations, and more recently from ship, aircraft and satellite surveys (e.g. Vestine *et al.*, 1947a, b; Alldredge and Stearns, 1974; IGRF Evaluation Group of Japan, 1975).

In practice, many more than 48 independent measurements are needed to determine ψ up to degree 6 reliably, and the measurements should ideally be uniformly distributed over the earth's surface. Usually a method of weighted least squares is used to determine these coefficients from the measurements. The reliability with which these coefficients have actually been determined is difficult to assess, as has been pointed out explicitly in the analysis of the statistical problem by Wells (1973). The Gauss coefficients for the 1980 epoch up to degree 8 are given in Table 2.1.

2.2.4 Interpretation of Terms in the Spherical Harmonic Analyses

Individual terms in (2.13) can be interpreted in terms of "sources" at the earth's centre. The g_1^0 term is given by:

$$\frac{g_1^0}{\mu_0} P_1^0 \frac{a^3}{r^2} = \left(\frac{g_1^0 4\pi a^3}{\mu_0}\right) \frac{\cos\theta}{4\pi r^2} \tag{2.16}$$

This term is the potential associated with a geocentric dipole with strength $(g_1^0 4\pi a^3/\mu_0)$ oriented along the z axis (in the +z direction; Fig. 2.3 gives the relationship between the axes). Thus the magnitude of g_1^0 in Table 2.1 is linearly related to the strength of the axial geocentric dipole, which points downwards because the sign of g_1^0 is negative. The g_1^1 term is:

$$\left(\frac{g_1^1 4\pi a^3}{\mu_0}\right)\left(\frac{\cos\phi \sin\theta}{4\pi r^2}\right) \tag{2.17}$$

If γ is the angle between the x and r axes in Fig. 2.3, then it is easy to show that:

$$\cos\gamma = \cos\phi \sin\theta \tag{2.18}$$

Therefore, the g_1^1 term given above is that of a geocentric dipole oriented in the +x direction. Similarly, it is found that the h_1^1 term $(h_1^1 4\pi a^3/\mu_0)(\sin\phi \sin\theta/4\pi r^2)$ corresponds to a geocentric dipole oriented along the +y axis. The magnitude and direction of the geocentric dipole, p, can be determined by the usual vector addition

$$p = \frac{4\pi a^3}{\mu_0}(g_1^{0\,2} + g_1^{1\,2} + h_1^{1\,2})^{1/2}$$

and is found to be tilted at roughly $11\frac{1}{2}°$ to the rotation axis. Approximately 90% of the earth's magnetic field at the earth's surface can be described by this tilted dipole (Table 2.1; see also Vestine, 1967). Similarly, the terms involving $l = 2$ (r^3 in the potential) represent the geocentric quadrupole terms, $l = 3$ (r^4 in the potential) represent the geocentric octupole terms and so forth. The g_1^0, g_1^1 and h_1^1 terms collectively represent the *dipole field*, while the remaining terms collectively represent the *non-dipole field*. Sometimes the dipole and quadrupole terms are combined to give an estimate of the "best" eccentric dipole field (see next section).

The coefficients given in Table 2.1 indicate that the series (2.13) converges (Fig. 2.6). Even when higher order terms are included the series converges out to degree $l = 10$ to 12, beyond which the errors in the determination of the coefficients are typically comparable in size to the coefficients themselves (Kolesova and Kropachev, 1973).

The lowest degree terms correspond to the largest wavelength features of

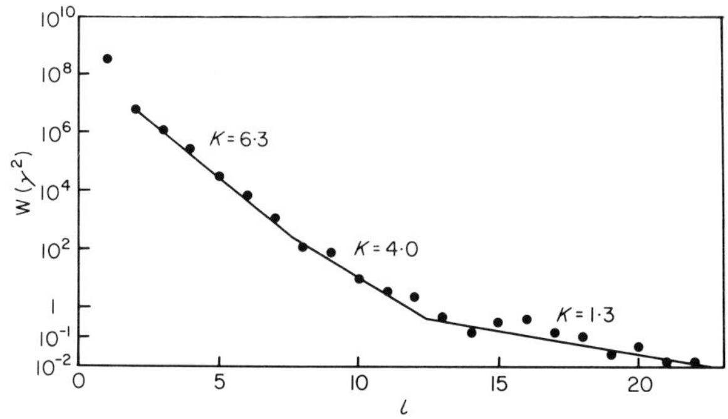

Fig. 2.6. Normalized power in spherical harmonics of the internal geomagnetic field as computed by Cain *et al.* (1974). The power is computed from the equation

$$W = \frac{1}{2l+1} \sum_m \{(g_l^m)^2 + (h_l^m)^2\}$$

for each l. Over several values of l the power varies according to $W \simeq W_0 K^{-l}$. For $l = 2$ to 7, $K = 6.3$, for $l = 8$ to 12, $K = 4.0$ and for $l = 13$ to 22, $K = 1.3$.

the field, as can be appreciated by considering the zonal harmonics. The g_1^0 term has a $\cos\theta$ dependence, so that the wavelength of the associated magnetic feature is the earth's entire circumference, $2\pi a$. The g_2^0 term has a (πa) wavelength, the g_3^0 term a $(2\pi a/3)$ wavelength, and the g_l^0 term a $(2\pi a/l)$ wavelength. As a rule of thumb, the depth of the source of a particular magnetic field feature is expected to be less than the wavelength of that feature. This follows from (2.1). For example, the wavelength of the magnetic field at a distance R from a linear electric current source is $2\pi R$. The field is circular about that source and the depth to the source is $1/(2\pi)$ of the wavelength of the field.

The wavelengths for the internal magnetic-field sources that have harmonics of degree $l = 8$ or less are usually attributed to sources that lie in the earth's core. Crustal magnetic sources (associated mostly with remanent magnetization, although a few percent can also be due to induced magnetization) generally have maximum associated wavelengths of a few tens or occasionally a few hundreds of kilometres. Such wavelengths are believed to be too short to be represented in the truncated spherical harmonic series used to describe the earth's magnetic field up to $l = 8$. This is of course a physical argument that is mathematically non-unique. It is conceivable that some long wavelength features could be associated with crustal remanence, as might occur say from continental-sized magnetic anomalies. A review of the observed long-wavelength crustal magnetic anomalies has been given by Cain (1971, 1975). The interpretation of terms with $l = 8$ to $l = 22$ is controversial (Cain, 1975; Harrison and Carle, 1982; Schure and Parker,

1982). That these intermediate wavelength features have changed with time between recent magnetic epochs suggests either that they are partly of core origin, or that there has not been a complete removal of external magnetic field effects in the analysis.

The decreased power in intermediate wavelength features, less than 4000 km or so ($l > 10$) and larger than a few hundred kilometres, probably reflects the lack of magnetic sources in most of the mantle. This is to be expected because only the uppermost mantle can have temperatures low enough to permit permanently magnetized material. There are induced electric currents in the mantle associated with variations in the external magnetic field, but these currents are not expected to make substantial contributions to the apparent "main" field when averaged over long times (for further discussion on potential erroneous effects stemming from external sources see Alldredge and Stearns, 1974; Lowes, 1975, 1976).

It is controversial whether the series of terms represented in Table 2.1 exhibits convergence at the core–mantle interface (see Lowes, 1974). The problem in assessing convergence there is that the shorter wavelength terms (higher order harmonics) are magnified more by the downward continuation process (because the negative power of r increases with increase in the degree of the harmonic), and so small errors associated with the estimation of those terms are also greatly magnified. In addition, the downward continuation process is valid only if the contributions from the crustal and mantle sources to the field that is continued downwards are negligible. Using MAGSAT data Benton *et al.* (1982) have shown from the sensitivity of locations and intersections of key geomagnetic indicators (specifically curves where $H_R = 0$, $\partial \boldsymbol{H}_R/\partial\theta = 0$, $\partial \boldsymbol{H}_R/\partial\phi = 0$; see §9.1) that spherical harmonic expansions of the geomagnetic field should be truncated at or below degree 8 if they are to be used to infer sources in the core.

2.3 Uniqueness and Other Mathematical Problems

2.3.1 Application of Laplace's Equation to the Earth

Laplace's equation is clearly not satisfied throughout the earth's interior, because the analysis of the last section indicates that there are strong internal sources. The question arises as to whether Laplace's equation is even satisfied at the earth's surface, because electric currents (such as lightning) do sometimes cross the earth's surface.

In §2.2 it was assumed that $\nabla \times \boldsymbol{H} = \boldsymbol{0}$ in order to make use of Laplace's equation. Therefore, the line integral of the horizontal component, $\boldsymbol{h}$, of the field around any closed curve, L, on the earth's surface must satisfy:

$$\oint \boldsymbol{h} \cdot \mathrm{d}\boldsymbol{L} = 0 \tag{2.19}$$

Analysis by several workers (Schmidt, 1939; Chapman, 1942; Vestine *et al.*, 1947a, b; Vestine, 1967) of charted magnetic data indicate that the above condition is "reasonably well satisfied" for the earth. By "reasonably well" is meant that although there are some small electric currents crossing the earth's surface even in fair weather conditions, the lowest degree coefficients given in Table 2.1 are not affected much by these currents. That is, for most magnetic epochs the contoured magnetic data satisfy (2.19) for almost all closed curves L on the earth's surface. Departures are so small that they could be due to various errors involved in the construction of the charts. Therefore, the use of Laplace's equation at the earth's surface in §2.2 is regarded as a valid procedure.

2.3.2 Parseval's Theorem

It is possible that errors can arise out of using a truncated series of spherical harmonic terms to represent the geopotential. An extremely important concept associated with orthogonal functions is sometimes referred to as *Parseval's approximation in the mean*. To make this concept clear and to simplify the algebra, consider the orthonormal set of functions $u_l(x)$ defined in terms of the Legendre polynominals, $P_l(x)$, by:

$$u_l(x) = \sqrt{\frac{2l+1}{2}}\, P_l(x) \tag{2.20}$$

where $x = \cos\theta$ as before.

It is easy to show that these are orthonormal functions, satisfying the following relation:

$$\int_0^1 u_l(x) u_{l'}(x)\, \mathrm{d}x = \delta_{ll'} \tag{2.21}$$

By using $u_l(x)$, consideration is restricted to one of azimuthal symmetry. However, the theory given here is essentially the same for spherical harmonics, although the algebra is somewhat more complicated.

Suppose values of the magnetic potential are expanded in a *truncated* series of $u_l(x)$:

$$S_N = \sum_{l=0}^{N} A_l u_l(x) \tag{2.22}$$

Now S_N affords an approximation of the potential (say) $\psi(x)$ (which in this example is assumed to have azimuthal symmetry) over the sphere. Of course $\psi(x)$ can be precisely represented by an infinite series of $u_l(x)$:

$$\psi(x) = \sum_{l=0}^{\infty} a_l u_l(x) \tag{2.23}$$

The question arises as to what relationship, if any, do the A_l of (2.22) have to the a_l of (2.23)?

The quantity

$$\varepsilon_m \equiv \int_0^1 |\psi(x) - S_N|^2 \, dx \tag{2.24}$$

is a positive number which measures the departure of S_N from ψ on the sphere. The choice of A_l which corresponds to the smallest possible value of ε_m respresents the *best approximation of* $\psi(x)$ *in the sense of least squares*. This best choice is achieved when $A_l = a_l$. To do this (2.24) is expanded making use of (2.22) and (2.23) to obtain:

$$\varepsilon_m = \int_0^1 \psi^2 \, dx - \sum a_l^2 + \sum |A_l - a_l|^2 \tag{2.25}$$

which clearly has its smallest value for $A_l = a_l$.

In brief, the approximation S_N *to* ψ *which is obtained by deleting from the orthogonal series for* $\psi(x)$ *those terms for which* $l > N$ *is the best approximation attainable in the sense of least squares.*

2.3.3 Uniqueness of Source

It has been argued that magnetic sources represented by harmonics of degree of $l = 8$ and less probably originate primarily in the core. However, it is important to realize that the spherical harmonic procedure is a mathematical technique for representing the field by hypothetical sources placed precisely at the earth's centre. It is highly unlikely that the actual sources of the earth's magnetic field lie in the solid inner core of the earth (see Chapter 7), so that a spherical harmonic analysis does not uniquely give the sources. As a corollary, the individual terms in a spherical harmonic analysis should not be considered as representing separate "real" physical entities.

The ambiguity in a spherical harmonic analysis can be illustrated mathematically by considering a simple example of a dipole source displaced from the earth's centre a distance d along the z axis. If the dipole were at the centre, it could be represented by the g_1^0 term in the spherical harmonic expansion:

$$\frac{p \cos \theta}{4\pi r^2}$$

where $p = (g_1^0 4\pi a^3/\mu_0)$ has been substituted for the dipole strength. The potential for the offset dipole is:

$$\frac{p \cos \theta}{4\pi(r^2 + d^2 - 2rd \cos \theta)} = \left(\frac{p \cos \theta}{4\pi r^2}\right)\left(\frac{1}{1 + d/r^2 - 2d/r \cos \theta}\right)$$

The second term on the right side can be expanded in a Taylor series for $d < r$ to obtain the potential, ψ_d:

$$\psi_d = \frac{p}{4\pi r^2} \cos \theta + \frac{2pd}{4\pi r^3} (P_2^0) + \dots \tag{2.26}$$

That is, an offset dipole will be represented in a spherical harmonic expansion by an axial geocentric dipole of the same strength plus an axial geocentric quadrupole of strength $(2pd)$, and so forth. The larger d, the larger the number of zonal harmonics required to approximate that offset dipole. Thus, an offset dipole in the core is mathematically indistinguishable from a dipole plus a series of higher degree multiple fields at the earth's centre. This not only indicates the non-uniqueness of spherical harmonic analyses (*or any other analyses*), but it also emphasizes that separate terms in a spherical harmonic analysis do *not* represent separate "real" magnetic sources. It is experimentally impossible to determine the magnetic sources uniquely in the core no matter how many measurements of the field are made at the earth's surface, even in the ideal case when those measurements are perfect estimates of the mean magnetic field! Of course one can add sufficient "physical constraints" to improve and constrain the variability of allowed solutions, but this increased resolution is directly dependent on the applicability of the "physical constraints".

2.3.4 Non-spherical Harmonic Representation of the Earth's Magnetic Field

Because of certain mathematical advantages (such as illustrated by Parseval's theorem when dealing with convergence problems), it seems desirable to express magnetic field data in terms of spherical harmonic coefficients. However, such representations are not of direct physical significance, as is clearly seen by the fact that all internal sources are placed at the earth's centre in a spherical harmonic analysis, while the primary location of the actual internal sources is most probably in the outer core. Ultimately, one hopes to be able to construct models that have more physical significance than spherical harmonic analyses descriptions.

Most alternative descriptions have used magnetic dipoles as elementary sources of the internal magnetic field, as pioneered by McNish (1940) and Lowes and Runcorn (1951; also see Lowes, 1955). Subsequently several models consisting of one centred axial dipole with several eccentric radial dipoles have been advocated (Alldredge and Hurwitz, 1964; Pudovkin and Kolesova, 1968; Pudovkin *et al.*, 1968; Alldredge and Stearns, 1969; Zidarov and Bochev, 1965, 1969). Most of the above workers attempted to explain the observed geomagnetic secular variation in terms of their models. Much of the earlier work and some of the later work constrains the eccentric dipoles to lie close to the core–mantle boundary, based on the argument that the sources giving rise to the observed secular variation cannot lie too deep in the core because of the electromagnetic screening by the electrically conducting core (see Chapter 9 for further discussion). Similar arguments have been advanced by Cox (1968) and Harrison and Ramirez (1975) who use dipoles to model some changes in the magnetic field as observed in the palaeomagnetic record. Some of the later models have tended to relax the condition that all the dipoles be radial and even in some cases to allow the magnitude of the dipoles to change with time (Bochev, 1965, 1969, 1975; see also Zidarov, 1969, 1970, 1974).

A few workers have adopted circular current loops as more physical representations of the earth's magnetic field, as pioneered by Zidarov and Petrova (1974). Recent work by Peddie (1979) considers models that constrain circular current loops to the core–mantle boundary (up to seven loops) as well as models for unconstrained current loops (up to four loops).

None of the models given above is very satisfactory in providing a physical picture of the internal sources. This is partly manifested in the work by Peddie (1979) which constrains the current loops to lie at the core–mantle interface, a constraint which results in intersecting current loops. Moreover, it is questionable whether any given model will ever be completely satisfactory because of uniqueness problems, as can be appreciated by recalling the uniqueness problems associated with modelling a dipole displaced from the earth's centre discussed in §2.3.3. In addition, modern dynamo theory (Chapter 8) suggests that "real" sources are probably far more numerous and complex than suggested by any of the above modelling schemes.

In view of these difficulties and because of the mathematical advantages already noted, spherical harmonic representations are used predominantly in this book. At present, there is no evidence that alternative models to spherical harmonic analyses are as accurate or as efficient in modelling the earth's magnetic field of core origin and its secular variation. Nevertheless, this should not be misconstrued as a reason for suspending research to construct more viable physical models such as undertaken by Peddie (1979).

The use of spherical harmonics to represent crustal magnetic sources is a different matter. Crustal sources are close to the observer, even when satellites such as POGO and MAGSAT are used. Thus a very large number of spherical harmonic terms are needed to represent the crustal fields. These fields typically vary significantly over short distances and rapid convergences of the spherical harmonic series should not be expected. Thus alternative representations of the fields of crustal origin are probably desirable. Discussion of such problems with respect to MAGSAT data can be found in Benton *et al.* (1982), Galliher and Mayhew (1982), Langel and Estes (1982) and Schmitz *et al.* (1982).

2.3.5 Magnetic Annihilator

Backus and Gilbert (1967, 1968, 1970) have developed a method to optimize the information that can be obtained from non-unique problems in geophysics. The method requires that there exists a viable solution to the *forward problem*. In the case of geomagnetism this would mean that the hydromagnetic dynamo problem could be solved to the extent that the observed properties of the earth's magnetic field could be obtained from the theory. Because this has not yet been done, it might be supposed that the application of techniques like those developed by Backus and Gilbert would be inappropriate to geomagnetism. Although such is often the case, there are situations in which the theory may be applicable and there is at least one case where it has actually been applied. For example, inverse theory might be applicable to the source problem of the last subsection when (say) dipole sources are assumed to exist in the outer core and to give rise to the observed instantaneous field. A particular distribution of such dipole sources that produce the observed field would then be taken as a solution of the forward problem. One then would use the inverse theory methods to investigate the statistical variations associated with the set of all allowable solutions, commonly referred to as the *inverse* problem. Such an approach has not yet been tried on this problem, but it has been attempted on the related problem which deals with the inversion of small wavelength magnetic anomaly data to obtain crustal magnetic sources (Parker and Huestis, 1974).

The direct application of the Backus–Gilbert inverse technique to the magnetic anomaly problem resulted in such wide limits for acceptable models, that additional modifications of the techniques and certain geophysically reasonable constraints were needed to obtain useful results (Parker and Huestis, 1974). The Parker–Huestis approach will not be extensively discussed here other than to introduce the concept of the annihilator, which is useful in dealing with non-uniqueness problems. Following Parker (1977) it is convenient to assume for the purposes of illustration that the problem is linear, although it turns out that all inverse problems in geomagnetism are essentially non-linear. Methods for treating non-linear problems are beyond the scope of this section, but they have been discussed by Sabatier (1977) and Parker (1977).

Suppose $F(z)$ represents the unknown source function to be determined. Information concerning this source function is represented by data obtained at the surface of the earth that can (say) be represented by the function $D(x)$. We assume that $D(x)$ and $F(z)$ are linearly related, or:

$$D(x) = \int G(x, z)F(z)\,\mathrm{d}z \tag{2.27}$$

where $G(x, z)$ is a kernel that describes the mapping of $F(z)$ into $D(x)$. If there is no function $f(z)$, such that

$$\int G(x, z)f(z)\,\mathrm{d}z = 0 \tag{2.28}$$

then $F(z)$ in (2.27) can be shown to be unique. A solution of the forward problem in such a case is sufficient. However, usually there are non-trivial solutions of (2.28). There is no way to learn about the $f(z)$ from the data, because the substitution of $F(z) + af(z)$, where a is any constant, for $F(z)$ in (2.27) does not change $D(x)$. The set of all functions $f(z)$ that satisfies (2.28) is called the *annihilator* (Parker, 1977). No information about the annihilator can be obtained from the data. The annihilator is a manifestation of the non-uniqueness of the problem that cannot be removed, no matter how perfect and complete are the data. A simple example of a magnetic annihilator in the flat-earth model often used in magnetic anomaly studies is a uniformly magnetized horizontal layer (Parker and Huestis, 1974). No magnetic field exists external to such a layer.

Equation (2.27) can also be used to illustrate the concept of mathematical stability. Parker (1977) points out that if $G(x, z)$ greatly smooths $F(z)$, then the inversion of $D(x)$ to obtain $F(z)$ will be unstable. This means that small changes in $D(x)$ can result in large changes in $F(z)$. Unfortunately, one of the best known examples of an unstable problem is the inversion of a potential field that satisfies Laplace's equation (Bullard and Cooper, 1948). That is, small changes in the magnetic scalar potential at the earth's surface can lead to very large changes in potential at the core–mantle interface (as previously noted in §2.2.3).

The conclusion is that there is a large degree of non-uniqueness in the magnetic source problem that can never be resolved no matter how perfect and complete are the data. In addition, small errors in measurements at the earth's surface will be greatly magnified by downward continuation to the core–mantle interface. These facts only serve to strengthen the view that spherical harmonic expansions are to be preferred, when possible, for representation of the magnetic data. The above illustrates that it will probably never be possible to define the magnetic sources and source regions in the core very well. In particular, it is doubtful if the process of downward continuation of the magnetic potential to the core–mantle boundary will ever show convincingly that the core is turbulent or not, or show that the sources of the non-dipole field lie at the core–mantle boundary or deeper in the core (see Chapters 8 and 9). Of course, this conclusion does not mean that there are not other ways of answering such questions.

2.4 Geomagnetic Secular Variation: Time Variation of the Internal Sources

2.4.1 Overview

It has been known for over 300 years, since the discovery of Gellibrand in 1635, that the geomagnetic field changes with time. For example, the declination at London, England is known to have changed gradually from $11\frac{1}{2}$°E in 1576 to 24°W in 1823 before turning eastward again (Fig. 2.7).

Variations in the geomagnetic field observed at the earth's surface occur on time scales ranging from milliseconds to millions of years. The short-term variations arise mostly from currents flowing in the ionosphere giving rise to magnetic storms and the more or less regular daily variation (see §2.5.3). These rapid fluctuations are superimposed upon a much slower change with periods of years to millions of years. These slower changes are generally referred to as the *geomagnetic secular variation*.

Although the cut-off between internal and external sources in terms of period is probably not as distinct as is sometimes argued, almost all geomagnetists agree that periods of a few tens of years or longer are of internal origin while periods less than a year or so are of external origin. The reason that shorter periods of internal origin are not seen at the earth's surface can be traced to screening effects of the lowermost electrically conducting mantle, as is discussed in the next subsection. Variations in the external magnetic field are discussed in §2.5.

2.4.2 Internal Cut-off Period and Mantle Conductivity

Because the mantle is an electrical conductor (technically it is really a semi-conductor) there will be a minimum period for a magnetic field fluctuation of

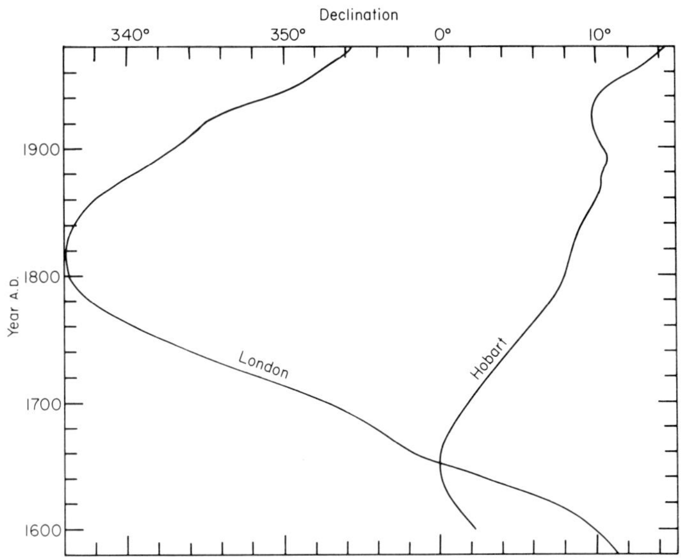

Fig. 2.7. Variation in declination at London, England (51.5°N) and at Hobart, Tasmania (42.9°S) from observatory measurements. The earliest measurement in the Tasmanian region was made by Abel Tasman at sea in 1642 in the vicinity of the present location of Hobart. Pre-observatory data have been derived also by interpolation from isogonic charts.

core origin, that can be observed at the earth's surface. This follows readily from screening depth calculations obtained from Maxwell's equations. From (2.1)

$$\nabla \times \boldsymbol{H} = \sigma \boldsymbol{E} + \varepsilon \frac{\partial \boldsymbol{E}}{\partial t} \tag{2.29}$$

where $\boldsymbol{E}$ is the electric field, σ is the electrical conductivity and ε is the electric permittivity. Taking the curl of (2.29) and using another of Maxwell's equations, $\nabla \times \boldsymbol{E} = \mu(\partial \boldsymbol{H}/\partial t)$, the wave equation for $\boldsymbol{H}$ is obtained:

$$\nabla^2 \boldsymbol{H} = \sigma\mu \frac{\partial \boldsymbol{H}}{\partial t} + \varepsilon\mu \frac{\partial^2 \boldsymbol{H}}{\partial t^2} \tag{2.30}$$

The first term on the right side of (2.30) represents a damping term; electromagnetic waves described by (2.30) (and a similar one for $\boldsymbol{E}$) exhibit exponential damping with distance x. This can be seen by showing that the damped plane wave, $\boldsymbol{H} = \boldsymbol{H}_0 e^{-k_1 x} e^{i(k_2 x - \omega t)}$, is a solution of (2.30) by direct substitution, where k_1 and k_2 are the real and imaginary parts of the wave number and ω is the angular frequency.

The problem in using the above theory is that either one has to know the mantle conductivity to analyse the geomagnetic spectrum or one has to know the geomagnetic spectrum to obtain the mantle conductivity. Runcorn (1955) noted that for a flat-earth model, in which the mantle is approximated by a plane of constant conductivity, σ_0, and thickness L, the amplitude of the spectrum at the surface has the form:

$$\boldsymbol{H} = \boldsymbol{H}_0 \exp\left(-L/(\mu_0 \sigma_0 \omega)^{1/2}\right)$$

Therefore, by assuming that the secular variation at the core–mantle interface could be described by a white spectrum, Runcorn could estimate the mean mantle conductivity. McDonald (1957) further improved on this procedure by considering a spherical model and by allowing for geometrical spreading in a mantle in which the conductivity was allowed to vary according to a power law of the radius. His analysis depended heavily on mathematical techniques developed by Lahiri and Price (1939) to investigate upper mantle conductivity. Subsequently, Currie (1966, 1967, 1968) and Banks (1969) used time series analysis methods to estimate a conductivity near 3×10^2 Sm^{-1} for the lower mantle. Stacey (1977a) argues that there may be a very high conducting region at the base of the mantle near 10^3 Sm^{-1}.

Currie (1968) further showed that the internal field cut-off period was near four years (3.7 years by his analysis, although the resolution does not give confidence in the second significant figure). That is, shorter periods are screened out by an electrically conducting mantle and for periods up to four years, the geomagnetic spectrum originates from outside the earth. Using the method to separate sources of internal and external origin described in §2.2.1, Malin and Hodder (1982) have shown that a 1969–1970 sudden pulse (jerk) in the geomagnetic secular variation is definitely of internal origin. They argue that the period of this pulse is less than two years. However, Backus (1983) demonstrates that short periods observed at the earth's surface may not reflect the true periods of the core sources. This follows because both filtering and mixing of the source frequencies occur. Backus suggests a period near 13 years is responsible for the 1969–1970 pulse or jerk.

The interpretation of the spectrum with periods between four and 30 years is controversial. For example, Currie (1973) showed the tendency for spectral lines from several observations to cluster around a period of 21.4 years. Currie suggested that this variation is of external origin and due to the 22-year sunspot cycle (see Chapter 12). In contrast, Alldredge (1977) argues that the 21.4 years variation is of internal origin on the basis that there is no good evidence for synchronous world-wide variation with a common period. It should be noted that these uncertainties in distinguishing between internal and external sources will be reflected in corresponding uncertainties in estimates of lower mantle conductivity. In any case, there appears to be a consensus that most periods in the geomagnetic spectrum that are longer than a few years are predominantly of internal origin.

The above analysis was only possible because Lahiri and Price (1939) (see also reviews by Rikitake, 1966 and Price, 1967) had already developed and utilized methods to determine the upper mantle conductivity distribution from "short period" electromagnetic fluctuations of external origin. They found, as supported by subsequent investigations, that the electrical conductivity of the upper mantle is more than five orders of magnitude less than the conductivity estimated for the lower mantle (McDonald, 1957; Price, 1967; Currie, 1968). Therefore, the conductivity of the upper mantle will have a far less serious effect on damping the internally produced secular variations than that of the lower mantle. The techniques pioneered by Lahiri and Price play a very important role in geomagnetism and geophysics generally and as a class provide one of the few geophysical tools besides seismology for probing the earth's mantle. Nevertheless, because the average conductivity of the upper mantle is so much smaller than the lower mantle, the determinations of its detailed conductivity variation, albeit of great geophysical importance, do not yield valuable information on the properties of the earth's internal magnetic field. Because it is the earth's internal magnetic field that is of prime concern in this book, discussion is not taken further regarding magnetotelluric and geomagnetic depth sounding methods for determining the crustal and upper mantle conductivity. However, some relevant results using this technique are referred to in Chapter 11 regarding the properties of the Moon's interior.

2.4.3 Westward Drift of the Non-dipole Field

The largest changes in the direction of the earth's magnetic field during the past 150 years are associated with the non-dipole part of the field. This can readily be seen as a westward shift of various isoporic foci (maxima or minima) on isomagnetic charts between different magnetic epochs. Figure 2.8 compares the vertical component of the non-dipole field for 1829 and 1980. Yukutake and Tachinaka (1968) noted there were two main types of non-dipole anomalies; there were those that clearly drifted westwards (such as the central African anomaly in Fig. 2.8) and those that have remained stationary and just increased or decreased in magnitude (such as the Mongolian anomaly which increased by 50% over the 150 years).

Bullard *et al.* (1950) made use of data tabulated by Vestine *et al.* (1947a, b) for the time period spanned by the 1907 epoch to the 1945 epoch and determined the average velocity of the non-dipole field to be 0.18° per year westward, the so-called *westward drift of the non-dipole field*. Thus, a whole

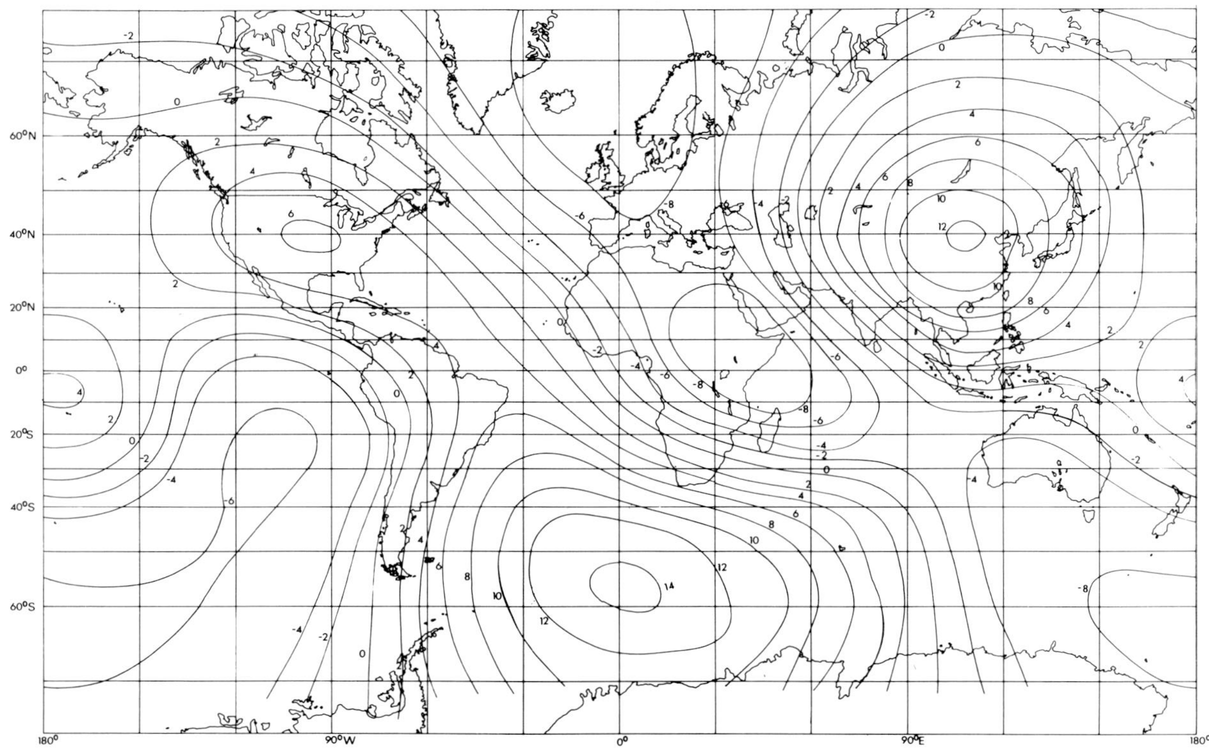

60°N
40°N
20°N
0°
20°S
40°S
60°S
180°
90°W
0°
90°E
180°

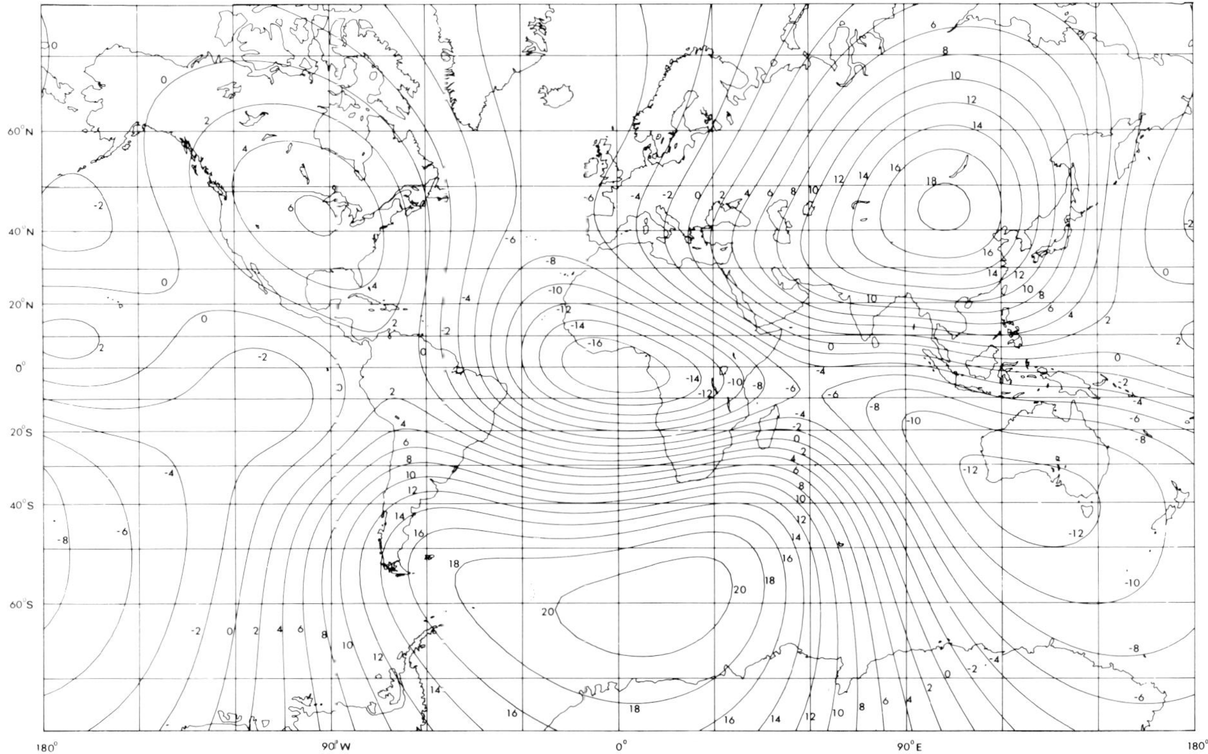

Fig. 2.8. The vertical component of the non-dipole field for 1929 (top) (from Yukutake and Tachinaka, 1968) and 1980 (bottom) (courtesy of D. R. Barraclough and the Geomagnetism Unit, Institute of Geological Sciences, Edinburgh, Scotland). Contours are labelled in units of 2000 nT.

circuit of the earth would take 2000 years, a period sometimes erroneously used to represent "sufficient time" to average out secular variation in palaeomagnetism.

The methods for determining the westward drift of the non-dipole field are relatively simple. For example, one can substitute $\phi_0 + \partial\phi/\partial t$ for ϕ in (2.13), and then determine the value of $\partial\phi/\partial t$ that minimizes residuals (numerically calculated at grid points by a computer) for a given time interval (such as the 37.5 year interval used by Bullard *et al.*, 1950). Yukutake (1962) and Rikitake (1966) discuss alternative methods for estimating the westward drift of the non-dipole field; all methods agree in giving values near 0.2° per year.

2.4.4 Secular Variation not Explained by the Westward Drift

The drift rate of particular features such as foci on isoporic charts is estimated to be near 0.3° per year, significantly larger than the westward drift of the non-dipole field (Burlatskaya *et al.*, 1965, 1970a, b; Burlatskaya, 1972; Yukutake, 1967, 1970; Yukutake and Tachinaka, 1969). This apparent paradox can be resolved by recognizing that although the westward drift of the non-dipole field is the most dominant secular variation feature, a considerable amount of the non-dipole field's variation must be explained by mechanisms other than westward drift.

Yukutake and Tachinaka (1969) and Yukutake (1970, 1979) suggest that the solution of this problem lies in separating the earth's non-dipole field into *drifting* and *standing parts*. Their analyses indicate that the standing and drifting parts of the non-dipole field are approximately the same size and intensity (Fig. 2.9). The drifting field consists mainly of low harmonics ($l \leq 3$) whereas the standing field has a more complicated distribution. The drifting part of the non-dipole field moves westward at a rate slightly greater than 0.3° per year. The inclusion of the standing part of the non-dipole field in the drift rate lowers the average rate to near 0.2° per year. This is argued to be the main reason why the rate estimated from the non-dipole field is smaller than that of the secular variation. Of course, the separation of the non-dipole field into standing and drifting parts will automatically improve on the "spherical harmonic fit". The main unresolved question is whether the division into standing and drifting parts is physically meaningful.

James (1971, 1974) argues that a substantial part of the secular variation still cannot be explained by the above analyses. James correctly points out that neither standing nor westward-drifting non-dipole fields can explain large changes in the axisymmetric part of the non-dipole field as characterized by zonal harmonics. For example, g_2^0 shows very large changes with time, including a reversal in sign since the time of Gauss (Vestine, 1967). Either some north–south movement of the non-dipole field is required as argued for by James, or the amplitude of the standing or westward-drifting part of the non-dipole field must vary with time. In summary, the dominant feature of secular variation of the non-dipole field during historical times is its westward drift. Nevertheless, roughly half of the secular variation of the non-dipole field cannot be explained by this westward drift.

2.4.5 Variations of the Dipole Field with Time

The dipole field is also changing with time. The intensity of the dipole field has decreased at the rate of about 5% since the time of Gauss' analysis in 1835 (Leaton and Malin, 1967; Vestine, 1967; McDonald and Gunst, 1968) (Fig. 2.10a). Indeed, Leaton and Malin (1967) and McDonald and Gunst (1968) have speculated on the demise of the main dipole around A.D. 3700 to 4000 if the present trends continue. The dipole axis, however, as represented by the position of the North Geomagnetic Pole has hardly changed its position since the analysis of Gauss (Bullard *et al.*, 1950). Over the past 150 years there appears to have been a slow westward change of near 0.05° per year in azimuth angle, but no progressive motion in polar angle (Nagata, 1965; McDonald and Guest, 1968).

Veinberg and Shibaev (1970) have produced a catalogue of values of geomagnetic declination (D), inclination (I) and horizontal intensity (H) reduced to 12 epochs between 1550 and 1940 inclusive, and to positions whose latitude and longitude are multiples of 10°. Prior to 1800 there are no historical intensity measurements and there are very few inclination values before 1750. Analyses of these old historic measurements using spherical harmonics have been carried out by Benkova *et al.* (1970, 1974), Adam *et al.* (1970, b), Yukutake (1971), Braginskiy and Kulanin (1971), Braginskiy (1972) and Barraclough (1974). Without intensity data it is only possible to determine relative magnitudes of the Gauss coefficients in a spherical harmonic expansion. The coefficients are normalized to g_1^0.

When there are only declination values available the method due to Bauer (1894) can be used and has been discussed in detail by Benkova *et al.* (1970). For any point on the earth's surface, the relation

$$X \sin D = Y \cos D$$

can be rewritten by expanding X and Y in spherical harmonics. If there are N values of D a series of N inhomogeneous equations result and may be solved to obtain relative values of the coefficients (see Barraclough, 1974). If values of inclination are available at points for which declination data also exist, then Bauer (1894) proposed the relation

$$Z \cos I = H \sin I = \frac{X}{\cos D} \sin I$$

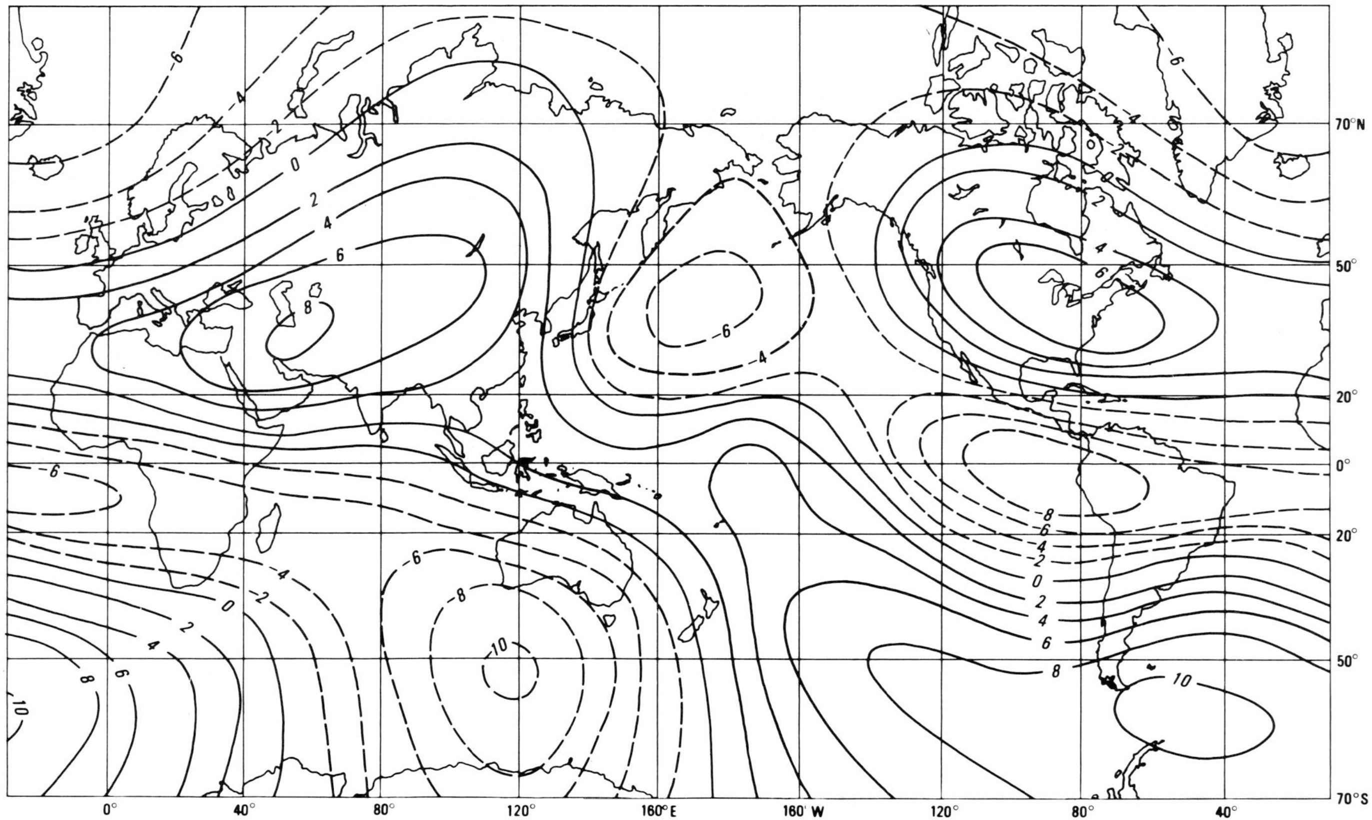

70°N
50°
20°
0°
20°
50°
70°S
0°
40°
80°
120°
160°E
160°W
120°
80°
40°

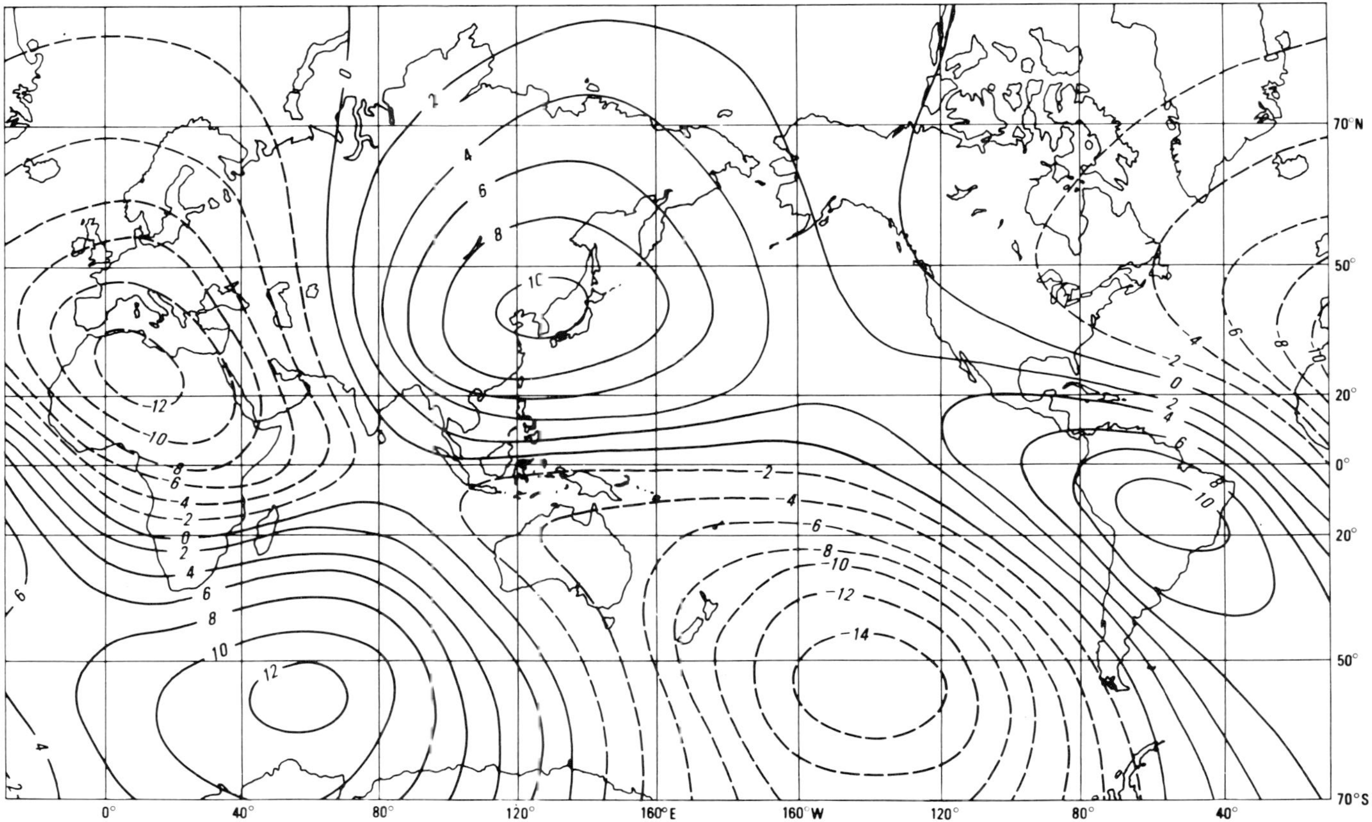

Fig. 2.9. The vertical component of the non-dipole field for 1900 subdivided into its standing (top) and drifting (bottom) parts after Yukutake and Tachinaka (1969). Contours are labelled in units of 1000 nT.

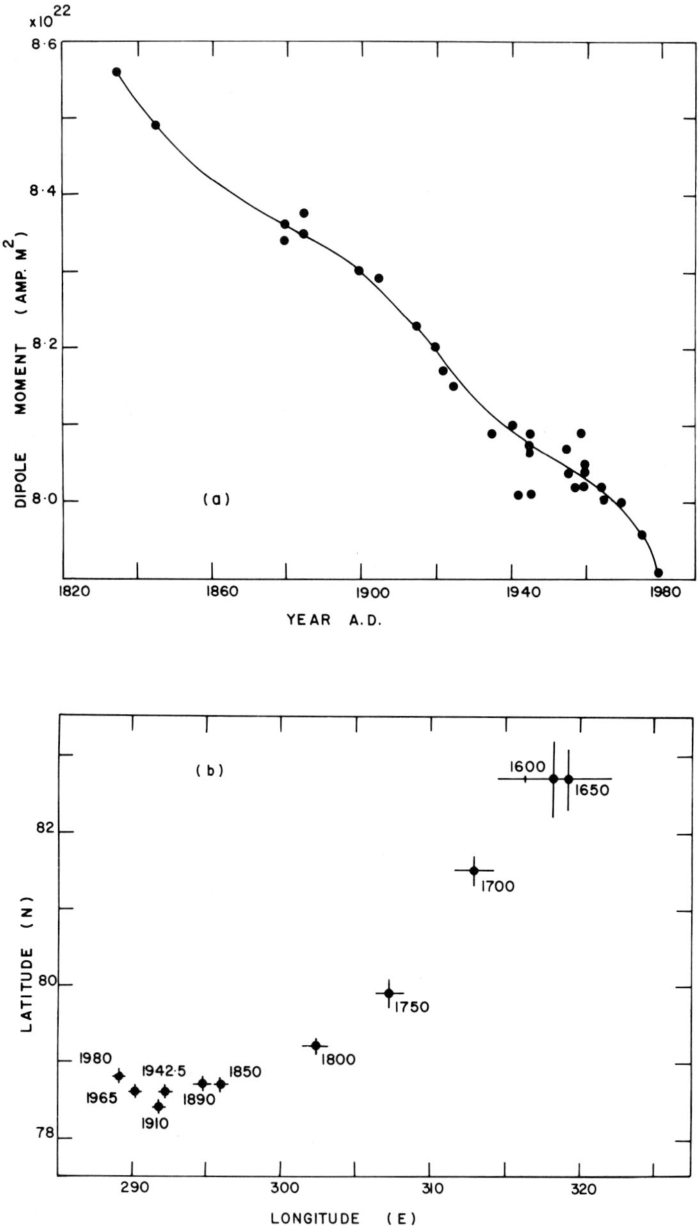

Fig. 2.10. Variations of the dipole field with time. (a) Variation of the dipole moment from successive spherical harmonic analyses since Gauss (1838). (b) Variation of the dipole axis since A.D. 1600 as represented by the change in position of North Geomagnetic Pole. Spherical harmonic analyses of Barraclough (1974) and model fields for 1965 and 1980.

be used in addition. An alternative method used by Braginskiy and Kulanin (1971), Yukutake (1971), Braginskiy (1972), Benkova *et al.* (1974) and Barraclough (1974) determines values of g_1^0 by extrapolating back in time from values determined since the time of Gauss. Barraclough (1974) fitted a straight line to values of g_1^0 from 170 spherical harmonic models of the field between 1829 and 1970 to derive the relation

$$g_1^0(t) = -31110.3 + 15.46(t - 1914)$$

where t is the epoch in years A.D.

Barraclough (1974) has analysed the field for epochs since 1600 using all of the above methods. The positions of the North Geomagnetic Pole since 1600 are plotted in Fig. 2.10b. They show that the dipole field has drifted westwards at about 0.08° per year since 1600 and has changed latitude at the very much slower rate of 0.01° per year. More recently Yukutake (1979) suggests that the *equatorial* dipole field be separated into two components drifting in opposite directions. Braginskiy (1972) made a similar suggestion but the amplitudes and velocities of the westward and eastward drifting components differ considerably in the two analyses. The main feature however is quite clear and this is the westward drift of the equatorial dipole.

2.5 The External Magnetic Field

2.5.1 The Magnetosphere

The earth is immersed in an expanding solar atmosphere which consists mainly of supersonic transport of H^+ and $^4He^{2+}$ and a few heavier elements away from the sun. As is discussed further in Chapter 12, this *solar wind* interacts with the earth's magnetic field to produce the *magnetosphere* (Fig. 2.11).

The *magnetopause* separates the earth's environment from the interplanetary space where solar wind particles are found. By attaching a coordinate system to the solar wind, the earth moves through the solar wind at a typical velocity of 400 km s^{-1}. This is about ten times the local Alfven speed (defined in Chapter 9) and therefore the flow is "supersonic". This motion of the earth in the solar wind creates a shock front referred to as the *bow shock*. The bow shock is separated from the magnetopause by a region referred to as the *magnetosheath* (Fig. 2.11). The bow shock is usually regarded as the boundary of the magnetosphere.

As discussed briefly in Chapter 12, following the development of the frozen-in-field theorem in Chapter 7, the solar wind "drags" along the solar

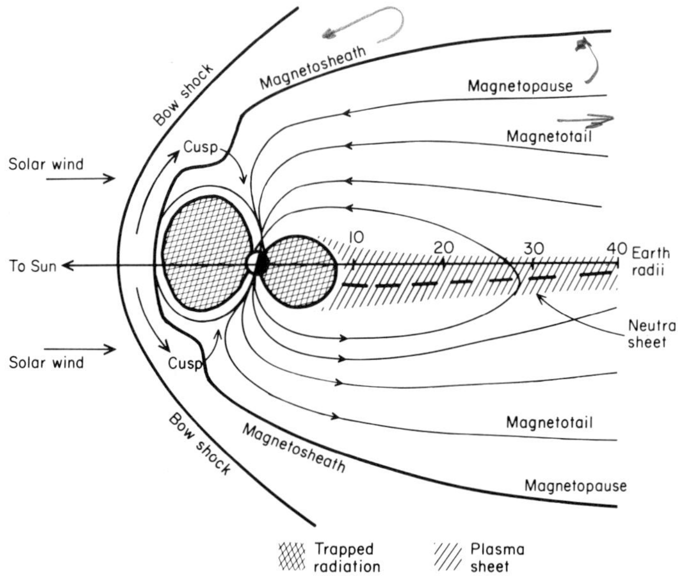

Fig. 2.11. The different regions of the magnetosphere in the noon–midnight meridian plane.

magnetic field. This produces an enhancement of the intensity of the magnetic field on the sun's side of the earth and a decrease of the intensity on the opposite side relative to that expected from the earth's field of internal origin, which can here be considered as essentially a dipole field. The interactions between the interplanetary and geomagnetic fields have been reviewed by several workers (e.g. Burch, 1974; Hill, 1974; Stern, 1977). The enhancement of the field on the day side is due to solar wind pressure which compresses the geomagnetic field there (Hill, 1974; Stern, 1977; see also Chapman and Ferraro, 1931; Parker, 1958).

The geomagnetic field on the dark side of the earth is highly distorted from a dipole, having been "stretched" by the solar wind. The magnetopause would form a teardrop shape if the geomagnetic field originating in the earth's core were the only source of internal pressure resisting the solar wind (Axford *et al*., 1965). However, there are observed dawn to dusk currents that are carried by particles in the *plasma sheet* (Fig. 2.11) that results in a much larger *geomagnetic tail*, or *magnetotail*, than was first predicted (Bird and Beard, 1972a, b; Stern, 1977).

Because the high latitude geomagnetic lines of force are mostly dipolar and are either compressed on the front side of the earth by the solar wind or swept away by the solar wind on the back side, a *cusp* geometry will occur (Fig. 2.11). This cusp (or cleft) is the region of the last line of force that crosses the front side of the earth and still goes to the geomagnetic tail. The cusp is one of the easiest places where solar wind particles could enter the earth's ionosphere.

Although it is still controversial, most workers now accept that the magnetosphere is probably "open" (Stern, 1977), following the pioneering work of Dungey (1961, 1963) and Axford *et al.* (1965). The term "open" means that some of the geomagnetic field lines connect with those of the interplanetary magnetic field. A schematic view of an open magnetosphere is shown in Fig. 2.12. Note that the topology of the magnetic field lines is such that there are always at least two points (A and B in Fig. 2.12) at which the intensity of the earth's magnetic field vanishes. At such points, sometimes referred to as *neutral points*, the very concept of magnetic field lines becomes vague; magnetic "field lines" can open up and form new connections, a process referred to as *magnetic merging* or *magnetic reconnection*. This process is also important in many dynamo models discussed in Chapters 7 and 8. Point A (Fig. 2.12) illustrates one way in which charged particles spiralling along a magnetic field line carried off the sun can find their way into the magnetosphere. Work by Axford and others (see Axford, 1969) suggests that the neutral point B (Fig. 2.12) is located inside the plasma sheet of the tail, in the region where the magnetic field is very weak and is therefore sometimes referred to as the *neutral sheet* (Speiser and Ness, 1967). Violent outbursts of particle accelerations observed on the night side of the earth, a kind of *magnetospheric substorm*, are believed to be associated with enhanced merging at this nightside neutral point B (Stern, 1977).

The solar wind is highly variable in time. Therefore its pressure on the geomagnetic field will vary with time and will cause the magnetopause to move around. This will invariably lead to changes in the magnetic field at the earth's surface. However, the tracing of the sources of disturbance of the magnetic field at the earth's surface is extremely difficult. This follows

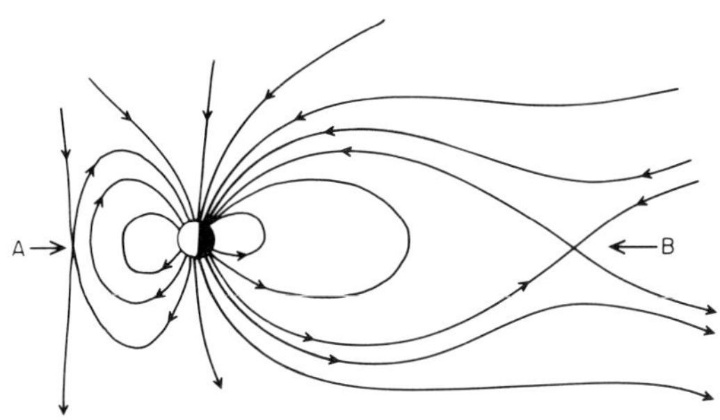

Fig. 2.12. Schematic view of an open magnetosphere with an extended tail. Points A and B are neutral points at which the earth's magnetic field vanishes. From Stern (1977).

because processes that occur in the above distant regions can induce other major disturbances closer to the earth in the ionosphere. Conversely, changes in ionospheric currents could lead to changes far above the ionosphere.

2.5.2 The Ionosphere

With the exception of thunderstorms (the movement of cumulonimbus clouds overhead can change the magnetic field at the earth's surface by 20 nT or so and lightning discharges can have even much larger effects), the region from the earth's surface up to about 50 km can be regarded as a "vacuum" from an electromagnetic point of view. Roughly between 50 km and 1500 km lies the ionosphere, subdivided into the *D region* (50–90 km), the *E region* (90–120 km) and the *F region* (120–500 km). The ionization originates from the interaction of solar ultraviolet radiation with various atmospheric constituents. The ionization is greatest in the F region in which typical midday electron densities will reach 10^5–10^6 cm^{-3}. A typical energy of an ionospheric electron is a few eV. The degree of ionization decreases as one moves from region F to D.

Particles in the ionosphere are affected by the rotation of the earth and by tidal forces from the moon and sun. Thermal effects are important in the first order analysis only in the bottom-most regions of the ionosphere. Neutral particles in regions D and below rotate in unison with the earth. The degree of ionization is so small in the D region that regional collisions with neutral particles result in synchronous rotation. Region D disappears at night. In region E the collisions are fewer than at the lower altitudes, the degree of ionization is larger so that there is a partial decoupling between the ionized particles and neutral particles. In the F region, it is usually assumed that, to a first approximation, there is a complete decoupling between the neutral and ionized particles. That is, in the F region, electromagnetic forces are much stronger than collisional forces and hence the motions of charged particles are controlled by processes that occur in the magnetosphere and solar winds.

Electromagnetic forces in the ionosphere can lead to large electric currents. Some of these currents have associated magnetic fields of magnitude up to 1000 nT at the earth's surface, as is further discussed in the next section.

2.5.3 Transient Magnetic Variations, Storms and Substorms

Geomagnetism has two main areas of interest. The first area covers the variations of the external magnetic field as recorded on magnetograms (recordings of the magnetic elements, such as intensity, declination, etc., with time) and the possible origins of those variations. However, in the second major area of emphasis which deals with the magnetic field of internal origin, the external variations are generally considered to be a nuisance, as they can only serve to reduce the resolution of the magnetic field spectrum of internal origin.

Even if there were no variation in the solar wind (which the reader is reminded is actually highly variable), there would be variations in the magnetic field due to tidal effects on the magnetosphere from the sun and moon and due to the fact that the dipole is tilted with respect to the general direction of the solar wind as the earth rotates. The degree of variation in the magnetic elements thus varies widely from day to day and from season to season. Furthermore there are considerable variations due to changes in sunspot activity, either due to the sudden occurrence of solar flares or else corresponding to the sunspot cycle (Chapter 11). At any station the days when the three main elements undergo smooth and regular variations are termed magnetically *quiet days*, and when these variations are more or less irregular the days are termed magnetically *disturbed days*. When extreme, a magnetic disturbance is called a *magnetic storm*. Except during magnetic storms it is evident from magnetic observatory records that there is a regular daily variation in the magnetic elements. The main part is the *solar daily variation*, $\boldsymbol{S}$, with a period of 24 h together with the much smaller *lunar daily variation*, $\boldsymbol{L}$, with a period near 25 h. The *disturbance variation*, $\boldsymbol{D}$, refers to the additional field(s) present during disturbed and extremely disturbed days. $\boldsymbol{S}$ and $\boldsymbol{D}$ are very easily recognized in the magnetic records but $\boldsymbol{L}$ can only be determined after careful analysis of a large data set.

The solar daily variation is seen in its pure form on extremely quiet days, and the mean taken over these days is referred to as the *solar quiet day variations*, $\boldsymbol{S}_q$. On normal days or days with only minor disturbance, there is in addition the *solar disturbance daily variation*, $\boldsymbol{S}_D$. The variation $\boldsymbol{S}_D$ is part of the $\boldsymbol{D}$ field generally and only in the absence of storms and substorms will $\boldsymbol{D} \simeq \boldsymbol{S}_D$. The disturbance field $\boldsymbol{D}$ is defined as the difference $\Delta \boldsymbol{F}$ between the total field at any instant and the average field (over a month or more) after allowing for the $\boldsymbol{S}_q$ and $\boldsymbol{L}$ fields, so that

$$\boldsymbol{D} = \Delta \boldsymbol{F} - \boldsymbol{S}_q - \boldsymbol{L}$$

The intensity of $\boldsymbol{S}_D$ varies with the intensity of the general disturbance. Both the $\boldsymbol{S}_q$ and $\boldsymbol{S}_D$ fields vary with latitude and local time (that is their time variation is roughly constant relative to the meridian plane containing the sun). However both the latitude and time variations for the $\boldsymbol{S}_q$ and $\boldsymbol{S}_D$ fields are themselves very different (Fig. 2.13). In particular, whereas the horizontal component of the $\boldsymbol{S}_q$ field changes sign at about 30° latitude, that of the $\boldsymbol{S}_D$ field changes sign at about 55° latitude. Also the time variation in $\boldsymbol{S}_D$ is independent of the intensity of the disturbance, ranging from the very slight disturbances present on average days to the very intense ones present during magnetic storms (Fig. 2.13). For general discussion of $\boldsymbol{S}_q$, $\boldsymbol{S}_D$ and $\boldsymbol{L}$ see Chapman and Bartels (1940, 1962) and Matsushita (1967). Both $\boldsymbol{S}_q$ and $\boldsymbol{L}$ most likely arise from the tidal motion of the air in the ionosphere in a horizontal direction across the lines of force of the earth's magnetic field so inducing electric currents. The solar atmospheric tide is probably caused by thermal rather than gravitational action of the sun on the atmosphere. For this reason $\boldsymbol{S}_q$ is very much larger than $\boldsymbol{L}$ which is probably due to gravitational tides.

Along the magnetic equator within a narrow band of about 2° latitude there is an enhancement of the horizontal component of $\boldsymbol{S}_q$ by about a factor of two. The narrow equatorial current system inferred from this observation flowing from west to east on the sunward hemisphere was named the *equatorial electrojet* by Chapman (1951). Details of this effect and detailed theory are given by Onwumechilli (1967). The cause of the equatorial electrojet lies in a special equatorial feature of the electrical conductivity of the ionosphere. When there are crossed electric and magnetic fields a Hall

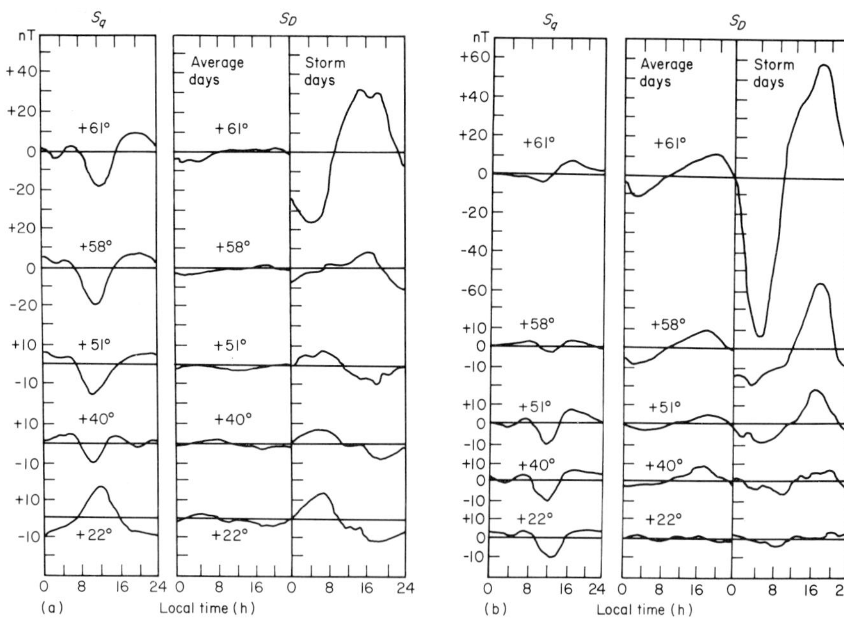

Fig. 2.13. Average S_q and S_D variations for various northern latitudes. In the southern hemisphere the vertical component variations would be reversed in sign. (a) Horizontal component; (b) vertical component. From Chapman and Bartels (1962).

current usually flows perpendicular to both the electric and magnetic fields, When the medium is bounded in the direction of the Hall current polarization results and opposes the flow. The result is enhanced conductivity of the medium. In the ionosphere the Hall current leads to polarization but this easily leaks away along the lines of force of the geomagnetic field, except in the region of the magnetic equator where the geomagnetic field is entirely horizontal. The resulting increased ionospheric conductivity is sufficient to account for the observed enhancement of $\boldsymbol{S}_q$.

Magnetic storms typically can be divided into three phases termed the initial, main and recovery phases. The initial phase may be gradual or be represented by an abrupt change called a *sudden commencement*. In addition to the $\boldsymbol{S}_D$ component (related to local time) there is a component related in time to the beginning of the storm termed the *storm-time variation*, $\boldsymbol{D}_{st}$, so that during magnetic storms

$$\boldsymbol{D} = \boldsymbol{D}_{st} + \boldsymbol{S}_D$$

There is also an irregular part $\boldsymbol{D}_i$ which only becomes significant at high latitudes. The main characteristic of magnetic storms is the reduction in the horizontal component of the geomagnetic field during the main phase (Fig. 2.14). The main phase of a storm is defined as beginning when the horizontal component of the $\boldsymbol{D}_{st}$ field decreases below its pre-sudden commencement value and ends when this reaches its maximum decrease. The initial and

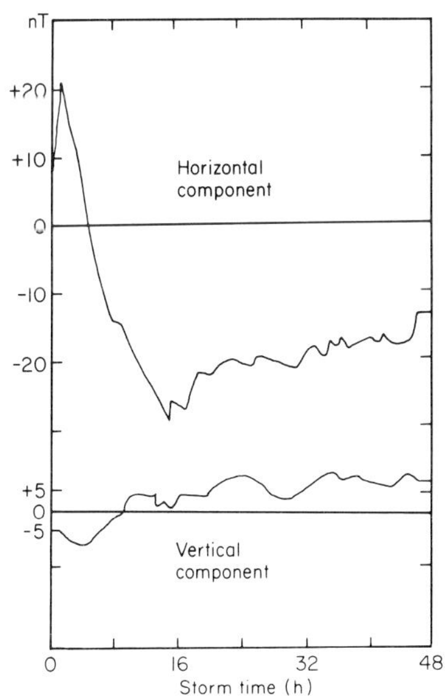

Fig. 2.14. Average D_{st} variation at 40°N. In the southern hemisphere the vertical component variations would be reversed in sign. From Chapman and Bartels (1962).

recovery phases are simply before and after the main phase. The magnitude of the reduction in the horizontal component of $\boldsymbol{D}_{st}$ during the main phase has a maximum at the equator and decreases to a minimum at about 60° latitude thereafter increasing rapidly towards the auroral zone. The actual observed changes in the horizontal component of the geomagnetic field H during a magnetic storm will depend very much on local time. The components $\boldsymbol{D}_{st}$ will have the form shown in Fig. 2.14, but the resulting geomagnetic field variation will depend on the phase of $\boldsymbol{S}_D$ at the beginning of the storm, in the limits either greatly enhancing the reduction in H or greatly diminishing it (see Chapman and Bartels, 1940, 1962).

The theory of magnetic storms has been intensely debated since the work of Chapman and Ferraro (1931, 1932, 1933). In the 1960s a model was developed that has been the dominant view of magnetic storms until very recently (Akasofu and Chapman, 1967; Parker, 1967; Piddington, 1967; Cole, 1967; Akasofu, 1979a, b, c). The model was built around the concept that when a transient plasma stream arrives at the boundary of the magnetosphere with a speed substantially greater than that of the steady solar wind, the outer region of the magnetosphere experiences a strong impact. It was argued that this resulted in the sudden compression of the magnetic field and plasma in that region which was presumably transmitted inward as a hydromagnetic perturbation and observed as a sudden change in the magnetic field at the earth's surface (the sudden commencement of a magnetic storm). During the last decade it has become increasingly clear that this model is vastly oversimplified. For example, although few scientists doubt that some large solar flares (which produce large transient effects on the plasma sheet) are correlated with magnetic storms, other large solar flares are not (Perreault and Akasofu, 1978).

One of the variables missed in the earlier magnetic storm models is the direction of the magnetic field carried along with the solar wind. Smith *et al.* (1978) showed that the magnetic field carried by the solar wind even changes sign in the vicinity of the earth due to a "warped" current sheet (see Chapter 12 for a more detailed discussion). The "reconnection" of the magnetic field carried by the solar wind and the earth's magnetic field depends vertically on the sign of the magnetic field (see Fig. 12.4). If reconnection occurs, then it is much easier to transfer electric charges between the solar wind and the earth's magnetosphere. Perreault and Akasofu (1978) and Akasofu (1979a, b) suggest that the energy coupling between the solar wind and the magnetosphere is expressed by:

$$\varepsilon = vB^2F(\theta)l_0^2 \tag{2.31}$$

where v = solar wind speed, B = magnitude of the interplanetary magnetic field (IMF), $F(\theta)$ is a function of θ, where $\theta = \tan^{-1}(|B_y/B_z|)$ for $B_z > 0$ and

$\theta = 180° - \tan^{-1}(|B_y/B_z|)$ for $B_z < 0$. Perreault and Akasofu argue that the first order approximation of $F(\theta)$ is $\sin^4\theta/2$. l_0 is a constant that is roughly 7 earth radii.

Although it is likely that the precise form of (2.31) is not yet well-known, this equation marks an important departure from the older and less acceptable models of magnetic storms. In any case, there is a consensus that a *geomagnetic storm* (measured at the earth's surface) is the magnetic manifestation of a *magnetospheric storm*. The critical variables so far identified are the velocity of the solar wind and the magnitude and the *direction* to the IMF.

The concept of the *magnetospheric substorm* was introduced with the thought that many geomagnetic phenomena were correlated and possibly tied to a single major cause. Magnetospheric substorms can be viewed as mini-magnetospheric storms. A magnetospheric storm can be viewed as a non-linear superposition of intense magnetospheric substorms (Akasofu, 1979c). Because of this, it has been suggested (Akasofu, 1979b) that (2.31) is also applicable to use for correlations between solar activity and substorms.

As an example consider one source of a substorm, that of auroral phenomena which occur in the higher latitude regions in the ionosphere; the maximum variation occurs in a geomagnetic latitude belt of 65° to 70°. The soft green light typically associated with aurora comes from the forbidden emission of oxygen precipitated by the bombardment of electrons from the outer radiation belts. Auroral events are usually accompanied by large-scale electric fields. These fields move charges around in a manner to produce large ionospheric currents. The strongest of these currents runs from dawn to dusk, peaks around midnight and is referred to as the *auroral electrojet* (Stern, 1977). The current can be so strong that magnetic fields at the earth's surface underneath it can change as much as 1000 nT, although typical variations are more like 200 and 300 nT. The deviations in the magnetic field observed at the earth's surface from auroral substorms resemble "bay-like" features and are frequently referred to as *magnetic bays*. Some of the common periods and amplitudes associated with geomagnetic phenomena are listed in Table 2.2.

Thus, there are magnetic substorms (describing magnetic disturbances), ionospheric substorms (describing ionospheric disturbances), auroral substorms, micropulsation substorms, particle substorms, etc. The implication is that the various substorms are local manifestations of some general disturbance as would be indicated by the use of (2.31) for both storms and substorms. However, it takes only a little reflection to realize how complicated and difficult it is in practice to try to trace back a local disturbance such as an auroral substorm to some "primary cause".

It now appears that to understand the origin of a geomagnetic storm that

TABLE 2.2
Periods and average amplitudes of various geomagnetic phenomena as observed in mid-latitudes

Phenomenon	Period	(amplitude in nT)
Micropulsations	1 ms–3 min	($\lesssim 1$)
Magnetospheric substorms and bays	1–2 h	(~ 10)
Solar quiet day variation, $\boldsymbol{S}_q$	24 h	(~ 20)
Solar disturbance daily variation, $\boldsymbol{S}_D$	24 h	(~ 5–20)
Lunar daily variation, $\boldsymbol{L}$	25 h	(~ 1)
Storm-time variation, $\boldsymbol{D}_{st}$		
(a) initial phase	~ 4 h	(~ 15)
(b) main phase	~ 8 h	(~ 35)
(c) recovery phase	~ 60 h	(~ 35)
External magnetic field	$\lesssim 4$ yr	
Internal magnetic field	$\gtrsim 4$ yr	
Sunspot cycle variation	11 and 22 yr	
Westward drift	2000 yr	

one must understand first the origins of magnetospheric substorms. During substorms a group of ring current particles are injected into the inner magnetosphere, forming a mini-ring current. A large ring current can be built up when intense injection occurs very frequently (Akasofu, 1979c). The geomagnetic storm is a manifestation of this ring current. "Magnetospheric dynamos" play a critical role in the more detailed processes involved. Unlike the earth dynamo processes considered in Chapters 7 to 10, it is often necessary to retain the electric displacement current in the work on magnetospheric dynamo processes. These processes locally intensify magnetic fields and on occasion non-linearly combine to produce magnetospheric storms.

Chapter Three

Fundamentals of Palaeomagnetism

3.1 Rock Magnetism

3.1.1 Types of Magnetization Acquired by Rocks

When a rock forms it usually acquires a magnetization parallel to the ambient magnetic field (normally presumed to be the earth's magnetic field) and this is referred to as a *primary magnetization*. This primary magnetization then provides information about the direction and intensity of the magnetic field in which the rock formed. However, there are numerous pitfalls that await the unwary; first, in sorting out the primary magnetization from the *secondary magnetization* (acquired subsequent to formation) and secondly, in extrapolating the properties of the primary magnetization to those of the earth's magnetic field. For more detailed texts on rock magnetism readers are referred to Fuller (1970) and Stacey and Banerjee (1974).

The most common magnetic minerals in rocks are either *antiferromagnetic* or *ferrimagnetic*. In such substances negative exchange energy dominates the magnetic ordering and the adjacent antiparallel magnetic moments are either equal in magnitude (antiferromagnetic) or unequal in magnitude (ferrimagnetic). The temperature at which the exchange energy decreases sufficiently so that the magnetic ordering disappears is called the *Néel temperature*. A ferrimagnetic substance exhibits spontaneous magnetization (and may therefore have remanence) and in this case the Néel temperature is usually referred to as the *Curie temperature* as in normal ferromagnetic substances like iron.

When an igneous rock cools from a temperature above the Curie temperature of its magnetic minerals in an external magnetic field (such as the earth's magnetic field), it acquires a remanent magnetization referred to as *thermoremanent magnetization* or *TRM*. Technically speaking a *remanent magnetization* or *RM* is the net magnetization present in zero magnetic field. In a magnetically isotropic rock the TRM is aligned parallel (or in rare cases antiparallel) to the external field in which the rock cooled. Although essentially all magnetic minerals exhibit magnetic anisotropy (that is there are preferred directions of magnetization in any magnetic mineral grain), by far the majority of igneous rocks are magnetically isotropic because the magnetic minerals they contain are randomly oriented. When the TRM is produced antiparallel to the external field the rock is said to have *self-reversed*. Self-reversal was predicted on theoretical grounds by Néel (1955), but in practice appears to be rare in nature. The first discovered self-reversing rock was studied extensively by Uyeda (1958). The numerous occurrences of rocks with magnetization almost exactly opposed to the direction of the present earth's magnetic field are now known to be due almost entirely to the fact that the field has reversed its polarity many times in the past. The evidence relating to this and a full discussion of self-reversal mechanisms in rocks is given in Chapter 5.

Unfortunately when palaeomagnetists measure the remanent magnetization of an igneous rock it is often not a pure TRM. For this reason the

TABLE 3.1
Some common types of remanent magnetizations

Thermoremanent Magnetization (TRM): That RM acquired by a sample during cooling from a temperature above the Curie temperature in an external magnetic field (usually in a weak field such as that of the earth)

Chemical Remanent Magnetization (CRM): That RM acquired by a sample during a chemical change in an external magnetic field

Viscous Remanent Magnetization (VRM): That RM acquired over a long time in an external magnetic field

Isothermal Remanent Magnetization (IRM): That RM acquired in a very short time at one temperature (usually room temperature) in an external magnetic field (that is usually strong)

Anhysteretic Remanent Magnetization (ARM): That RM acquired in a weak steady magnetic field, when an alternating magnetic field is decreased from some large value to zero. (This definition differs from that used in magnetic tape research in which the intensity of the steady field is also decreased to zero with the alternating field)

Natural Remanent Magnetization (NRM): That RM acquired by a sample under natural conditions

Depositional Remanent Magnetization (DRM): That RM acquired by sediments when grains settle out of water in the presence of an external magnetic field

Post-depositional Remanent Magnetization (post-DRM): That RM acquired by physical processes that cause translation or rotation of sedimentary grains after deposition

magnetization first measured in the laboratory is referred to as the *natural remanent magnetization* or *NRM*. The NRM may consist of primary and secondary magnetizations, the latter arising from subsequent chemical alteration, relaxation effects, etc. Well over 20 different kinds of remanent magnetizations have been proposed. The definitions of the most common forms of RM are given in Table 3.1. Of these palaeomagnetists are most interested in the primary magnetization which is usually a TRM in igneous rocks, *depositional remanent magnetization* (*DRM*), *post-depositional RM* or occasionally *chemical remanent magnetization* (CRM) in sedimentary rocks.

3.1.2 Magnetic Hysteresis

Some application of the origin and properties of remanent magnetization in rocks may be obtained by considering magnetic hysteresis. Consider an oversimplified example of uniformly magnetized (single domain) grains that are identical and non-interacting (the magnetic field produced by the magnetization of any grain is negligibly small at the site of any other grain). Assume that the grain possesses some uniaxial magnetic *anisotropy energy*, E_{an} per unit volume where

$$E_{an} = K \sin^2 \theta \tag{3.1}$$

K is defined as the uniaxial anisotropy constant and θ is the angle the magnetization makes with the z axis in a cartesian co-ordinate system (x, y, z). z is an *easy axis of magnetization* and the x–y plane is a hard anisotropy plane. The anisotropy energy is lowest for $\theta = 0$ and largest for $\theta = \pi/2$. The physical origins of magnetic anisotropy are discussed briefly in §3.1.3, but anisotropy of some type is a necessary, but not sufficient, condition for a grain to possess a remanent magnetization.

There will be an extra energy associated with a grain having magnetic moment $\boldsymbol{M}$ in an external field $\boldsymbol{H}$ when $\boldsymbol{M}$ and $\boldsymbol{H}$ are not parallel. From Fig. 3.1 the energy originating from the interaction of the grain's magnetization and the external field is given by:

$$E = K \sin^2 \theta - M_s H \cos(\theta - \alpha) \tag{3.2}$$

where M_s is the saturation magnetization (the highest a grain can have) and is used because it is assumed that all the grains are uniformly magnetized. The grain magnetization is $M = M_s V$, where V is the grain volume. Equilibrium is obtained by setting $\mathrm{d}E/\mathrm{d}\theta = 0$ and there is a minimum energy state if $\mathrm{d}^2E/\mathrm{d}\theta^2 < 0$. Consider the situation when $\boldsymbol{H}$ is along the positive z axis $(\alpha = 0)$. It is easily seen that $\theta = 0$ is a minimum value. Suppose all grains are magnetized parallel to $+z$ near absolute zero temperature. This can be

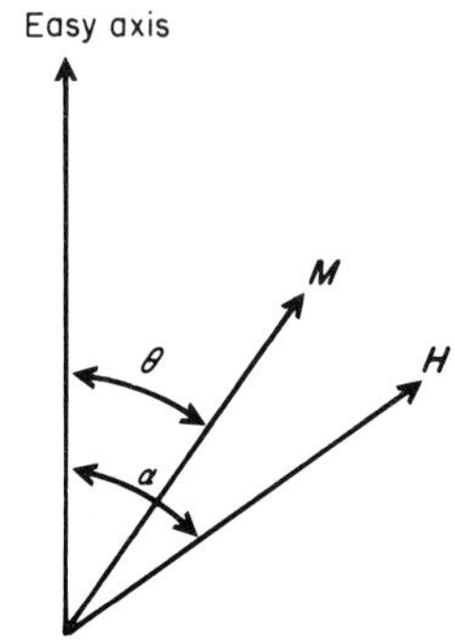

Fig. 3.1. The geometry of the easy axis of magnetization (vertical axis), the magnetization direction (M) and the magnetic field (H).

achieved by increasing $\boldsymbol{H}$ enough so that the magnetization is "saturated". If $\boldsymbol{H}$ is now reduced to zero, the magnetization will retain its saturation value, M_s. That is, the saturation isothermal remanent magnetization, M_{rs}, will equal M_s (Fig. 3.2). If $\boldsymbol{H}$ is now increased in the opposite direction ($-z$; $\alpha = 180°$ in (3.2)) then it is easily shown that an energy minimum still exists at $\theta = 0$ for $H < 2K/M_s$. However, for $H > 2K/M_s$, the $\theta = 0$ direction becomes a maximum energy direction and the magnetization will change to $\theta = 180°$. This produces half of the rectangular hysteresis loop shown in Fig. 3.2a. The origin of the other half is obvious from consideration of the symmetry. However, if $\boldsymbol{H}$ is applied at right angles to z ($\alpha = 90°$ in (3.2)) the conditions for an energy minimum are:

$$\sin\theta = \frac{M_s H}{2K} \quad \text{for} \quad |H| \le \frac{2K}{M_s}$$

$$\theta = \pi/2 \quad \text{for} \quad |H| \ge \frac{2K}{M_s}$$

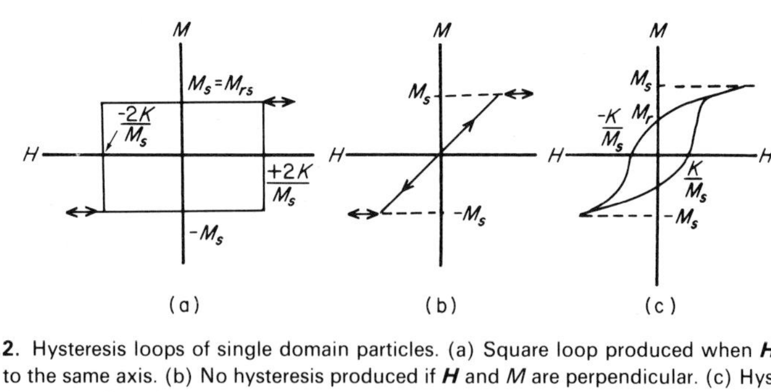

Fig. 3.2. Hysteresis loops of single domain particles. (a) Square loop produced when $\boldsymbol{H}$ and M are confined to the same axis. (b) No hysteresis produced if $\boldsymbol{H}$ and M are perpendicular. (c) Hysteresis loop produced by a set of randomly oriented grains.

The magnetization along the z axis is:

$$M = M_s \sin\theta = \frac{M_s^2 H}{2K} \quad \text{for} \quad |H| \leq \frac{2K}{M_s}$$

Therefore, there is no hysteresis and no remanence, as shown in Fig. 3.2b. Suppose one is dealing with an assembly of uniaxial single domain grains that have random orientations of their easy axes. Their integration over all possible angles $\boldsymbol{H}$ can make with the easy axis is required. This produces the hysteresis loop given in Fig. 3.2c. The M_{rs}/M_s value is around 0.5 and the coercivity H_c (the value of H that reduces the magnetization to zero) is K/M_s. M_{rs} is the saturation remanent magnetization.

TABLE 3.2

Some hysteresis loop parameters that are useful in distinguishing single domain (SD) from multidomain (MD) grains. M_s is the saturation magnetization. M_{rs} is the saturation remanent magnetization (that remaining after the sample has been exposed to a saturation field). H_c is the bulk coercivity (the magnetic field required to reduce an initially magnetized sample to zero net magnetization in the presence of the external field). H_{rc} is the remanent coercive force (the field required to reduce the remanence to zero, in zero external field)

	SD	MD
$\frac{M_{rs}}{M_s}$	Ideally 0.5 in an isotropic sample but typically between 0.3 and 0.5	Less than 0.1 (and often much less)
$\frac{H_{rc}}{H_c}$	Between 1 and 1.2	Variable but typically much higher (~5) than SD grains

Only small grains are uniformly magnetized. Larger grains are non-uniformly magnetized because they are typically multidomain grains (see §3.1.5), and parameters of their hysteresis loops are often substantially different from single domain grains. Thus hysteresis parameters are often useful for determining the magnetic character of a sample and for distinguishing grains with single domain assemblages from grains with multidomain assemblages (see Table 3.2), as was pioneered by Stoner and Wohlfarth (1948) and extended to geophysics by Wasilewski (1973), Fuller (1974), Day *et al.* (1976, 1977) and Day (1977). The method is not without its shortcomings however as is discussed more fully by Day (1977) and Levi and Merrill (1978). Other methods are therefore often used to supplement the hysteresis loop analysis for characterizing the state of remanence (see Lowrie and Fuller (1971); Johnson *et al.* (1975a); Levi and Merrill (1978)).

3.1.3 The Demagnetizing Field and Magnetic Anisotropy

Once a sample has been magnetized it produces its own internal magnetic field that tends to oppose the magnetization. This arises from the shape of the body because it is more easily magnetized in some directions than others. Because the demagnetizing field of non-uniformly magnetized grains is complicated, the discussion will be restricted to the consideration of uniformly magnetized grains (Merrill, 1977, 1981). The demagnetizing field $\boldsymbol{H}_d$ is uniform only in simple shapes such as ellipsoids and is usually expressed in the form

$$\boldsymbol{H}_d = \bar{\boldsymbol{N}} \cdot \boldsymbol{M}_s$$

where $\bar{\boldsymbol{N}}$ is the demagnetization tensor. In many cases such as when the magnetization is along one of the major axes of an ellipsoid, $\bar{\boldsymbol{N}}$ reduces to a scalar N, referred to as the *demagnetizing factor*. Consider the case when rectangular co-ordinates are aligned along the axes of an ellipsoidal grain that is uniformly magnetized. If N_x, N_y and N_z are the appropriate demagnetizing factors for the ellipsoid then it can be shown (Chikazumi, 1964) that

$$N_x + N_y + N_z = 1$$

Thus for a sphere $N_x = N_y = N_z = 1/3$, for an infinite cylinder $N_z = 0$ and $N_x = N_y = 1/2$ (where z lies along the cylinder axis) and for a thin disc $N_z = 1$ and $N_x = N_y = 0$ (where z lies perpendicular to the disc).

Because the demagnetizing field depends on the direction of magnetization for all shapes that are not spheres, it is easier to magnetize samples along those directions in which the demagnetizing field is smallest. For example it is easier to magnetize the infinite cylinder above along the z axis than perpendicular to it. This is one way in which magnetic anisotropy can originate and is referred to as *shape anisotropy*. Consider a grain in the shape of a prolate spheroid with major axis in the z direction and let it have magnetization M_s at an angle θ to z. The energy associated with the demagnetizing field is called the *magnetostatic energy* and for the spheroid example, is given by (Cullity, 1972)

$$E_{ms} = \tfrac{1}{2}M_s^2 N_z + \tfrac{1}{2}(N_x - N_z)M_s^2 \sin^2\theta \tag{3.3}$$

The angle dependent term has the same form as (3.1) and this defines the shape anisotropy constant K_s as:

$$K_s = \tfrac{1}{2}(N_x - N_z)M_s^2 \tag{3.4}$$

For a spherical grain $N_x = N_z$ and $K_s = 0$. The shape anisotropy constant is thus determined by the axial ratio of the grain and its magnetization. Recall that the coercivity of a uniaxial grain is given by:

$$H_c = \frac{2K}{M_s} \tag{3.5}$$

so that the coercivity of a grain due to its shape anisotropy is given by

$$H_c(\text{shape}) = (N_x - N_z)M_s \tag{3.6}$$

Besides shape anisotropy there are other forms including *magnetostrictive* (or *stress*) *anisotropy*, *magnetocrystalline anisotropy* and occasionally *exchange anisotropy*. Exchange anisotropy exists when there is exchange coupling between two different magnetic phases and will not be discussed further. Readers interested in learning more about anisotropy are referred to books such as Chikazumi (1964), Morrish (1965), Cullity (1972) and Stacey and Banerjee (1974).

Magnetocrystalline anisotropy is the only anisotropy that exists in an idealized infinitely large crystal free of defects and therefore it is sometimes referred to as an intrinsic anisotropy. Although in principle its origin is understood theoretically (it originates from three sources: spin-spin interaction, relativistic spin-orbit coupling, and ionic or atomic distortions), in practice it must be determined experimentally. It is described by a magnetocrystalline anisotropy tensor, $\bar{\boldsymbol{K}}$. for a uniaxial crystal the coercivity is simply related to the first magnetocrystalline anisotropy constant K_1 and from (3.5):

$$H_c(\text{magnetocrystalline}) = \frac{2K_1}{M_S} \tag{3.7}$$

Magnetostrictive (or stress) anisotropy originates from internal stress, such as would exist around dislocations. Like magnetocrystalline anisotropy its origin is understood in principle (see Mathis, 1965) but in practice experimental values must be used. Even an ideal crystal undergoes some strain when it is magnetized. Similarly any internal stress $\bar{\boldsymbol{\sigma}}$ can be resolved into a strain $\bar{\boldsymbol{\varepsilon}}$ not associated with the magnetization and a strain $\bar{\boldsymbol{\lambda}}$ directly associated with the magnetization. $\bar{\boldsymbol{\lambda}}$ is referred to as the magnetostrictive anisotropy tensor and is determined experimentally (Chikazumi, 1964). When the magnetostriction is isotropic and there is an angle θ between M_s and the internal stress σ there is a *magnetoelastic energy*, E_{me} given by (Cullity, 1972)

$$E_{me} = \tfrac{3}{2}\lambda\sigma \sin^2 \theta \tag{3.8}$$

From (3.1) and (3.5) the corresponding coercivity due to magnetostriction is

$$H_c(\text{stress}) = \frac{3\lambda\sigma}{M_s} \tag{3.9}$$

3.1.4 Single Domain Theory for TRM

In 1949 Néel presented a theory for an assembly of identical non-interacting uniformly magnetized grains with uniaxial anisotropy that remains at the foundations of rock magnetism. Although the theory has some defects (which are not discussed here), it does provide valuable insight into many of the observed phenonomena in rock magnetism.

For simplicity consider the case when the applied external field H is along the easy anisotropy axis of the grains. At all times it is assumed that the grains are uniformly magnetized (including during changes in direction). Under these assumptions there are two equilibrium magnetic states for the magnetization in the grain. The lowest state has energy $-VM_sH$ and the highest state has energy $+VM_sH$. Using Boltzman statistics for n identical particles the number of grains in the upper, n_u, and lower, n_l, states is

$$n_u = \frac{n\,e^{-(VM_sH)}/kT}{e^{+(VM_sH)}/kT + e^{-(VM_sH)}/kT} \tag{3.10}$$

$$n_l = \frac{n\,e^{+(VM_sH)}/kT}{e^{+(VM_sH)}/kT + e^{-(VM_sH)}/kT} \tag{3.11}$$

where k is the Boltzman constant and T is temperature. The magnetization is given by:

$$M = n_l VM_s - n_u VM_s = nVM_s \tanh\left(\frac{VM_sH}{kT}\right) \tag{3.12}$$

where (3.10) and (3.11) have been used for n_l and n_u. Equilibrium conditions have been invoked by using Boltzman statistics (small relaxation times) and, in analogy with paramagnetism (which considers magnetic moments associated with individual atoms, not grains), the grains are called *superparamagnetic*.

At this point in the theory Néel (1949) made a crucial assumption, he assumed that there is a *blocking temperature* above which the grain assembly is in equilibrium, but below which the magnetization is locked into the sample. Therefore the room-temperature value for TRM, M_T, can be obtained from (3.12) and is:

$$M_T = nVM_s \tanh\left(\frac{VM_sH}{kT}\right)_{\text{blocking}} \tag{3.13}$$

where "blocking" refers to the fact that the parameters in the argument of the hyperbolic tangent are to be evaluated at the blocking temperature.

An important second contribution from Néel was the explanation as to why the TRM might sometimes be stable over long time periods. He pointed out that dn_l/dt would be negative once the external field was removed due to thermal fluctuations causing some of the magnetic moments to pass over the anisotropy barrier which has an energy height of $VM_sH_c/2$. dn_l/dt is proportional to $(n_u - n_l)$, because it will increase for moments traversing the barrier from the initially higher energy state and will decrease in the opposite case, so that:

$$\frac{dn_l}{dt} = \frac{1}{2\tau}(n_u - n_l)$$

$$\frac{dn_u}{dt} = \frac{1}{2\tau}(n_l - n_u)$$

where 2τ is the constant of proportionality. Subtracting the second equation from the first, using (3.12), and integrating:

$$M = M_0\,e^{-t/\tau} \tag{3.14}$$

where M_0 is the initial remanence and τ is now seen to be a relaxation time. To obtain information on the geomagnetic field at some time before present, t_p, it is necessary to have rocks that contain some magnetic grains which have relaxation times greater than t_p.

From Boltzman statistics, $1/\tau$ must be proportional to the probability of transversing the anisotropy barrier or:

$$\tau = \frac{1}{f}\exp\frac{VM_sH_c}{2kT} \tag{3.15}$$

where f has the dimensons of frequency. Both Néel (1949) and Brown (1959) derived different relationships for f, but both are similar in that f only varies slowly as a function of temperature. For most magnetic minerals in rocks f can be taken to be of the order of 10^8 to 10^{10} s. In the case when a magnetic field H is present and along the easy axis, then the relaxation time of remanence towards and away from that field are respectively (Néel, 1949):

$$\tau_- = \frac{1}{f}\exp\frac{VM_s(H_c - H)^2}{2kTH_c}$$

$$\tau_+ = \frac{1}{f}\exp\frac{VM_s(H_c + H)^2}{2kTH_c}$$

It is readily seen by substituting reasonable values for elongated single domain magnetite into (3.15) and using the above value for f, that a sample with a relaxation time of 1 s at 550°C (30° below magnetite's Curie temperature) will have a relaxation time longer than the age of the earth at room temperature. These results indicate the reasonableness of the blocking temperature concept at the first order approximation level and indicate why it is possible for rocks to retain a record of the ancient field directions for millions and even billions of years.

Taking f to be constant and selecting some relaxation time, τ_0 (say 1 year $\approx 3 \times 10^7$ s) as indicating blocking, then the blocking temperature T_b, can be obtained from (3.15) and is:

$$T_b = \frac{VM_sH_c}{2K \log_e (f\tau_0)}$$

where V, M_s and H_c are to be evaluated at T_b. Because H_c, V and M_s can vary considerably between individual grains in an actual rock sample, one typically finds a spectrum of blocking temperatures present. Therefore, there is a range of temperatures over which the TRM is "locked into" the rock. Crudely, for a mineral such as magnetite half of the TRM would be stabilized during cooling in the first 150° or so below its 580° Curie point. Many of the over-simplified assumptions in the single domain theory given above can, and have been, relaxed. Various contributions to single domain theory made since Néel's (1949) fundamental work have been reviewed by Stacey and Banerjee (1974).

3.1.5 Pseudo-single Domain and Multidomain Grains

Although the theory in §3.1.4 is widely referred to and well respected, it has never been completely verified even for single domain particles (see Dunlop and West, 1969). In addition, it has long been recognized that larger grains are not uniformly magnetized. In the case of the mineral magnetite (Fe_3O_4), the size of equant grains that are uniformly magnetized is roughly one order of magnitude smaller than can be optically observed (Butler and Banerjee, 1975). Because larger magnetic grains are common in rocks, it is important to try to understand the origin and properties of their magnetization.

The reason grains become non-uniformly magnetized can be traced to their demagnetizing fields discussed in §3.1.3. A uniformly magnetized grain with magnetic moment M_sV (where V is the grain volume) has a magnetostatic energy (energy of self demagnetization) of $\frac{1}{2}V(\boldsymbol{M}_s \cdot \boldsymbol{H}_d)$. This energy is directly proportional to the grain volume. As V becomes larger and larger (and depending on the grain shape), the magnetostatic energy becomes too large for the grain to be uniformly magnetized, and subdivision occurs because the energy is decreased. For large grains a multidomain (MD) configuration is realized (Kittel, 1949) as is illustrated in Fig. 3.3. The magnetization in each domain is in a uniform direction. A narrow transition zone (a few tens to a few hundreds of angstroms wide, depending on such factors as mineralogy, temperature, etc.), referred to as a *domain wall*, separates adjacent domains.

The actual domain configuration in a real situation is far more complicated than shown in Fig. 3.3 and depends on the details of the magnetocrystalline anisotropy, the distributions of internal defects, shape, pressure, temperature, and even the magnetic history of the grain (Stacey and Banerjee, 1974; Levi and Merrill, 1978). The magnetic moments in the domains will generally lie along easy magnetocrystalline anisotropy directions. On the other hand, the magnetic moments in the walls typically have a component of magnetization directed in a hard magnetocrystalline anisotropy direction. This leads to a domain wall energy that would not be present if the grain were uniformly magnetized. It is the balancing of this domain wall energy with the magnetostatic energy that, in the first approximation, determines how large the grain volume must be to become multidomain (Kittel, 1949).

A remanence can exist in the multidomain grains shown in Fig. 3.3 if one or more of the domain walls is displaced from the "demagnetized" state shown. In some cases there are sufficient defects present that the wall cannot move back to its demagnetized central position when the external field is removed. In such cases the MD grain will exhibit a remanence. Other possible ways in which a remanence might exist in a MD grain are discussed further by Stacey and Banerjee (1974). Basically therefore, magnetizations can exist in a wide range of grains ranging from single domain size to multidomain size. The transition from single domain size to multidomain size for a given mineral will clearly depend on many factors, such as mineralogy (because the saturation magnetization, magnetic anisotropies and exchange energy are all mineralogically dependent), shape (which affects the demagnetization field) and temperature. (The role of many of these factors with respect to magnetite are discussed by Butler and Banerjee, 1975.)

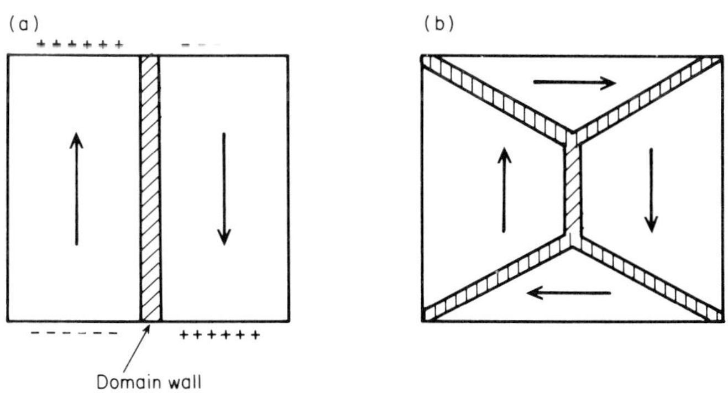

Fig. 3.3. Two hypothetical multidomain states. (a) A two-domain grain separated by a 180° domain wall. The magnetostatic energy of this grain is roughly one-half that of a uniformly magnetized grain of the same size and shape, but there is an additional domain wall energy. (b) A four-domain grain with closure domains (four 90° walls and one 180° wall). The magnetostatic energy is now zero, but there is an increase in domain wall energy compared with that in (a).

There is considerable controversy over the nature of the magnetic state in grains that have volumes slightly larger than that expected theoretically for single domain behaviour. The single domain theory of §3.1.4 predicts that the relaxation time, τ, varies exponentially with V. On the other hand, τ is found experimentally to decrease with V for truly multidomain material, so that a relaxation time versus volume curve will have a maximum (all other factors constant), as shown in Fig. 3.4. The region that separates the single domain

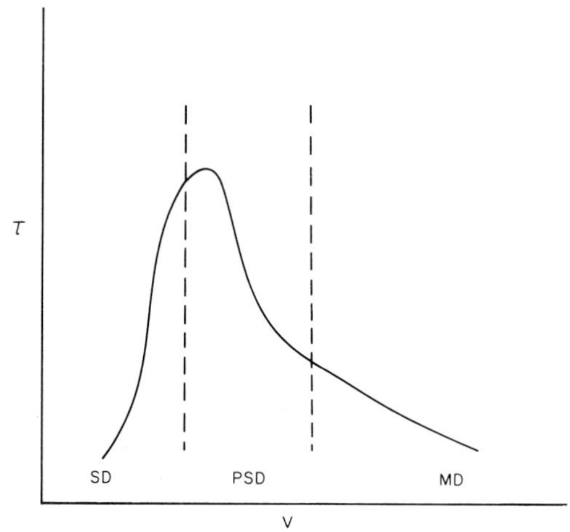

Fig. 3.4. Hypothetical change in relaxation time, τ, as a function of volume, V, all other factors held constant. The shape of the curve is not well known (see Banerjee, 1977; Levi and Merrill, 1978), but it appears to have a single maximum. SD = single domain, PSD = pseudo single domain, MD = multidomain.

(SD) from the multidomain (MD) configuration is sometimes referred to as the *pseudo-single domain* (PSD) region. The origin of the remanence in this region is still controversial and is very important because some of the PSD grains appear to have very large relaxation times and contribute significantly to the remanence. Schmidt (1973) supposes that PSD behaviour is nothing more than MD behaviour in grains with only two domains or so, and although *some* remanence will certainly arise this way, it is generally agreed that this cannot account wholly for observed PSD behaviour (see Fuller, 1970).

A considerable number of investigations, led principally by Dunlop (1973, 1977; also see Dunlop *et al.*, 1974) and Stacey and Banerjee (1974; see also Banerjee, 1977) have been made into PSD behaviour (see also a special symposium on TRM published in the *J. Geomag. Geoelect.* **29**, 1977). The most fashionable model at present (which evolved primarily from the work of

Dunlop, Stacey and Banerjee) is that small PSD grains consist of part, or all, of a domain wall. A second model for PSD behaviour involves small closure domains that are argued to act independent of the main domain structure (Stacey and Banerjee, 1974; also see Banerjee, 1977). In both models a theory similar to Néel's SD theory is utilized, although problems with this theory still exist (see Day, 1977; Banerjee, 1977). Similarly there is still no satisfactory MD theory (see Néel, 1955; Verhoogen, 1959; Stacey, 1963; Fuller, 1970; Lowrie and Fuller, 1971; Schmidt, 1973, 1976; Dunlop and Waddington, 1975; Merrill, 1977, 1981; Day, 1977), although most rock magnetists now believe that very large MD grains do not usually contribute much to the *stable* remanence in rocks.

3.2 Magnetic Mineralogy

3.2.1 Properties of Magnetic Minerals

The magnetic minerals that can contribute significantly to the remanence in rocks include iron-titanium oxides, iron-manganese oxides and oxyhydroxites, iron sulphides, and iron and nickel-cobalt alloys. The purpose here is simply to give a few examples of these minerals and properties so that those not previously familiar with rock magnetism can gain some appreciation of the complexities and variations that arise from mineralogy. More detailed magnetic mineralogy can be found in reviews by Nagata (1961), Stacey and Banerjee (1974) and Haggerty (1976). A list of the properties of many magnetic minerals is given in Table 3.3.

For simplicity discussion is restricted to the Fe-Ti-oxides given in the ternary system in Fig. 3.5. The members of the solid solution series $x\mathrm{Fe_3O_4}$ $(1-x)\mathrm{Fe_2TiO_4}$ are referred to as *titanomagnetites* and those of the series $x\mathrm{Fe_2O_3}(1-x)\mathrm{FeTiO_3}$ are referred to as *titanohematites*. That titanomagnetite forms a *solid solution series* means that $\mathrm{Fe_2TiO_4}$ and $\mathrm{Fe_3O_4}$ are mutually soluble in each other. The Curie temperature of ferrimagnetic magnetite, an inverse spinel which has a saturation magnetization value of 480×10^3 A m^{-1} (480 emu cm^{-3}), is 580°C and to the first order approximation, the Curie temperature decreases linearly with mole percent ulvöspinel in solid solution to a value near −150°C for antiferromagnetic ulvöspinel (Nagata, 1961). Similarly, the amount of ilmenite also nearly linearly decreases the Curie point in the titanohematite series from 680°C for hematite (the Curie temperature of hematite is still controversial; see Fuller, 1970) to near −223°C for ilmenite (Nagata, 1961). Although at high temperatures both the titanomagnetites and the titanohematites form solid solution series, at lower temperatures *exsolution* occurs (unmixing of phrases; so that in the titanomagnetites exsolution would result in an ulvöspinel-rich phase and a magnetite-rich phase). The temperature of the *solvus* (the maximum temperature for exsolution) for both the titanohematites and titanomagnetites are controversial. For further discussion see reviews by Merrill (1975) and Haggerty (1976, 1978).

In addition to exsolution, oxidation commonly occurs after crystallization, but still at high temperatures above the Curie temperature, in terrestrial magnetic minerals. Fundamental work on the mineralogy associated with this oxidation was first carried out by Buddington and Lindsley (1964) and has continued for many years (see Lindsley, 1976). This work assumes thermodynamic equilibrium conditions, and needs slight adjustments for the non-equilibrium conditions that seem to exist in nature (Sato and Wright, 1966; Grommé *et al.*, 1969). In addition, low temperature oxidation (below 200°C or so) of titanomagnetites produces titanomaghemites, *cation deficient spinels* (vacancies sit on some of the lattice sites occupied before oxidation by cations), the compositions of which lie in the field between titanomagnetites and titanohematites in Fig. 3.5 and have been characterized magnetically by Readman and O'Reilly (1970, 1972), Ozima and Sakamoto (1971), Nishitani (1979) and Keefer and Shive (1981). Palaeomagnetists often use a useful empirical classification scheme for high temperature oxidation developed by

TABLE 3.3

Magnetic properties of some common minerals (M_s = saturation magnetization in units of 10^3 A m^{-1}; T_c = Curie (or Néel) temperature)

Mineral	Composition	Magnetic state	M_s	T_c (°C)
Magnetite	Fe_3O_4	Ferrimagnetic	476	580
Ulvöspinel	Fe_2TiO_4	Antiferromagnetic		−153
Hematite	αFe_2O_3 (hexagonal)	Antiferromagnetic with a parasitic ferromagnetism	~2.2	~680[a]
Ilmenite	$FeTiO_3$	Antiferromagnetic		−233
Maghemite	γFe_2O_3 (cubic)	Ferrimagnetic	426	~600
Pyrrhotite	$Fe_{1-x}S (0 < x < 1/7)$	Ferrimagnetic	90	~320
Troilite	FeS	Antiferromagnetic		~305
Jacobsite	$MnFe_2O_4$	Ferrimagnetic	424	~300
Geothite	$\alpha FeOOH$	Antiferromagnetic with a parasitic ferromagnetism	~2(?)	~120
Iron	Fe	Ferromagnetic	1714	770
Cobalt	Co	Ferromagnetic	1422	1131
Nickel	Ni	Ferromagnetic	484	358
Awaruite	Ni_3Fe	Ferromagnetic	950	620
Wairauite	CoFe	Ferromagnetic	1936	986

[a] Estimates vary from 675° to 725°C. More precise experiments done this decade suggest 696°C rather than 680°C.

Wilson and Haggerty (1966) and Ade-Hall *et al.* (1971) and a low-temperature scheme developed by Haggerty (1976) and Johnson and Hall (1978).

Reviews of the magnetic effects associated with chemical changes have been given by Merrill (1975) and Henshaw and Merrill (1980) who show that sometimes chemical changes of magnetic minerals lead to pronounced magnetic changes and at other times little magnetic effects occur at all. Although many of these variations can be explained theoretically (Merrill, 1975; Henshaw and Merrill, 1980), often the variations depend on such subtle effects as minute exsolution, grain growth, etc., that they are sometimes difficult to detect in practice, as is clearly shown by the work of Haigh (1958), Kobayashi (1961), Strangway *et al.* (1968), Larson *et al.*

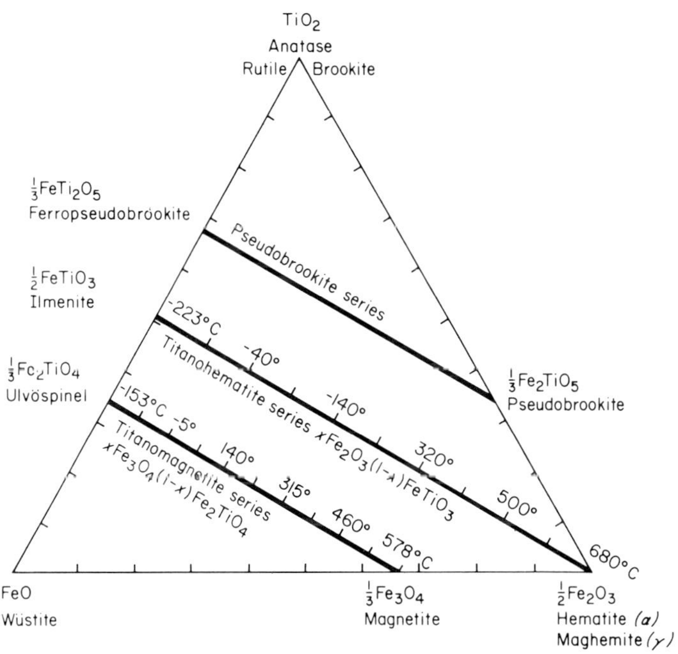

Fig. 3.5. FeO–TiO_2–Fe_2O_3 ternary system showing the three principal solid solution series found in igneous rocks. Members of the pseudobrookite series are all paramagnetic above liquid oxygen temperatures and therefore have little significance. Approximate Curie (or Nèel) temperatures for various values of mole function x are indicated for the titanomagnetite and titanohematite series. In the titanomagnetite series M_s decreases with decreasing x but in a complicated way depending on ordering of Fe^{2+}, Fe^{3+} and Ti^{4+} on two possible lattice sites. In the titanohematite series there is complex magnetic variation with x. For $1 \geq x \gtrsim 0.5$ the Ti ions are disordered and the magnetization weak ($\sim 2 \times 10^3$ A m^{-1}); for $0.5 \gtrsim x \gtrsim 0.2$ the Ti ions are ordered and the material ferrimagnetic with a maximum magnetization near 20×10^3 A m^{-1}; for $0.2 \gtrsim x > 0$ the material is antiferromagnetic with ferrimagnetic clusters.

(1969), Evans and McElhinny (1969) and Evans and Wayman (1974). There is little doubt that some remagnetization due to chemical changes occurs in nature, but the extent of this occurrence is very controversial.

The titanohematite system well illustrates the complexities of magnetic minerals in rocks. The higher temperature end member hexagonal (rhombohedral) hematite (commonly found in red sandstones), is basically antiferromagnetic on which a weak or "parasitic" ferromagnetism is imposed. Above $-15°C$ the magnetic moments lie in planes (basal planes) perpendicular to the C axis (Nagata, 1961). The coupling between moments in adjacent basal planes is essentially antiferromagnetic. However, a slight canting of the spins in the basal planes results in a small moment (weak ferromagnetism) that is perpendicular to the basal spins and is referred to as the spin canted moment (Dzyaloshinski, 1958). In addition, there are defect moments parallel to the spin directions that probably originate from partial ordering of the defects on every other basal plane (a second source of weak ferromagnetism). The saturation magnetization is somewhat variable for hematite, but is approximately 2.2×10^3 A m^{-1} (2.2 emu cm^{-3}) at room temperature. Below $-15°C$ (the Morin transition) the easy axis of magnetization becomes the C axis, the spin-canted moment vanishes leaving only the defect moment, a fact exploited by Fuller (1970) and Dunlop (1970) to study the origin and properties of hematite's weak ferromagnetism.

Both the defect and the spin canted moments seem to persist when Ti is added to the system, although the Morin temperature decreases rapidly with small amounts of added Ti. Broadly speaking, around $x = 0.5$ in the $xFe_2O_3(1 - x)FeTiO_3$ system there is an ordering phenonomen. For $x \approx 0.5$ the Ti ions are preferentially ordered on alternating basal planes (the symmetry group is $3\bar{R}C$), while for $x < 0.5$ the Ti ions are disordered (symmetry group $3\bar{R}$). The room-temperature saturation magnetization increases nearly 50-fold on this ordering, as the ordered phase is ferrimagnetic. At the other end of the solid solution series, $FeTiO_3$ is antiferromagnetic. Therefore, the system varies from antiferromagnetism with a weak ferromagnetism to ferromagnetism and finally to antiferromagnetism as the Ti content in the titanohematite is increased (Ishikawa and Akimoto, 1958; Uyeda, 1958; Nagata, 1961; Ishikawa and Syono, 1963; Allan and Shive, 1974). The room-temperature saturation magnetization varies in titanohematites from zero to a value around 1/4 that of magnetite. Some palaeomagnetists have erroneously dismissed the magnetic role titanohematites can have in rocks, because they considered only the end members of the solid solution series.

3.2.2 Magnetic Properties of Rocks

As a rule of thumb intrusive igneous rocks usually contain larger grain-sizes for magnetic minerals than extrusives. Because the majority of these are often multidomain grains, intrusives (for example, gabbro) often exhibit a lower mean coercivity than the chemical equivalent in extrusive rocks (for example, basalt). Basic rocks (for example, gabbro) are generally more magnetic than acidic rocks (for example, granite). Rocks that cool very rapidly, such as submarine basalts, often exhibit skeletal magnetic grains, whose elongated shapes and larger internal stresses increase their relative coercivities over more slow-cooled samples. Intrusive rocks often contain titanomagnetites and titanohematites with high Curie temperatures (greater than 500°C) while fresh submarine basalts typically contain titanomagnetites with Curie temperatures around 200°C or lower (such rocks readily undergo low-temperature oxidation, a process that raises their Curie temperatures), as was predicted from theoretical considerations (Carmichael and Nicholls, 1967) and confirmed observationally (Irving, 1970; Irving *et al.*, 1970; Gromme *et al.*, 1969; Ozima and Ozima, 1971; Marshall and Cox, 1971; Lowrie *et al.*, 1973a, b; McElhinny, 1977; Johnson and Atwater, 1977; Johnson and Hall, 1978; Johnson, 1979). Acidic rocks are usually more oxidized than basic rocks, presumably because of variability in the water content (Haggerty, 1978).

NRM consists mostly of TRM in the majority of igneous rocks, even though CRM and VRM also often contribute significantly to NRM. Titanomagnetites and sometimes titanohematites are primarily responsible for carrying the primary remanence in most terrestrial igneous rocks. Although fine-grained pyrrhotite often occurs in the ground mass of many igneous rocks, it fortunately only occasionally contributes significantly to the remanence. A variety of other magnetic minerals also can contribute to the remanence in igneous rocks, such as jacobsites (Fe–Mn spinels) and geothite (αFeOOH). Pure iron or iron-nickel alloys are rare in most terrestrial igneous rocks, but dominate the magnetic properties of lunar samples (see Chapter 11). A possible exception to this is in the case of serpentinized ultramafics in which the Fe–Ni–S system in some cases plays the dominant role (Haggerty, 1978).

Post-depositional RM is thought to be responsible primarily for the NRM of deep-sea sediments, while either DRM or post-depositional RM is responsible primarily for the NRM in lake sediments (Irving and Major, 1964; King and Rees, 1966; Verosub, 1977; Barton *et al.*, 1980). Although the magnetic mineralogy of deep-sea sediments is not yet well known, the NRM is carried mainly by titanomagnetites even though contributions from pyrrhotite (in certain anoxic environments) and ferromanganese oxides and oxyhydroxides also contribute (Haggerty, 1970; Kobayashi and Nomura, 1974; Kent and Lowrie, 1974; Johnson *et al.*, 1975b; Henshaw and Merrill, 1980).

The magnetic mineralogy of terrestrial sediments is highly variable, because the sediments originate in part from pre-existing igneous, sedimentary and metamorphic rocks and also because magnetic minerals sometimes form in sediments by authigenesis and diagenesis. Although the origin of the magnetization of so-called "red-beds" (principally red sandstones and shales) is currently very controversial, few workers question that CRM in hematite typically contributes substantially to the remanence (Collinson, 1965; Roy and Park, 1972; Larson and Walker, 1975; Elston and Purucker, 1979; Butler, 1982). The controversy largely involves whether the CRM occurred rapidly soon after deposition, or whether it was graded and therefore the NRM would be unusable for palaeomagnetic studies. Although no attempt will be made to resolve this controversy here, no significant interpretations made in this book depend on the outcome.

Although magnetic properties have been generalized with respect to different rock types above, it needs to be emphasized that all too often exceptions occur. For example, it is quite possible in a given study to find that a particular granite is more strongly and stably magnetized than a particular basalt. Because subtle changes in such factors as grain size and oxidation state can have pronounced affects on the magnetic properties of rocks, it should not be expected that generalizations such as given above, can be applied to any given situation with much confidence. They are "statistical" arguments only, and deviations from the mean are typically large.

3.3 Palaeomagnetic Directions and Poles

3.3.1 Demagnetization Procedures and Consistency Checks

Although all magnetic minerals exhibit anisotropy, the majority of igneous rocks and sedimentary rock used in palaeomagnetism are not significantly magnetically anisotropic as a whole. On the other hand metamorphic rocks which are rarely used in palaeomagnetic studies often do exhibit magnetic anisotropy. The absence of magnetic anisotropy in rocks occurs because the orientations of individual grains in a sample are random, or nearly random. Of course, because all rocks are not magnetically isotropic, palaeomagnetists should determine in the laboratory whether the isotropic condition is satisfied in any given study.

The remanence acquired when a rock forms is called a primary magnetization and this will be parallel (or very rarely antiparallel) to the external field during its formation providing the rock is magnetically isotropic. This follows since the external field is the only parameter with a preferred direction left in the problem. Subsequent to formation secondary magnetization(s) (CRM, VRM, etc.) are often acquired. To distinguish between primary and secondary magnetizations palaeomagnetists use various demagnetization procedures on the NRM. The two most commonly employed techniques are Alternating Field (AF) demagnetization and thermal demagnetization.

In *AF demagnetization*, a sample is placed in an alternating field of amplitude H_0 that is slowly (relative to the period, t_a, of the alternating field) reduced to zero while the sample is tumbled as randomly as possible. This process randomizes the remanence due to those grains having relaxation times corresponding to coercivities $H_c \leq H_0$ (recall that τ is a function of H_c; see (3.15)). H_0 is then increased in steps, so that the directions of more and more stable (larger τ) remanence can be ascertained. If only one type of remanence contributes to the NRM, no change in direction will be observed on demagnetization. If two or more types of remanence are present and have different directions, then a directional change on demagnetization is to be expected since it would be coincidental if the relaxation time spectra of the two remanences were identical. Usually the most stable component of remanence is argued to be the primary magnetization. (This is somewhat oversimplified, as a variety of more sophisticated criteria have been used in recent years.) AF demagnetization then measures the stability of remanence in alternating fields, from which stability in time is inferred.

Thermal demagnetization measures stability of remanence as a function of temperature, from which stability in time is inferred. This is carried out by heating the sample up to some temperature, T_0, and cooling in a non-magnetic space. That remanence with blocking temperatures less than T_0 will be randomized (providing magnetic interactions between blocked remanence and unblocked remanence is negligible, as often seems to be the case). T_0 is stepwise increased up to the Curie temperature. Analogous to AF demagnetization, thermal demagnetization makes it possible to discern different remanence types, if they are present. The direction of the primary magnetization is often taken to be that of the remanence with the highest blocking temperatures, when more than one type of remanence is present. In addition, other demagnetization procedures including chemical demagnetization, low-temperature demagnetization and various combinations of the above are used in actual palaeomagnetic studies. A brief description of these is given by McElhinny (1973). The formal analysis of vectors from multi-component systems during demagnetization procedures has been given by Halls (1978), Hoffman and Day (1978), Kirschvink (1980) and is discussed by Dunlop (1979).

Of course it is still possible occasionally for the secondary magnetization to be more stable than the primary magnetization. In this case the so-called cleaning techniques described above will by themselves provide incorrect results. Sometimes more detailed rock magnetic procedures prove adequate to distinguish such exceptions, but inevitably sometimes palaeomagnetists end up with "bad" directions. However, there are often other criteria available that provide confidence in the palaeomagnetic data. An example of this is when different rock types of the same general region and of the same relative age give the same magnetic directions. It would be rather surprising if the magnetization of a sediment (typically DRM or post-DRM) agreed with that of an igneous rock (typically TRM) unless their magnetizations were both primary. Consistency of directions from different rock formations in the same general locality and of the same age provide considerable confidence for the reliability of palaeomagnetic directions. Also there are a variety of field tests, as illustrated in Fig. 3.6, that make it possible to place a minimum age on the magnetization, even supposing it may not be primary.

By using criteria such as those just described it is possible to obtain some idea of the reliability of most palaeomagnetic data. The main problem is that each palaeomagnetic study will involve different techniques and tests depending on the circumstances that arise. Some will demonstrate that the NRM is a primary remanence, while others will rely on less convincing arguments. There is no standard applicable to all studies because in many cases it might be impossible to do all that a given standard might require. Inevitably therefore a few erroneous directions arise, but the subject has now

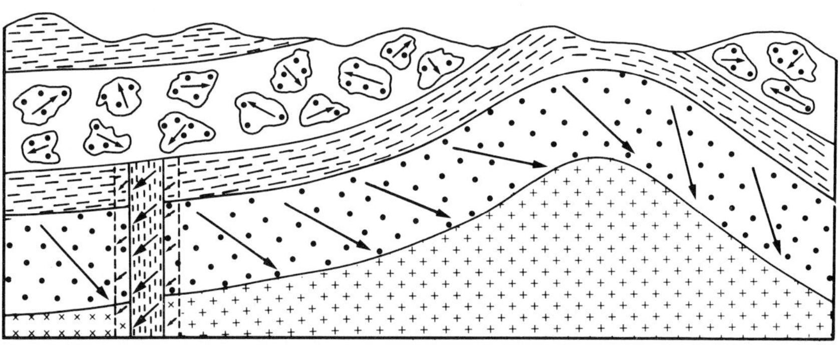

Fig. 3.6. Fold and Conglomerate test of Graham (1949) and the Baked Contact test for palaeomagnetic directions. (a) If the directions of magnetization of samples collected from different limbs of a fold converge after unfolding the beds, the magnetization pre-dates the folding and has been stable since. (b) If directions of magnetization in conglomerate pebbles derived from the beds under investigation are randomly oriented, the magnetization of the parent beds has been stable since formation of the conglomerate. (c) If directions of magnetization in the baked zone surrounding an intrusion are parallel to those observed in the intrustion, but differ from that of the unbaked country rock, the magnetization of the intrusion has been stable since formation.

advanced to such a level of sophistication that hopefully such errors will form only a minute fraction of the whole. The analysis of a mixed set of data on a global scale as is carried out in the next few chapters thus relies on its careful selection according to some minimum criteria of acceptability. Examples of some possible methods of selection are given in McElhinny (1973).

3.3.2 The Geocentric Axial Dipole Field Hypothesis

Suppose the direction of the primary magnetization of a rock at a particular location has been determined and that a reliable age for the time of formation of the rock is available. As discussed in Chapter 2, the instantaneous field is moderately complex at any time so the problem arises as to how a local measure of that field at some time in the past can be analysed. The problem is complicated by the fact that not all rocks record the field in the same way. Imagine a series of lava flows, each with a thickness of a few metres, extruded successively at random times by a volcano. Each lava flow will cool to the ambient temperature over time intervals that depend on the thickness but will be measured typically in years. Thus the palaeomagnetic direction measured in each lava will be an average of the local variations over at most a few years. Over such time intervals the field is essentially unchanged and so each lava flow effectively has recorded the instantaneous field. Thicker lavas, up to several tens or a hundred metres or more will cool much slower, so that the field recorded in the flow will be more complex. At any given horizon the instantaneous field will be that recorded when it cooled through the blocking temperatures. A series of samples collected across the flow will thus record changes over a few hundred or even a thousand years. The average magnetization of such thick flows, and even more so of thick intrusions that cool even slower, will be a *time-averaged* magnetic field over several hundred or thousand years.

Because the process of magnetization acquisition in sediments is usually either depositional or chemical, the time span covered by a typical specimen of thickness 2.5 cm, as is used in the laboratory, may already be more than a thousand years (deposition rates in the pelagic ocean environment are usually measured in millimetres per thousand years). Successive samples collected over great thicknesses could correspond to time spans of hundreds of thousands or millions of years. In these cases the field measured by just a thin specimen could be time-averaged as much as a very thick lava flow or intrusion. Palaeomagnetists therefore design their sampling schemes to attempt to measure the field in the past at as many different instances in time as they can. The average of these magnetizations, determined by finding the direction of the vector sum, is termed the *palaeomagnetic field* for that rock formation.

The palaeomagnetic field is thus a time-averaged field. From knowledge of the present field over historic times it is supposed that over periods of several thousands of years any locality would record the total variations in the field corresponding say to the westward drift. Over such periods of time, the time-averaged field would, to the first approximation, correspond not to a dipole inclined to the axis of rotation but to a dipole aligned along that axis. This hypothesis is termed the *geocentric axial dipole field hypothesis*. The problem that arises is that one is never sure of the extent to which the time-averaging procedure has been carried out, so that each palaeomagnetic direction will be different from every other such direction in the time factor. It might be supposed that over a sufficiently long enough time all palaeomagnetic directions for a given age at any locality would coincide. The basic question is what is "sufficient time"? This question will be posed in more detail in subsequent chapters.

For the moment it is sufficient to assume the assumption is valid and investigate the consequences. For a dipole of strength p the radial and tangential components of the field at some colatitude θ at the earth's surface (radius R) are respectively:

$$F_R = \frac{2\mu_0 p \cos\theta}{4\pi R^3}$$

and

$$F_\theta = \frac{\mu_0 p \sin\theta}{4\pi R^3} \tag{3.16}$$

where the field F is measured in teslas. Since the tangent of the magnetic inclination I is F_R/F_θ then:

$$\tan I = 2\cot\theta$$

or

$$\tan I = 2\tan\lambda \tag{3.17}$$

where λ is the latitude and corresponds to the geographic latitude and not the magnetic latitude when the geocentric axial dipole field hypothesis is applicable. There is no information about longitude.

Because the time-averaged inclination corresponds to the palaeolatitude on the geocentric axial dipole field assumption, and the time-averaged declination indicates the direction of a meridian, the position of the corresponding geographic pole on the earth's surface can be calculated (Fig. 3.7). The co-ordinates of this pole (λ^1, ϕ^1) determined from the site co-ordinates (λ, ϕ) and palaeomagnetic direction (D, I) are given by:

$$\sin\lambda^1 = \sin\lambda\cos\theta + \cos\lambda\sin\theta\cos D \quad (-90° \leq \lambda^1 \leq +90°) \tag{3.18}$$

$$\phi^1 = \phi + \beta \quad \text{when} \quad \cos\theta \geq \sin\lambda\sin\lambda^1$$

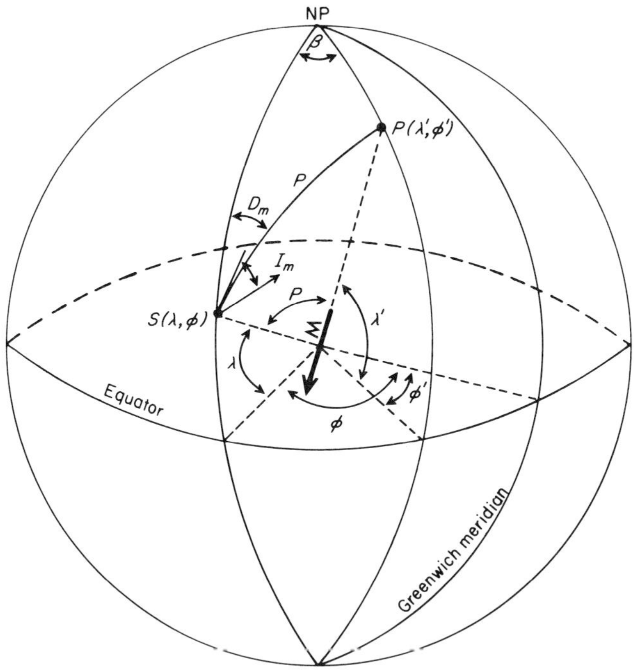

Fig. 3.7. Relationship to calculate the palaeomagnetic pole P with coordinates (λ^1, ϕ^1) from average directions of magnetization (D_m, I_m) observed at sampling site S with coordinates (λ, ϕ).

or

$$\phi^1 = \phi + 180 - \beta \text{ when } \cos\theta \le \sin\lambda \sin\lambda^1 \tag{3.19}$$

where

$$\sin\beta = \sin\theta \sin D/\cos\lambda^1 \; (-90° \le \beta \le 90°) \tag{3.20}$$

The palaeocolatitude θ is determined from (3.17). The pole position (λ^1, ϕ^1) calculated in this way is called the *Palaeomagnetic Pole*. The use of the term implies that "sufficient" time-averaging has been carried out. Alternatively any instantaneous palaeofield direction may be converted to a pole position according to (3.18), (3.19) and (3.20), in which case the pole is termed a *Virtual Geomagnetic Pole* (VGP). The VGP is the palaeomagnetic analogue of the geomagnetic poles of the present field. The palaeomagnetic pole may then be calculated alternatively by finding the average of many VGPs, corresponding to many palaeodirections. Table 3.4 summarizes definitions for the poles referred to in this book.

If the location of the rocks used for palaeomagnetic study has moved since its formation relative to the present geographic pole, the palaeomagnetic pole will depart from the present geographic pole. Indeed, the positions of the palaeomagnetic poles from all continents depart more and more from the present geographic pole the further one goes back in time. The path traced by successive positions of the palaeomagnetic poles with time plotted on the present longitude–latitude grid for a given continental block is termed the *apparent polar wander path* for that block. The comparison of such paths between different continental blocks provides the evidence that continental drift has occurred. The matching of two paths over any time span then provides a method for the determination of the past relative positions of two blocks. An example is given in Fig. 3.8. The pole paths for Europe and North America diverge over the past 350 million years. If various segments of the apparent polar wander paths are made to overlap, it is possible to find the relative locations of Europe and North America in the past (Graham *et al.*, 1964). This shows that these two continents were previously joined together and have subsequently separated to form what is now the North Atlantic Ocean (see Fig. 3.8). Maps showing possible positions of the continents over the past 1000 Ma are given in Irving (1977, 1979), Morel and Irving (1978) and Smith *et al.* (1981). A full discussion of palaeomagnetic methods and their relation to continental drift and plate tectonics is given by McElhinny (1973).

The close correspondence of palaeomagnetic poles of a given age from widely separated regions of continental extent provides the key evidence that,

TABLE 3.4
Summary of poles used in geomagnetism and palaeomagnetism

North (south) magnetic pole	Point on the earth's surface where the magnetic inclination is observed to be $+90°$ ($-90°$). The poles are not exactly opposite one another and for epoch 1980 lie at 77.3°N, 258.2°E and 65.6°S, 139.4°E
Geomagnetic north (south) pole	Point where the axis of the calculated best fitting dipole cuts the surface of the earth in tne northern (southern) hemisphere. The poles lie opposite one another and for epoch 1980 are calculated to lie at 78.8°N, 289.1°E and 78.8°S, 109.1°E
Virtual geomagnetic pole (VGP)	The position of the equivalent geomagnetic pole calculated from a spot reading of the palaeomagnetic field. It represents only an instant in time, just as the present geomagnetic poles are instantaneous
Palaeomagnetic pole	The average palaeomagnetic field over periods sufficiently long so as to give an estimate of the geographic pole. Averages over times of 10^4 to 10^5 years are estimated to be sufficient. The pole may be calculated from the average palaeomagnetic field or from the average of the corresponding VGPs

to a first approximation, the geocentric dipole assumption is valid. In some instances a palaeomeridian may pass lengthwise across a continent so that the latitude variation is maximized. In such a case the geocentric dipole assumption may be tested explicitly as has been done for Africa by McElhinny and Brock (1975). However, these tests do not establish the *axial* nature of the palaeomagnetic field. For this it is necessary to invoke palaeoclimatic evidence. Past climate and temperature distribution can be expected to have an equator to pole variation dependent essentially on the axis of rotation. If the palaeolatitudes of various palaeoclimatic indicators determined from palaeomagnetic data also show such an equator to pole distribution, then the *axial* nature of the palaeomagnetic field is strongly supported (Irving and Gaskell, 1962; Irving and Briden, 1962; Briden and Irving, 1964; Drewry *et*

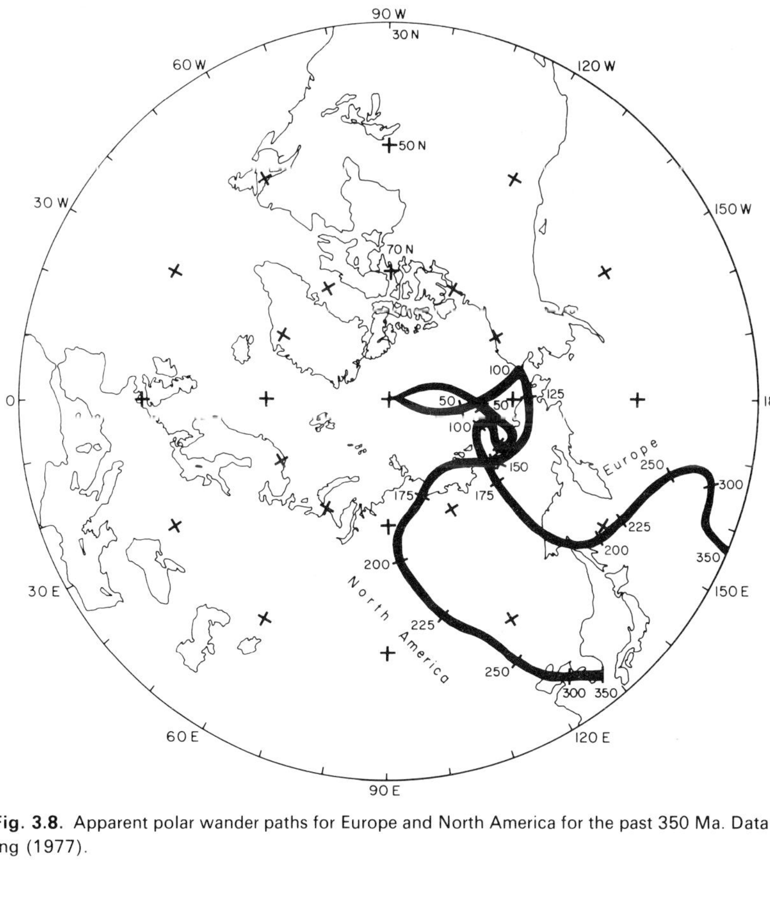

Fig. 3.8. Apparent polar wander paths for Europe and North America for the past 350 Ma. Data from Irving (1977).

al., 1974; see also summary by McElhinny, 1973). Further discussion of all these points is given in §6.1 and §6.2.

3.3.3 Statistical Methods in Palaeomagnetism

Statistics play an important role in palaeomagnetism. At a given sampling site errors may be introduced from orientation errors, local magnetic anomalies etc., and these may be further compounded in the laboratory because scatter may be introduced during procedures to remove secondary components. All of these errors produce a certain amount of scatter in the magnetizations between samples collected at a site and are referred to as the *within-site scatter*. Because sampling procedures are designed so that sites represent as far as is possible single points in time, the variation in magnetizations between sites will tend to represent variations produced by the secular variation. The *between-site scatter* is thus often used as a parameter to investigate changes in secular variation in the past. Because the statistical analysis of directional data is not normally dealt with in standard texts on statistics, an introduction to Fisher statistics, most commonly used in palaeomagnetism, is given here. It is assumed that readers are familiar with standard statistical techniques for normal distributions such as chi-square analyses, F-tests, etc.

Analogous to a normal distribution Fisher (1953) suggested that the distribution of vectors on a unit sphere has a probability density given by:

$$P(\theta) = c \exp(\kappa \cos \theta) \tag{3.21}$$

where $c = \kappa/4\pi \sinh \kappa$ from the requirement that $\int P(\theta)\, d\theta = 1$ when integration is carried out over the entire sphere. θ is the angle between the direction of a particular vector and the mean, defined as the direction of the vector sum of a set of unit vectors. κ is referred to as *Fisher's precision parameter* and describes the degree of scatter present. For $\kappa = 0$ the vectors are uniformly distributed over the sphere, and in the limit $\kappa \to \infty$, no scatter at all is observed.

For κ large, $\cos \theta \approx 1 - \theta^2/2$ and $P(\theta) \approx \exp^{-S\theta^2}$ where S is a constant. This is a normal distribution, so that in the case of small scatter in which the spherical surface can be approximated by a plane, the Fisher distribution reduces to a normal distribution. In that case κ is the *invariance* or reciprocal of the variance in all directions. Many of the discussions involving normal (Gaussian) distributions in texts on statistics are thus relevant to Fisher statistics.

In palaeomagnetism the direction of magnetization of a rock sample is specified by the declination, D, measured clockwise from true north, and the inclination I, measured positively downwards from the horizontal. This direction may be specified by its three direction cosines, as follows:

$$\begin{aligned}&\text{North component} \quad l = \cos D \cos I\\ &\text{East component} \quad m = \sin D \cos I\\ &\text{Down component} \quad n = \sin I\end{aligned}$$

The direction cosines (X, Y, Z) of the resultant of N such directions of magnetization are proportional to the sum of the separate direction cosines, and are given by:

$$X = \frac{1}{R}\sum_{i=1}^{N} l_i; \quad Y = \frac{1}{R}\sum_{i=1}^{N} m_i; \quad Z = \frac{1}{R}\sum_{i=1}^{N} n_i \tag{3.22}$$

where R is the vector sum of the individual unit vectors, given by:

$$R = \left[\sum l_i^2 + \sum m_i^2 + \sum n_i^2\right]^{1/2} \tag{3.23}$$

The mean declination, D_R, and inclination, I_R, are given by

$$\tan D_R = \frac{\sum m_i}{\sum n_i}; \quad \sin I_R = \frac{1}{R}\sum n_i \tag{3.24}$$

Fisher (1953) showed that the best estimate k of κ for $k \geq 3$ is given by

$$k = \frac{N-1}{N-R} \tag{3.25}$$

An angular standard deviation S can be defined as in the case of a normal distribution where

$$S^2 - \frac{1}{N-1}\sum_{i=1}^{N} \theta_i^2 \tag{3.26}$$

where θ_i is the angle individual vectors make with the mean. When θ_i is small, $\cos\theta_i = 1 - \theta_i^2/2$ and (3.26) may then be written as:

$$S^2 \approx \frac{1}{N-1}\sum_{i=1}^{N}(2i - 2\cos\theta_i) = 2\frac{(N-R)}{(N-1)} = \frac{2}{k}\ \text{radians} \tag{3.27}$$

In the limit of small scatter the *angular standard deviation* or *angular dispersion*, S is thus simply related to Fisher's precision parameter and becomes

$$S = \frac{81}{\sqrt{k}}\ \text{degrees} \tag{3.28}$$

This is of particular importance in palaeosecular variation discussed in §6.4.

Statistics in palaeomagnetism are of the hierachical type. At the lowest levels sample directions are averaged, then site directions are averaged to determine the palaeomagnetic pole. Suppose N samples are collected at each of B sites, then following standard statistical analysis, the total angular dispersion, S_T, is related to the between-site dispersion, S_B, and the within-site dispersion, S_W, by the relation:

$$S_T^2 = S_B^2 + S_W^2/N \tag{3.29}$$

It follows from (3.29) that the corresponding precision parameters are related by:

$$\frac{1}{k_T} = \frac{1}{k_B} + \frac{1}{k_W N} \tag{3.30}$$

Often in the case of lava flows the within-site scatter is so small that only $N = 4$ samples per site need be taken to make the last terms of (3.29) or (3.30) much smaller than the first (see §6.4).

The true mean direction of a set of N unit vectors for $k \geq 3$ lines within a circular cone about the resultant vector R with semi-angle α at the probability level $(1 - P)$ where:

$$\cos \alpha_{1-P} = 1 - \frac{N - R}{R}\left\{\left(\frac{1}{P}\right)^{1/(N-1)} - 1\right\} \tag{3.31}$$

In palaeomagnetism it is common to use $P = 0.05$ and give values for α_{95}, commonly referred to as the 95 percent circle of confidence. If a mean direction D_m, I_m is converted to a palaeomagnetic pole the α_{95} circle of confidence converts to an oval about the pole because of the dipole transformation (3.17). The errors in the mean direction $\mathrm{d}I_m$ and $\mathrm{d}D_m$ are given by (Irving, 1956):

$$\alpha_{95} = \mathrm{d}I_m = \mathrm{d}D_m \cos I_m \tag{3.32}$$

The error $\mathrm{d}I_m$ corresponds to an error $\mathrm{d}\theta$ in the colatitude given by:

$$\mathrm{d}\theta = \tfrac{1}{2}\alpha_{95}(1 + 3\cos^2\theta) \tag{3.33}$$

and an error $\mathrm{d}m$ in a direction perpendicular to the meridian given by:

$$\mathrm{d}m = \alpha_{95}\frac{\sin\theta}{\cos I_m} \tag{3.34}$$

Thus a 95 percent circle of confidence (α_{95}) of directions will transform to an oval of 95 percent confidence ($\mathrm{d}\theta$, $\mathrm{d}m$) for the palaeomagnetic pole. The converse is also true. If a set of VGP have mean with α_{95}, the corresponding mean direction at any locality will be specified with an oval of confidence.

The implications of the above are that because of the dipole transformation a set of directions having a Fisher distribution cannot in general transform into a set of VGPs having a Fisher distribution. The VGPs will have an oval distribution. Conversely a set of VGPs with Fisher distribution will transform to a set of directions at any locality with an oval distribution. Random errors that occur at the site level due to experimental errors of one sort or another can be expected to produce a Fisher distribution of directions within-site. The between-site distribution will depend on the properties of the magnetic field variations and depending on their nature will tend to produce either a Fisher distribution of directions or of poles in each case. In practice the result will be a combination of Fisher and oval distributions whose relative contributions might change with time and position on the earth's surface. In spite of this Fisher statistics are extremely useful in describing the data, particularly if the statistics are calculated using VGP as is discussed in detail in Chapter 6.

3.4 Palaeointensity Methods

3.4.1 The Problem

There are now numerous palaeointensity techniques that, it is argued, give reliable estimates for the ancient field strengths of the earth and moon. It is found experimentally for virtually all rocks that in fields of the order of the earth's magnetic field (10^{-4} T) the TRM, M_T, is proportional to the inducing field H, so that

$$M_T = C_1 H \tag{3.35}$$

where C_1 is a constant of proportionality that depends on the magnetic properties of the rock. Equation 3.35 follows from single domain theory by expanding the hyperbolic tangent of (3.13) and dropping the higher order terms. The linearity of M_T and H is expected therefore only for small inducing fields and this is what is observed experimentally.

Let M_L denote a TRM produced in the laboratory in a field H_L, then:

$$M_L = C_2 H_L \tag{3.36}$$

From (3.35) and (3.36) then,

$$\frac{M_T}{M_L} = \frac{C_1}{C_2}\frac{H}{H_L} \tag{3.37}$$

If the rock's magnetic properties have not been altered since formation then $C_1 = C_2$ and it is clearly a simple matter to determine H, the ancient field strength. Unfortunately in practice this is seldom the case and often chemical changes occur in the laboratory when the rock is heated. Coe (1967a, b) has shown that only a very small percentage of historically erupted lava flows (cases in which the correct field strength is known in advance) are suitable for palaeointensity studies. It might be expected that this small percentage would decrease even more for older rocks, because they are much more likely to have picked up unwanted remanences.

The basic idea behind all *reliable* palaeointensity techniques is to develop a method by which several *independent* estimates of the ancient field strength can be obtained from the *same* sample. Consistency between such estimates provides some confidence regarding reliability. Unfortunately many techniques do not do this. For example, several palaeointensity estimates have been determined by equating the ratio C_1/C_2 to the ratio of saturation magnetization of the sample before and after heating. The rock has undergone chemical change on heating and such a "correction factor" is used to take account of it. This technique is becoming increasingly fashionable but is often very dangerous in practice because it typically involves elimination of some or all of the consistency checks. Such methods have even been used to obtain palaeointensity estimates from Precambrian rocks, ignoring the high probability that such rocks cannot have a pure unaltered TRM.

The reader is warned that there are numerous poor palaeointensity values that have been published, making it difficult for those not familiar with palaeomagnetism to decide what is really known. Before accepting a palaeointensity value one should determine what *consistency checks* have been used to see whether or not they are acceptable. Two basic palaeointensity techniques are now described that illustrate what is meant by consistency checks. There are many other reliable techniques but all of them use similar principles to the ones described below.

3.4.2 The Modified Thelliers' Method

The classical method for palaeointensity determination was developed by the Thelliers for studies of baked archaeologial material, but it applies equally well to any material whose magnetization was acquired as a TRM (see review by Thellier and Thellier, 1959a, b). The methods used by most workers are in fact modified versions of the original technique and these, with the associated problems, have been discussed by Coe (1967a, b), Smith (1967a, b, c), Levi (1975), Domen (1977) and Kono (1978). All these techniques rely on Thellier's law of additivity, which effectively states that a *partial* TRM, $M_{T_i}^{T_{i+1}}$ acquired in any temperature interval (T_i, T_{i+1}), is independent of the remanence acquired in any other temperature interval. This is expressed mathematically (see Nagata, 1961) as follows:

$$M_{T_0}^{T} = \sum_{T_0}^{T} M_{T_i}^{T_{i+1}} \tag{3.38}$$

That is, the total TRM is equal to the sum of the partial TRMs. Equation 3.35 will usually hold for each temperature interval, although the constant C_1 will have to be replaced by C_i, because the constant of proportionality generally varies from temperature interval to temperature interval. Therefore (3.37) will also be applicable for each temperature interval, allowing one to obtain several separate estimates of the ancient field strength. It is these separate estimates that are the consistency checks in Thelliers' method.

There are several ways of doing this in practice. One common way is to heat the sample from room temperature to T_i and cool it in a non-magnetic space after which the remaining remanence is measured. Subsequently, the sample is heated a second time to T_i and cooled in a known field and the remanence is again measured. The remanence lost in the first heating and cooling cycle is then plotted against the remanence gained in the second (Fig. 3.9). These steps are then repeated to successively higher temperature, ideally producing the straight line as shown in Fig. 3.9. From the slope of this line the ancient field strength can be estimated. Deviations from a straight line are referred to as non-ideal Thellier behaviour (see Coe (1967a) for possible origins of this behaviour) and indicate that one of the following assumptions

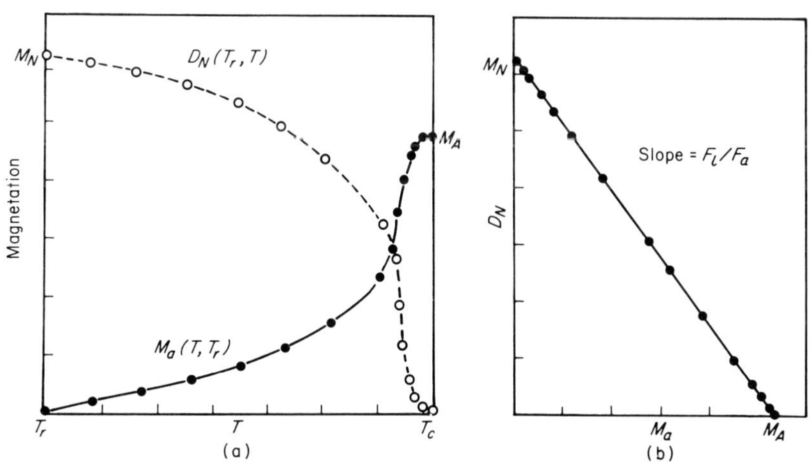

Fig. 3.9. (a) Ideal example of an NRM demagnetization curve $D_n(T_r, T)$ and a TRM acquisition curve $M_a(T, T_r)$ derived during the modified Thellier's method for palaeointensity determination. (b) The corresponding ideal NRM–TRM line obtained after T has been eliminated. The slope is the ratio of the laboratory inducing field (F_l) to the palaeofield intensity (F_a). After Coe (1967a).

has not been satisfied: (1) the NRM is a pure unaltered TRM; (2) the TRM is linearly proportional to the inducing field (this can be checked in the laboratory); (3) Thellier's additivity law is satisfied (deviations are typically small); (4) the physical–chemical properties of the magnetic minerals have remained unaltered since the acquisition of the initial TRM (that is the constants C_i have remained invariant). Typically, one or more of these assumptions are not met and in practice only a small percentage of rocks can be used for palaeointensity estimates. In the case of archaeological material, a common problem is the effect of weathering causing chemical changes. In certain circumstances this can be modelled to provide field strength estimates under non-ideal behaviour (Barbetti and McElhinny, 1976; Barbetti *et al.*, 1977). Recently Coe *et al.* (1978) have described statistical procedures for determining the quality of palaeointensity data.

3.4.3 The Comparative Coercive Force Method

An attempt to overcome the problem that most rocks contain secondary magnetization was made by van Zijl *et al.* (1962). If the initial palaeomagnetic study showed that the secondary magnetization was very easily removed by AF demagnetization in peak fields of about 20 mT, it was supposed that (3.37) would apply if both NRM and total TRM were demagnetized in that field. However, there was no consistency check in this method and Smith (1967a) extended this so that the coercive force spectrum was compared for NRM and TRM. The NRM is first demagnetized using increasing steps of peak alternating field. After completed demagnetization the sample is given a TRM which is subsequently demagnetized in the same way. A palaeointensity estimate is obtained from the linear part of the NRM–TRM plot shown in Fig. 3.10a. Thus the coercive force spectrum obtained on AF demagnetization is used to obtain the palaeointensity estimate, rather than the blocking temperature spectrum as used in Thelliers' method. The plot might deviate from a straight line at low fields because of the removal of secondary components from the NRM but not the TRM. The existence of a comparable portion of the coercive force spectra suggests this portion has remained unchanged as a result of laboratory heating.

Unfortunately there are ways in which rocks might be altered on heating and yet still maintain a comparable part of their NRM and TRM spectra. So although there is a consistency check of sorts, it is not necessarily a rigorous one. McElhinny and Evans (1968) improved on this by suggesting that, in addition, the coercive force spectrum of saturation IRM be compared before and after heating through AF demagnetization (Fig. 3.10b). Those parts of the spectrum that were not altered by the heating may then be used in the palaeointensity estimate. An independent consistency check was thus introduced. However saturation IRM is not a particularly good analogue of TRM and this consistency check is too stringent and unrealistic in practice.

It has been shown that ARM often has an identical AF demagnetization spectrum to TRM (unlike IRM). Using this observation Shaw (1974) introduced a consistency check for the comparative coercive force method using ARM. The coercive force spectrum of ARM is compared before and after heating through AF demagnetization (Fig. 3.10b). This determines if any chemical changes during heating have affected parts of the coercive force spectrum, and thus which parts of the AF demagnetization curve may be used in the palaeointensity estimate. A problem that has not yet been fully investigated is whether or not the AF demagnetization spectrum is really sensitive enough to distinguish CRM from TRM. This factor is the main question associated with the method. Some of the causes of non-ideal behaviour in Shaw's method have been examined by Kono (1978). Its main advantage is that it is very much less tedious than the Thelliers' method. The basic problem associated with either method is finding suitable samples that will give good results because often high failure rates are involved. This

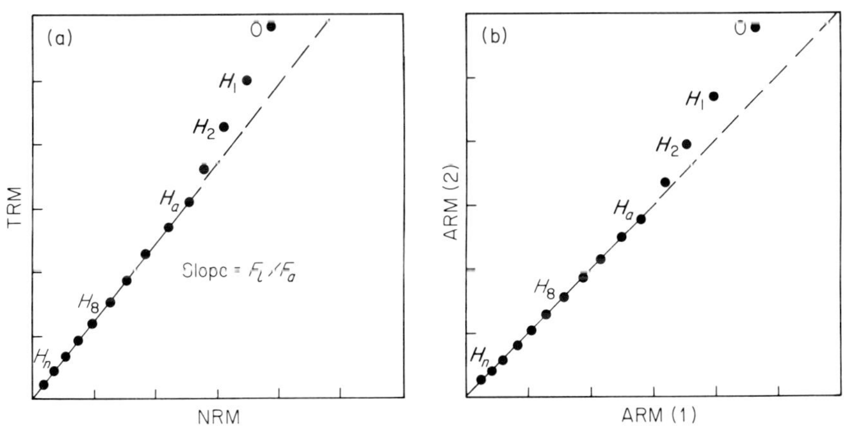

Fig. 3.10. (a) Hypothetical NRM–TRM plot at various AF demagnetization fields ($\tilde{H}_n$) obtained using the comparative coercive force method for palaeointensity determination. The slope of the straight line section at fields $\tilde{H}_n > \tilde{H}_a$ is the ratio of the laboratory inducing field for the TRM (F_l) to the palaeofield intensity (F_a). For $\tilde{H}_n < H_a$ the points do not fall on the straight line due to removal of secondary components during demagnetization of NRM but not of TRM, or else due to alteration of the coercive force spectrum in this region. (b) Hypothetical plot of ARM(2) determined after heating against ARM(1) determined before heating at various AF demagnetization fields (H_n), using Shaw's (1974) consistency test. Except in the region $H_n < H_a$ the plot is a line of gradient 1.0. For $H_n < H_a$ the points depart from the line because this part of the coercive force spectrum has been altered by the heating. A similar plot results from the McElhinny and Evans (1968) consistency test using saturation IRM. A plot of IRM(2) after heating against IRM(1) before heating should also give a line of slope 1.0.

problem has been overcome by Senanayaka and McElhinny (1981). They have noted that basic lava flows exhibiting high temperature oxidation of their magnetic minerals (occurring during initial cooling above the Curie temperature) are also those most resistant to AF demagnetization. Because oxidation is the most common form of chemical change that takes place during the laboratory heating, this process is naturally inhibited in the initially more oxidized samples. Such samples prove to have a high success rate using the Thelliers' method and their extended coercive force spectrum is also the most desirable characteristic for use in Shaw's method. Senanayake and McElhinny (1981) have shown that samples with these characteristics are quickly and easily selected from a large suite by virtue of their distinctive low temperature susceptibility variation. The safest and most efficient technique is to select samples as above, test whether a good Thelliers' method can be obtained, then use the quicker Shaw's method to provide back-up data (Senanayake *et al.*, 1982).

Stephenson and Collinson (1974) have developed a method to avoid laboratory heating altogether and so eliminate the problem of chemical changes associated with the heating. They use the fact that, apart from often having an identical AF demagnetization spectrum to TRM, ARM is often a linear function of the inducting field and satisfies an additivity law analogous to Thelliers' for TRM (Rimbert, 1959; Patton and Fitch, 1962; Levi and Merrill, 1976). Thus the ratio of ARM lost after AF demagnetization in some peak field H_p to the TRM lost after demagnetization in H_p should be constant irrespective of the value of H_p. The constancy of this ratio provides the necessary consistency check. The problem is how to relate the ARM intensity to the TRM intensity, assuming this has remained unchanged since formation. Unfortunately Levi and Merrill (1976) and Bailey and Dunlop (1977) have shown that the ARM to TRM ratio varies by an order of magnitude depending on grain size (and also probably on mineralogy). This ARM method may still however prove usable in spite of this problem but to be satisfactory it will have to be augmented by rock magnetic measurements such as hysteresis loop parameters (§3.1.2) from which estimates of the magnetic grain size distribution in a sample can be obtained.

Both the modified Thelliers' method and Shaw's variation of the comparative coercive force method appear to be the most reliable methods at present available for determining palaeointensities. These techniques are based on the premise that if the NRM is not an analtered TRM or if physical–chemical changes have occurred subsequent to initial cooling, then the blocking temperature distribution (Thelliers' method) or the AF demagnetization spectrum (comparative coercive force method) will be different for NRM and laboratory produced TRM. It is worth pointing out that thermal demagnetization and AF demagnetization do not affect the identical magnetic regions in a rock. Although rare, it is possible for a rock to exhibit high stability with respect to thermal demagnetization and low stability with respect to AF demagnetization or vice versa (see Levi and Merrill (1978) for further discussion). Thus rocks most suitable for analysis by Shaw's method might not be suitable for Thelliers' method and vice versa. Clearly when both methods can be used on adjacent samples and the palaeointensity estimates agree, there must be considerable confidence in the value obtained.

3.4.4 Dipole Moments

For a geocentric dipole of moment p the field intensity F observed at the surface of the earth, radius R, at magnetic colatitude θ is given from (3.16) by:

$$F = \frac{\mu_0 p}{4\pi R^3}(1 + 3\cos^2\theta)^{1/2} \tag{3.39}$$

where the field F is measured in teslas. Substituting from (3.17) the above can be expressed in terms of magnetic inclination I as:

$$F = \frac{2\mu_0 p}{4\pi R^3}(1 + 3\cos^2 I)^{-1/2} \tag{3.40}$$

Thus palaeointensity measurements are essentially a function of latitude or inclination. To enable results from different sampling localities to be compared Thellier arbitrarily referred all values to a palaeoisocline of 65° using (3.40). This is acceptable for archaeological results from a limited region such as Europe and the Mediterranean region from which the Thelliers have obtained results (Thellier and Thellier, 1959) but is not suitable for global comparisons or extension to the geological time scale. To enable results from different sampling localities at different latitudes to be compared it is more convenient to calculate an equivalent dipole moment p as in (3.39) or (3.40). This is analogous to the calculation of virtual geomagnetic pole positions (VGPs) from palaeodirectional data. Such a calculated dipole moment is called a *Virtual Dipole Moment* (VDM) (Smith, 1967a) by analogy.

In many archaeomagnetic studies palaeointensities are determined from broken pieces of earthenware (potsherds) and other baked material whose exact position when originally fired is not known. The magnetic inclination at the site at the time of firing thus may not be known. In such cases (3.39) or (3.40) have often been used with the magnetic colatitude θ or the present value of the magnetic inclination, I. The dipole moment calculated in this way is referred to as a *Reduced Dipole Moment* (RDM) (Smith, 1967b). However, even on the archaeological time scale the magnetic field at any place has varied widely so it is more appropriate to use the geocentric axial dipole assumption and insert the geographic colatitude in (3.39) to calculate a *Virtual Axial Dipole Moment* (VADM) (Barbetti, 1977).

The advantage of the VDM calculation is that no scatter is introduced by wobble of the main dipole because the magnetic colatitude determined is independent of the orientation of the dipole relative to the earth's axis of rotation. Calculation of the VDM is equally appropriate to the archaeological or geological time scales as long as the ancient magnetic inclination has been determined. McFadden and McElhinny (1982) have suggested that the observed VDMs for any geological epoch are derived from a set of *True Dipole Moments* (TDMs) plus the effect of non-dipole components and errors in the palaeointensity determinations. Assuming the non-dipole field strength is proportional to the TDM (see §4.1 and §6.3 for further discussion), the distribution of TDMs can be calculated from the observed VDMs. The peak in the calculated TDM distribution is not the same as that found by simply calculating the average VDM, but has more physical significance. The peak value of the TDM is the preferred value about which the dipole moment fluctuates. This value is thus the intensity analogue of the Palaeomagnetic Pole of directional data about which VGPs fluctuate and is referred to as the *Palaeomagnetic Dipole Moment* (PDM). The PDM is thus the time-averaged value of the earth's dipole moment (McFadden and McElhinny, 1982). Table 3.5 summarizes the various dipole moments referred to in palaeointensity studies. Further discussion follows in §6.3.

TABLE 3.5
Summary of various dipole moments used in palaeointensity studies

Reduced Dipole Moment (RDM)	Dipole axis assumed to be that determined from the *present* magnetic inclination at the site. Useful only for comparing archaeological data for the past few hundred years
Virtual Axial Dipole Moment (VADM)	Dipole axis assumed to be the axis of rotation (geographic axis). Used where no knowledge of magnetic inclination is available
Virtual Dipole Moment (VDM)	Dipole axis corresponds to that determined by the measured magnetic inclination (I) at the site. Analogous to the VGP for directional data. Dipole wobble does not introduce scatter in VDMs
True Dipole Moment (TDM)	Observed VDMs for any epoch are derived from a set of TDMs plus the effect of non-dipole components and errors in palaeointensity determinations. Only the *distribution* of TDMs can be calculated under given statistical assumptions
Palaeomagnetic Dipole Moment (PDM)	The time-averaged value of the earth's dipole moment, analogous to the Palaeomagnetic Pole of directional data. Determined as the peak value in the distribution of TDMs

Chapter Four

The Recent Geomagnetic Field: Palaeomagnetic Observations

4.1 Archaeomagnetic Results

4.1.1 Evidence for the Westward Drift

Pottery, and more usefully, bricks from pottery kilns and ancient fireplaces, whose last dates of firing can be estimated from carbon-14 contents of ashes, have a thermoremanent magnetization dating from their last cooling. The hearths of ancient fireplaces especially can provide measurements of declination and inclination at the time of their last firing, providing they have not since been disturbed. Bricks will often provide inclination data assuming they were fired horizontally, but all such material may be used for determining the ancient field strength typically through the use of Thelliers' techniques (§3.4.2). The most extensive archaeomagnetic studies to obtain directional data have been carried out by Soviet workers (Burlatskaya *et al.*, 1965, 1969, 1970a, b, 1975; Nachasova, 1972; Nodia and Chelidze, 1972; Rusakov and Zagniy, 1973). However, a high proportion of these results give inclination data only. The most detailed work has been carried out in Bulgaria and Yugoslavia by Kovacheva and Veljovich (1977) and Kovacheva (1980) covering the past 8000 years. Detailed studies have also been made in Japan mainly for the last 2000 years (Kawai *et al.*, 1965; Hirooka, 1971). Less detailed work has been carried out in North America (du Bois, 1974), France (Thellier, 1966), England (Cook and Belshé, 1958; Aitken and Weaver, 1962; Aitken *et al.*, 1963; Aitken and Hawley, 1966, 1967; Aitken, 1970) and Australia (Barbetti, 1977). Dated lava flows have been used for studying detailed variations in the geomagnetic field at Mount Etna (Chevallier, 1925; Tanguy and Wilson, 1973), in the Western USA (Champion, 1980) and on Hawaii (Doell and Cox, 1963; Holcomb *et al.*, 1980) and Japan (Yukutake, 1961).

Figure 4.1 compares the variations in inclination over the past 2000 years for the south-east European region (Bulgaria, Moldavia and the Ukraine) with that derived for south-west Japan. The curves have been derived as 50-year means using the available data for the two regions. These are the two most heavily sampled regions for archaeomagnetic data and they are separated by 110° of longitude at 35–40°N. There are two maxima and two minima in the inclination curves that occur between the present and about A.D. 600. These occur in Japan on average 350–400 years ahead of those in south-east Europe, corresponding to an average westward drift of about 0.3° longitude per year for the past 1400 years. Inclination is the most sensitive feature to consider. If isoporic foci are drifting westwards then maxima and minima will occur as they pass over the regions in turn.

Yukutake (1967) plotted the occurrences of the maxima and minima in declination and inclination for different regions of the northern hemisphere

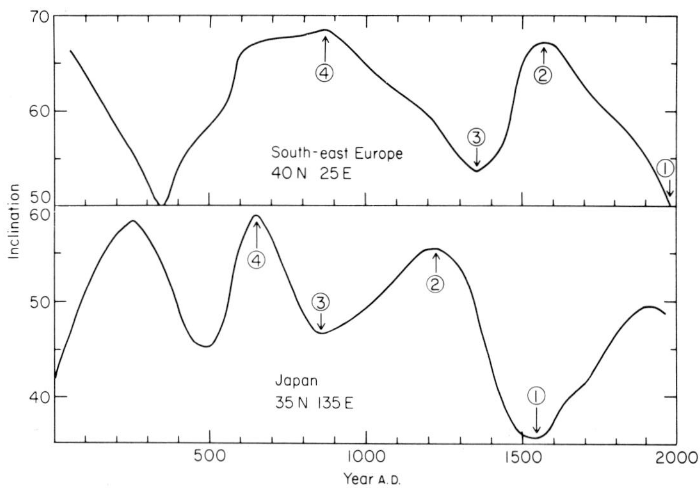

Fig. 4.1. Variations in magnetic inclination versus time for south-east Europe (upper figure) and Japan (lower figure), two regions separated in longitude by 110° but of similar latitude. Features indicated by the numbers 1–4 in the upper curve appear to be shifted back in time on the lower curve. This shift can be explained by an average westward drift near 0.3° longitude per year for the past 1400 years.

as a function of longitude. An updated version of his inclination analysis is shown in Fig. 4.2 using also data acquired over the past 15 years. The timing of the maxima and minima in each region depends critically on the amount of data from which a complete inclination curve is drawn. The largest variable in Fig. 4.2 is therefore the precise date of each maximum and minimum and these become less well-defined the older the date. However, the westward drift of the first three maxima and minima in the inclination curves is clearly demonstrated for almost the past 2000 years. The precision of the values of the rate of westward drift from each trend depends on how well defined each point is, so that it would be wrong to put too much emphasis on the varying rates obtained. Using the trends for the first two maxima and minima from Fig. 4.2 an average value of 0.38 ± 0.06° longitude per year is obtained (95% error) from the four values. This is similar to that deduced by Yukutake (1967) from analyses of declination and inclination maxima and minima. It is of course about twice the value of 0.2° per year deduced by Bullard *et al.*

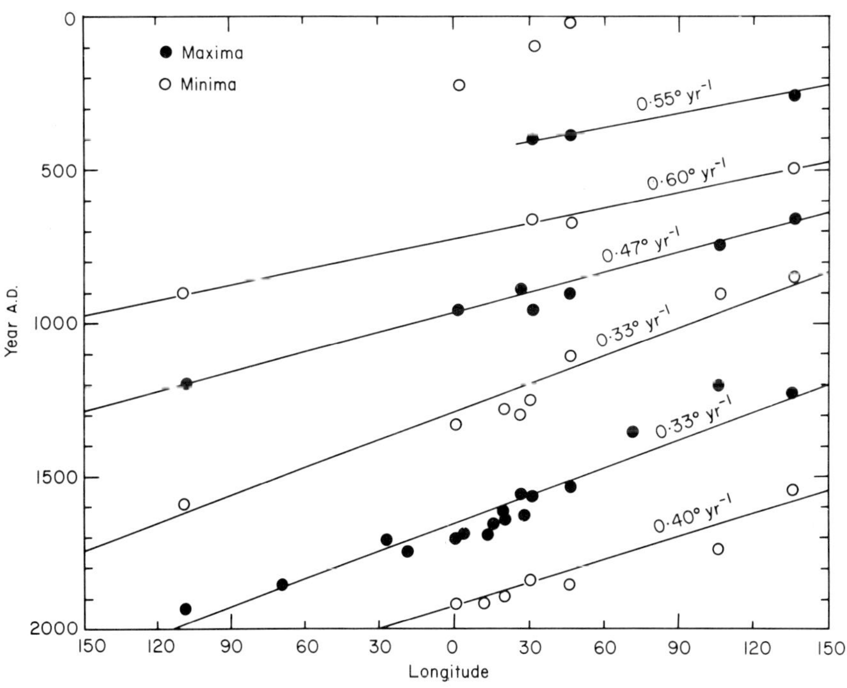

Fig. 4.2. Plot of various maxima (solid circles) and minima (open circles) in inclination as a function of longitude and age. Data come from mid-latitude regions in the northern hemisphere. Straight line fits to the data provide estimates for westward drift rates of the maxima and minima. All these estimates are significantly higher than the 0.2° per year drift rate of the non-dipole field obtained by Bullard *et al.* (1950). Updated from Yukutake (1967).

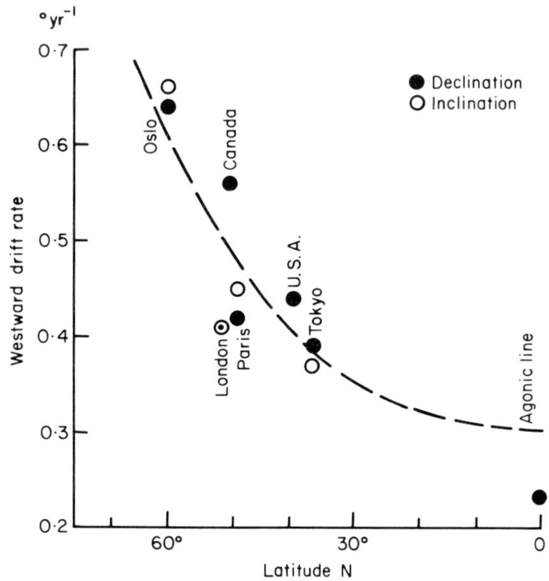

Fig. 4.3. Average westward drift rate estimates as a function of latitude for the northern hemisphere. Maxima and minima in inclination and declination data, such as given in Fig. 4.2 for mid-latitudes, have been used to construct this figure. The dotted line gives the distribution of angular velocity when the field rotates at a constant linear velocity of 0.058 cm s^{-1} at the surface of the core. (After Yukutake, 1967.)

(1950) for the westward drift of the non-dipole field. It was this observation that led Yukutake and Tachinaka (1969) to propose that the non-dipole field consisted of standing and drifting parts. The drifting field rotates at the faster rate and when the two are combined the lower value is then observed (§2.4.5).

Historical records also suggest the westward drift velocity is a function of latitude. Yukutake (1967) plotted the drift rates for features of declination and inclination for the equator, using the variation in longitude of the intersection of the west agonic line with the equator determined by Bauer (1895), and those for higher latitudes determined from Tokyo, the USA, Paris, London, Canada and Oslo. The latitude variation is illustrated in Fig. 4.3. There is a clear latitude dependence with values of around 0.4° per year associated with the mid-latitude regions from which the data of Fig. 4.2 were derived. The dashed curve in Fig. 4.3 gives the distribution of angular velocity when the field rotates at a constant linear velocity of 0.058 cm s^{-1} at the surface of the core.

4.1.2 Motion of the Dipole Axis

Spherical harmonic analyses of historical data back to A.D. 1600 have indicated a very slow movement of the dipole axis over the past 400 years (see §2.4.5). To attempt to extend this motion back in time using archaeomagnetic data, it is necessary to use some averaging technique that smooths out the variations in the non-dipole field so as to be able to see the main trends in the dipole field. Cox and Doell (1960) observed that the average of VGPs calculated from observatory data around the world plots very close to the present geomagnetic pole. Unfortunately archaeomagnetic data are not evenly place around the world, but are concentrated in the European region. Barbetti (1977) suggested that the effects of non-dipole field variations could best be averaged out if VGPs were averaged over 100-year intervals for a limited number of regions of the earth's surface. The gross movement of the dipole axis could be obtained by further averaging these 100-year interval means over a 500-year interval and then finding the regional mean for the whole globe.

The suggestion made by Barbetti (1977) has been used in detail by Champion (1980). He divided the world into seven arbitrary regions based on the number and location of available archaeomagnetic directional data. The regions chosen were Western North America, Central America, Northern Europe (Iceland, Great Britain and Poland), Southern Europe (Sicily, Bulgaria, the Ukraine and Soviet Georgia), East Africa, Australia and Japan. Since that analysis further results have become available from the island of Hawaii (Holcomb *et al.*, 1984). Champion's (1980) analysis has been extended to include Hawaii as one of eight regions and the 100-year means are listed in Table 4.1 and illustrated in Fig. 4.4. Several factors lead one to believe that these 100-year mean VGPs are close approximation to the successive positions of the North Geomagnetic Pole. The mean VGP calculated from the 1980 field values for the centre of each region lies close to the 1980 Geomagnetic Pole. Also the successive values for 1900 to 1600 lies close to, and have the same trend as, the positions of the Geomagnetic Pole calculated from historical observations (Barraclough, 1974; see §2.4.5).

Because the analysis is restricted to eight regions the 95% confidence limits (A_{95}) are never less than 5°. Furthermore the scatter in VGPs from epoch to epoch remains fairly constant with the precision parameter K at around 100. Successive positions of the North Geomagnetic Pole calculated in this way are not significantly different from one another, but the positions at the extreme locations on the path at 1700, 1300 and 900 years are different as shown by the error circles in Fig. 4.4. If 500-year averages are taken as suggested by Barbetti (1977), then the mean positions for each of the first three 500-year intervals are all significantly different from one another (Table 4.1). All of this strongly suggests that Fig. 4.4 displays the successive positions of the North Geomagnetic Pole at 100-year intervals over the past 2000 years. Interestingly the mean VGP for the 20 centuries lies close to the

TABLE 4.1

Estimated positions of the North Geomagnetic Pole (dipole axis) over the past 2000 years. 100-year means for eight possible regions, except 1980 which refer to present field values. *K*, Fisher's precision parameter; A_{95}, circle of 95% confidence

Age	No. regions	K	Lat. N	Long. E	A_{95}
1980	8	104.5	82.1	284.1	5.4
1900	8	98.8	82.3	288.2	5.6
1800	8	112.1	81.0	297.1	5.3
1700	7	108.3	81.1	307.1	5.8
1600	5	102.0	85.6	316.7	7.6
1500	6	99.3	86.3	301.5	6.8
1500–1900	8	177.5	82.4	299.3	4.2
1400	4	78.8	84.8	228.3	10.4
1300	4	197.8	83.2	189.1	6.6
1200	5	82.6	84.3	135.2	8.5
1100	4	100.1	85.3	110.0	9.2
1000	6	119.8	81.3	76.0	6.1
1000–1400	7	126.6	85.9	124.3	5.4
900	5	181.3	80.2	38.0	5.7
800	4	215.7	81.8	28.0	6.3
700	4	95.1	84.1	33.4	
600	5	56.3	85.6	6.6	10.3
500	4	145.2	86.1	343.5	7.7
500–900	7	147.6	83.8	13.8	5.0
400	4	45.8	86.0	316.3	13.7
300	4	56.2	88.9	131.9	12.4
200	4	59.5	87.7	160.3	12.0
100	4	123.9	87.7	143.9	8.3
0	3	80.0	86.4	121.4	13.9
0–400	6	109.3	88.9	102.7	6.4
20 centuries mean ($N = 20$)		180.7	88.6	358.1	2.4

North Geographic Pole suggesting that measurements over 2000 years are sufficient to average the geomagnetic axis to the axis of rotation. This observation needs to be viewed with considerable caution. It is not at all clear that the motion of the dipole axis over the past 2000 years as depicted in Fig. 4.4 can be regarded as a recurring feature, or that the average over the preceding 2000 years would also coincide with the geographic axis. The basic problem as to what is sufficient time to average palaeomagnetic results, so that the mean VGP coincides with the axis of rotation, will be discussed more fully in Chapter 6.

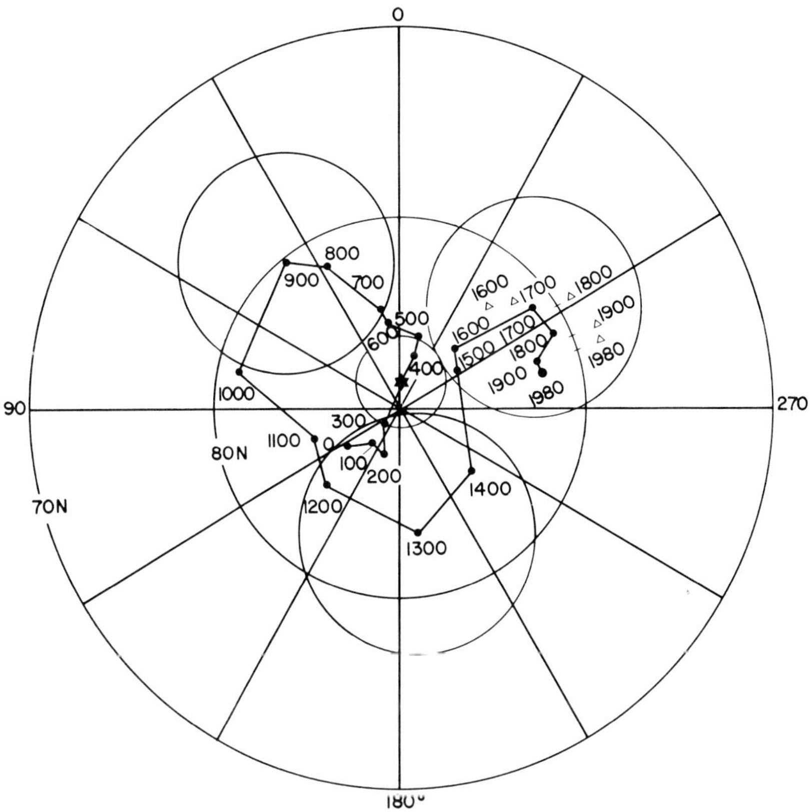

Fig. 4.4. The locations of the North Geomagnetic Pole (dipole axis) over the past 2000 years. 100-year means from eight regions are given along with the present field value (1980). Although successive values are not significantly different from each other, the extreme locations on the path are significantly different, as evidenced by 95% confidence circles around the 1700, 1300 and 900 year means. Triangles show poles obtained from spherical harmonic analysis of Barraclough (1974).

4.1.3 Variations in the Dipole Moment

Many more archaeomagnetic intensity determinations have been made compared with directional ones, but the distribution of sites is still very poor, grouped mostly in the northern hemisphere, and even then concentrated largely in Europe, the western USA and Japan (Barton *et al.*, 1979; McElhinny and Senanayake, 1982). Typically determinations have been made using the Thelliers' method on ancient hearths and pottery. Figure 4.5 shows the variations in intensity determined in south-east Australia from aboriginal fireplaces and other baked clays and sediments. The field strengths vary by a factor of two, and sometimes rapid changes occur over only a few

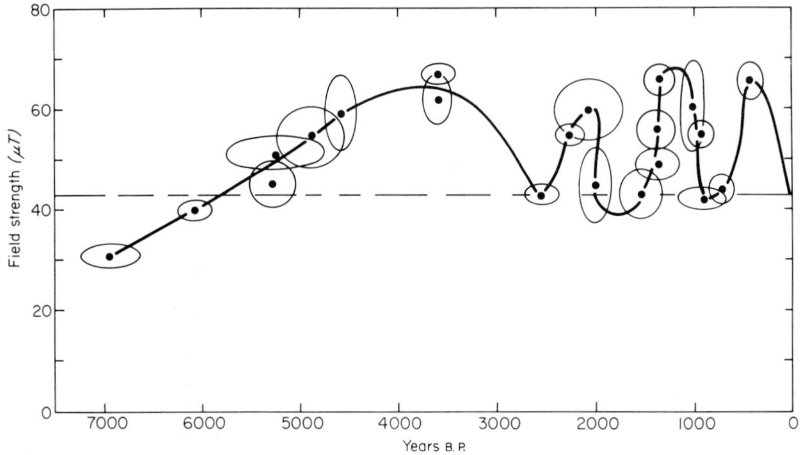

Fig. 4.5. Estimates of the intensity of the earth's magnetic field as a function of time in south-east Australia. Data are from aboriginal fireplaces and other baked clays and sediments. (After Barbetti, 1982.)

hundred years. These variations are argued to be too large to be due to experimental error. Instead they appear to be reasonable estimates of field strengths, and if so, accuracy of the dating indicates that the changes sometimes occur on the order of a hundred years or less. That such rapid changes in intensity at any locality might be a real feature of the past field is supported by results from other parts of the world, notably Egypt. In this case the intensity measurements have been made using a new technique applicable only to certain archaeological material.

The method of making adobe (sun-dried) mud bricks in Egypt is to fling mud into a wooden former, which is then removed, leaving the brick to dry in the sun. Typical drying times for such bricks are from three days to a week, depending on the weather. When a sample of mud is thrown into a container or stirred inside a container, the sample is always magnetized along the direction of the applied field (Games, 1977). It is argued that a mud brick, if kept abnormally wet over long periods but not stirred or mixed, will not change its direction of magnetization apart from the acquisition of VRM over the time interval. Games (1977) has carried out investigations into this magnetization acquisition and has discovered that the magnetization acquired by the mud brick in the high coercivity region (most likely in small micron-sized grains that are freer to rotate) seems to be caused by shear strain (squeezing) alone. This magnetization has been termed shear remanent magnetization (SRM) and Games (1977) has shown that it is not only an accurate recorder of the ancient field direction but that the magnitude is proportional to the ancient field strength.

Games (1977) has produced a simple piston device that reproduces the "throwing" procedures in the laboratory. A small sample of mud from an ancient brick can be reconstituted in the laboratory and its SRM acquisition measured. AF demagnetization of NRM and SRM are compared in the high coercive force region as in Shaw's method (§3.4.5). The AF demagnetization spectrum of ARM of the original brick and the laboratory produced SRM are also compared as in Shaw's method to make sure that the new SRM has the same characteristics as the original. The procedure for ancient field strength determination is similar to that outlined in Shaw's method, with SRM being substituted for TRM.

Results produced by this method from adobes in Egypt are illustrated in Fig. 4.6. Many of the bricks come from ancient tombs and other structures whose ages are known within a few years. Often pieces of pottery are found in the bricks and independent checks can be made of the intensity data using the Thelliers' method on this material. The agreement as shown in Fig. 4.6 is impressive. The data of Fig. 4.6 show that changes in field strength at any locality could be as large as a factor of three over only a few hundred years. Although such changes are recorded as variations in a dipole moment (§3.4.4), they are most likely due to changes in the non-dipole field. Figure 2.2c illustrates that the intensity of the present day field values varies by more than a factor of two at some latitudes (e.g. 30°S). Such wide variations in

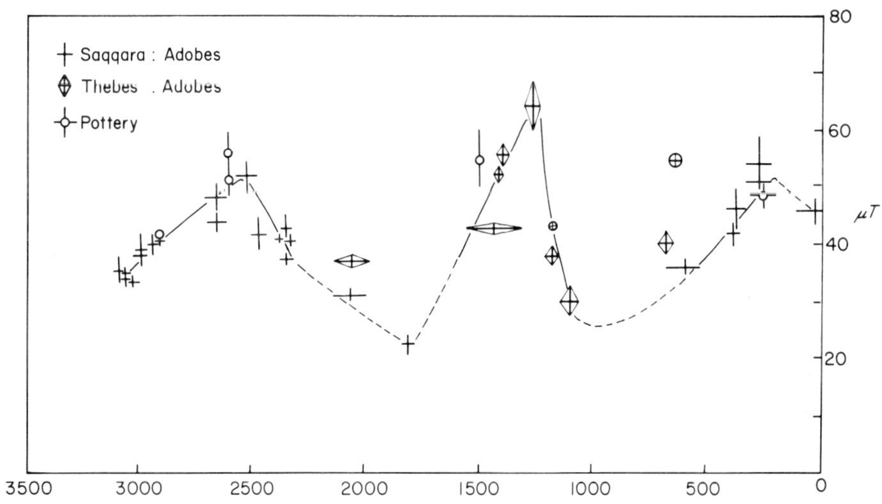

Fig. 4.6. Estimates of the intensity of the earth's magnetic field as a function of time in Egypt. Data are obtained from a new intensity technique to obtain estimates from mud bricks from two localities, Saqqara and Thebes (Games, 1977, 1980). Estimates using more conventional techniques on pottery are also shown. (After Games, 1980.)

intensity coupled with the poor distribution of sites pose problems for the analysis of global intensity data.

Barton *et al.* (1979) and Champion (1980) have analysed global archaeomagnetic data to determine possible time variations in the earth's dipole moment. Because most of the data were concentrated in the European region, it was difficult to judge the extent to which the data, when averaged, represent changes in the dipole moment. There appears to be an increase in intensity to a maximum about 2500 years ago, preceded by a minimum about 6500 years ago. Previously Bucha (1967, 1969) and Cox (1968) had suggested that this was indicative of variations in the dipole moment with simple periodicity of between 8000 and 9000 years with maximum and minimum respectively about 1.5 and 0.5 times the present dipole moment. Barton *et al.* (1979) argue strongly that inadequacies in the archaeomagnetic data are such that this periodicity should be regarded as highly tentative.

Barton *et al.* (1979) and Champion (1980) analysed about 600 intensity results from around the world. A much wider data set of 1167 world-wide results was used by McElhinny and Senanayake (1982) covering the past 12,000 years. In an attempt to average out non-dipole field variations it has been the practice to average results over 500- or 1000-year intervals (Cox, 1968; Barton *et al.*, 1979; Champion, 1980). Table 4.2 summarizes the analysis of McElhinny and Senanayake (1982) using 500-year averages back to 4000 years B.P. and 1000-year averages prior to that. The results are displayed in Fig. 4.7 together with the 95% error limits. The mean dipole moment for the last ten 1000-year intervals is 8.75×10^{22} A m^2 with an estimated standard deviation of 18.0% which may be attributed to dipole intensity fluctuations. For the past 8000 years the data from the European region are in general agreement with that from the rest of the world. There is a maximum about 2500 years ago and a minimum about 6500 years ago. Prior to 8000 years the European data have high values, a feature that encouraged some previous analysts to speculate on the completion of a sinusoidal cycle over about 10,000 years. However, the data from the rest of the world show continuing low values between 8000 and 12,000 years ago, so that the sinusoidal behaviour of the dipole moment is clearly not indicated. This conclusion is supported by the few data (14) that come from the period 15,000 to 50,000 years B.P. (McElhinny and Senanayake, 1982). The mean dipole moment for this interval is 4.44×10^{22} A m^2 with standard deviation 26.5% (see Fig. 4.9 and §4.1.5). This is significantly lower than the dipole moment for the past 1000 years and suggests there are changes in the dipole moment with time scales at least as large as 10^5 years.

Because there are contributions from rock magnetic and experimental errors and from the non-dipole field variation, the non-uniqueness problems in the determination of the variations of the dipole moment appear

TABLE 4.2
Global summary of dipole moments determined for the past 12,000 years. σ, standard deviation. From McElhinny and Senanayake (1982)

Years B.P.	N	Dipole moment (10^{22} A m^2)	σ	95% error
0–500	268	8.72	1.44	0.17
500–1000	187	10.30	1.89	0.27
1000–1500	205	10.90	2.01	0.27
1500–2000	131	10.94	2.18	0.37
2000–2500	89	11.10	2.63	0.54
2500–3000	60	11.28	2.49	0.63
3000–3500	43	9.64	2.85	0.85
3500–4000	17	9.21	1.90	0.90
4000–5000	34	8.87	2.20	0.74
5000–6000	44	7.20	1.94	0.57
6000–7000	36	6.73	2.00	0.65
7000–8000	18	7.08	1.44	0.66
8000–9000	15	8.61	2.32	1.17
9000–10000	14	8.26	2.39	1.25
(10000–11000)	5	6.76	1.34	1.17
(11000–12000)	2	8.36	(0.06)	(0.08)
Mean of 1000-year intervals	10	8.75	1.58	0.97

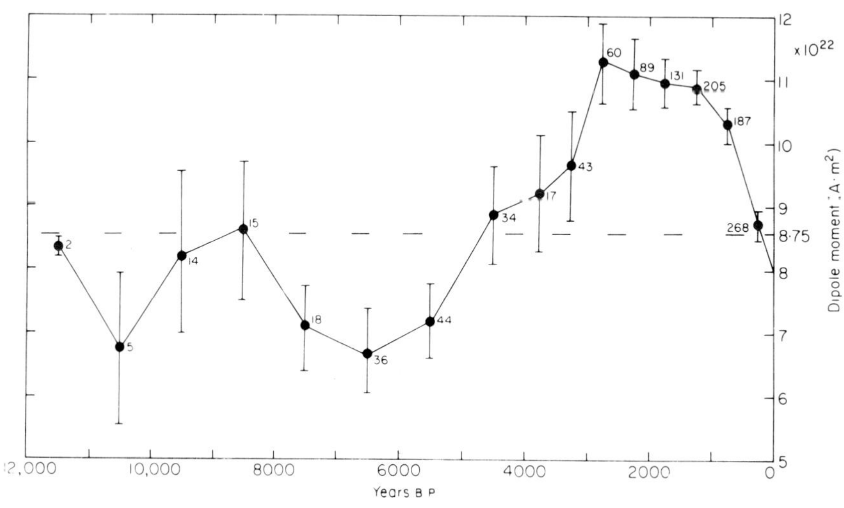

Fig. 4.7. Global Dipole Moment versus time estimates obtained from 500-year period averages. Data are taken from Table 4.2 and the error bars shown are for the 95% confidence level. (After McElhinny and Senanyake, 1982.)

formidable. The key to sorting out some of these problems lies in the observation from historical and palaeomagnetic data that the dipole field changes more slowly than the non-dipole field. In particular McElhinny and Senanayake (1982) utilize the observation that there is a fairly broad maximum in dipole moment between 1000 and 3000 B.P. (Fig. 4.7). By assuming a constant dipole moment in this interval, they obtained a first order estimate of the scatter due to other contributions. The estimated standard deviation of 472 values in this interval is 19.7%. If a similar analysis is done over the last 10,000 years (assuming constant dipole field in 1000-year intervals), the corresponding estimated standard deviation is 21.2%. Note that these last two numbers refer to the within-1000-year-interval standard deviation, whereas the figure given earlier (18.0%) refers to the between-interval standard deviation. Using the present field as a first order indicator of the magnitude of the non-dipole contributions, McElhinny and Senanayake (1982) obtain a value of 17.5% for the standard deviation of the non-dipole field contributions. This leaves a 10–12% estimate for the scatter attributed to rock magnetic and experimental errors.

The type of analysis given above makes it possible to make comparisons between the non-dipole field variation and the dipole field variation over the past 10,000 years or so. McElhinny and Senanayake (1982) have also done this for the 14 results between 15,000 and 50,000 years ago to find similar values to those given above. This led them to suggest that the variations in the non-dipole field always remain in the same proportion to the dipole field irrespective of its magnitude. This suggestion has been strongly supported by detailed statistical analyses of McFadden and McElhinny (1982). This is an important result and is discussed further in Chapters 6 and 10.

4.1.4 Deductions from Carbon-14 Variations

The isotope ^{14}C is one of several species of radionuclide produced by the interaction of cosmic rays with the earth's atmosphere. ^{14}C is produced mainly in the stratosphere by the (n, p) reaction with ^{14}N and it forms $^{14}CO_2$ which is rapidly and thoroughly mixed with ordinary carbon dioxide ($^{12}CO_2$ and $^{13}CO_2$) in the atmosphere. Carbon dioxide is continually exchanged between the atmosphere and the oceans which contain about 95% of the exchangeable carbon and where the mean residence time is of the order of 1000 years. Exchange between the oceanic and atmospheric reservoirs regulates the carbon dioxide content of the atmosphere. Carbon-14 dating is based on the fact that atmospheric CO_2, with its trace level of radiogenic ^{14}C, is absorbed and incorporated by living materials. After incorporation the level of ^{14}C gradually diminishes through radioactive decay with a half-life of 5730 years. Measurement of the residual ^{14}C allows an estimate of age to be made using the assumption that the atmospheric concentration in the past was constant.

Changes in atmospheric ^{14}C concentration, apart from recent artificial perturbations (e.g. fossil fuel and atomic bomb effects) can arise mainly from reservoir changes or from changes in the production rate of ^{14}C. Damon (1970) predicts that the effects of reservoir changes are small and that changes in the production rate are mainly due to changes in the high energy particle flux from the sun and to changes in the geomagnetic dipole moment. Observed variations in the atmospheric ^{14}C concentration on time-scales of less than 500 years are generally attributed to solar activity. The longer-term variations over several thousand years or more are generally attributed to variation in the geomagnetic dipole moment.

A charged particle moving in the plane normal to the lines of force of a uniform magnetic field B will describe a circle of radius R. It can be shown that the product BR is equal to the ratio of the relativistic momentum m to charge ze, where z is the charge number and e is the charge on the electron. The *magnetic rigidity* of the particle (Wolfendale, 1963) is defined by the quantity

$$\frac{mc}{ze} - cBR \text{ volts} \qquad (4.1)$$

where c is the velocity of light.

The effect of the geomagnetic field on the cosmic ray particle is such that certain classes of the trajectories of particles approaching the earth remain trapped along field lines (a slight leakage does occur but will be neglected in this discussion). These bounded trajectories therefore cannot be followed by incoming cosmic ray particles (see, for example, Rossi, 1964). At a given geomagnetic latitude particles with rigidity less than a certain value cannot reach the earth. Particles with a slightly greater rigidity may arrive at the earth's surface from the west (for positively charged particles), but not from the east. The critical value of rigidity with which positively charged particles may arrive at the earth's surface vertically or from the west, but not from the east, is known as the *vertical cut-off rigidity*, P_λ. Elsasser *et al.* (1956) give the following expression for P_λ as a function of geomagnetic dipole moment p and latitude λ:

$$P_\lambda = \frac{\cos^4 \lambda}{4} \cdot \frac{\mu_0 pc}{4\pi r^2} \text{ (volts)} \qquad (4.2)$$

where r is the radius of the earth. Taking the present value of the dipole moment p_0 as 8.0×10^{22} A m^2 and substituting values of μ_0, r and c, (4.2) reduces to

$$P_\lambda = 14.8 \frac{p}{p_0} \cos^4 \lambda \text{ (Gega volts)} \tag{4.3}$$

The above model of the modulation of the cosmic ray spectrum by the geomagnetic field has been used by several authors to calculate the ^{14}C production rate as a function of geomagnetic dipole moment (Elsasser *et al.*, 1956; Wada and Inoue, 1966; Bucha and Neustupny, 1967; Lingenfelter and Ramaty, 1970). Lingenfelter and Ramaty (1970) deduced the following relation for the ^{14}C production rate, Q:

$$Q = kp^{-1/2} \tag{4.4}$$

where k is a rather complex function derived from the model. Equation 4.4 is valid only for values of p close to its present value and the simple relationship breaks down for much smaller or much larger values. Applying this to the roughly sinusoidal variations in dipole moment deduced by Cox (1968) for the past 10,000 years, they concluded that the gross features of ^{14}C variations could indeed be understood in terms of the variations in geomagnetic moment.

The comparison of ages measured using the radiocarbon method with those deduced from dendrochronology has enabled the departures from the assumption of constant atmospheric ^{14}C concentration to be determined. Clark (1975) has listed estimates of the true age T_t corresponding to radiocarbon ages T_c back to 6500 years B.P. The relative atmospheric ^{14}C concentrations, C_A, can then be calculated from the expression

$$C_A = \exp\left\{\left(\frac{T_t}{5730} - \frac{T_c}{5568}\right) \ln 2\right\} \tag{4.5}$$

where the half-life of 5568 years is the one conventionally used to determine radiocarbon age. Assuming these ^{14}C variations are solely of geomagnetic origin, and that the production and decay of ^{14}C are instantaneously balanced so that the atmospheric concentration varies linearly with production, the geomagnetic dipole moment can be calculated (Barton *et al.*, 1979) from (4.4). The dipole moment calculated in this way is referred to as the *Radiocarbon Dipole Moment* (RCDM) and its deduced variation is shown in Fig. 4.8. The short-term variations in ^{14}C atmospheric concentration seen in Fig. 4.8 are probably due to changes in the solar high energy particle flux, and only the broad trends should be attributed to changes in dipole moment.

Comparison between the RCDM (Fig. 4.8) and the dipole moment obtained from archaeomagnetic data (Fig. 4.7) show the same general features, providing it is recognized that the former has been seriously affected by the large increase in fossil fuel burned by man, etc. These effects are

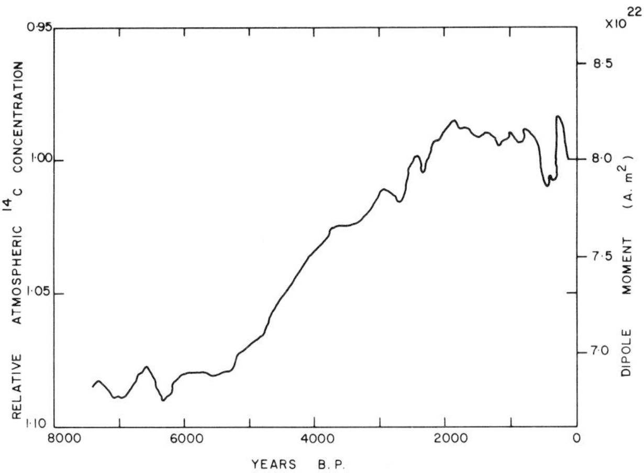

Fig. 4.8. Radiocarbon dipole moment (right vertical) axis versus age. The corresponding relative atmospheric ^{14}C concentration is shown on the left-hand side. Data for the last three or four hundred years have been seriously affected by artificial sources (fossil fuel burning by man). (After Barton *et al.*, 1979.)

essentially confined to the last three or four hundred years. (It should be noted that different time intervals are represented in Figs 4.7 and 4.8.) The broad trends depict the variations in dipole moment and a theoretical delay of about 1000 years (Houtermans, 1966; Houtermans *et al.*, 1973) can be expected between changes in ^{14}C concentration and changes in the dipole moment. The ^{14}C concentration minimum between 1000 and 2000 B.P. implies a broad maximum in dipole moment between about 2000 and 3000 B.P. as is observed in Fig. 4.7. The high ^{14}C concentrations between 5500 and 7500 B.P. imply a broad minimum in dipole moment between 6500 and 8500 B.P. as is also observed in Fig. 4.7. This provides confidence that the longer period changes in Figs 4.7 and 4.8 are due to variations in the dipole moment.

4.1.5 Dipole Moments Before 10,000 Years B.P.

Estimates of the geomagnetic dipole moment in the time range 50,000 to 10,000 years B.P. have been reviewed by Barbetti and Flude (1979b). The results at present suggest the geomagnetic dipole moment was very much lower than its present value during this time span (Fig. 4.9). The only exceptions are high values associated with the Lake Mungo geomagnetic excursion (§4.3.1) but these are probably not truly representative of the geomagnetic dipole moment at that time (see §4.3.3).

If the trends to lower values in Fig. 4.9 are representative of this time then similar trends should be seen in a comparison between ^{14}C dates and other dating methods. Barbetti (1980) has summarized ^{14}C and comparative thermo-luminescence or uranium series ($^{230}Th/^{234}U$) ages between 40,000 and 10,000 years B.P. Atmospheric ^{14}C concentrations deduced from these comparisons using (4.5) are almost all much higher than their present value implying a prolonged period of reduced dipole moment as suggested in Fig. 4.9. These results indicate that the quasi-cyclic geomagnetic moment variation suggested by several workers for the past 10,000 years (Cox, 1968; Bucha, 1969; Barton *et al.*, 1979) did not extend to previous epochs. Between 50,000 and 10,000 years B.P. the dipole moment was apparently very much weaker than the present value.

Between 20,000 and 30,000 years B.P. the 13 values of palaeointensity in Fig. 4.9 have an average value of 4.5×10^{22} A m^2, whereas the average over the past 10,000 years has been 8.6×10^{22} A m^2 (Table 3.2). One view would be that the most recent trends in dipole moment over the past 10,000 years represent a broad maximum and those between 20,000 and 30,000 years ago represent a broad minimum in a much longer-term variation with period around 50,000 years. This is consistent with an analysis of intensity variations in deep-sea sediments by Kent and Opdyke (1977). They suggested a possible secular variation period in palaeointensity of 43,000 years (see

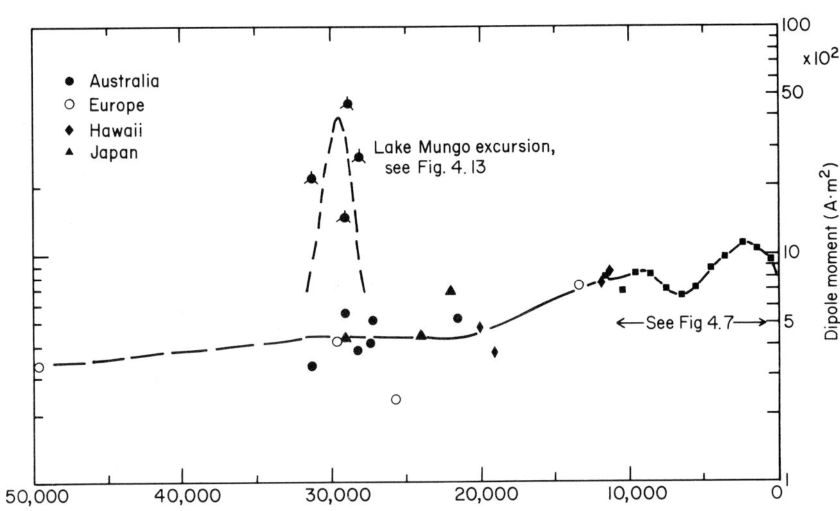

Fig. 4.9. Estimates of the dipole moment obtained from palaeointensity data for the last 50,000 years. Data for the last 12,000 years are shown in more detail in Fig. 4.7. Data for the Lake Mungo excursion are also shown and discussed in §4.3.3. (After McElhinny and Senanayake, 1982.)

§4.4.3) with amplitude about 30‰. However, global analysis of the dipole moment over the past five million years (McFadden and McElhinny, 1982) suggests a mean value of 8.7×10^{22} A m^2 for the normal field with standard deviation attributable to dipole fluctuations of 42‰ (§6.3.2). This suggests that the extended period of low dipole strength about 25,000 years ago are part of the expected variations over long time-scales. Extended time-spans at low dipole moment may contribute to the occurrence of other phenomena, such as geomagnetic excursions (§4.3.3).

4.2 Analysis of Recent Lake Sediments

4.2.1 Extension of Archaeomagnetic Results

Sediments deposited under suitably quiet conditions have been found to possess a stable remanent magnetization. In the case of slowly deposited deep-sea sediments this has provided a record of reversal history of the geomagnetic field (Chapter 5). Sediments deposited in lakes have provided a high resolution record of the geomagnetic field over long periods of time. The typical high sedimentation rate in lakes (of the order of 1 m per 1000 years) and their long life-span (of the order of 10,000 years), make them ideal not only for extending the archaeomagnetic record back in time but also because a continuous record is obtained.

The magnetization of lake sediments (and other types of sediments) is due to post-depositional DRM. Small magnetic particles (already magnetized from their previous history), slowly deposited in these lakes, are free to move in the water-filled interstices of a newly deposited sediment and their magnetic axes align (statistically) with the ambient magnetic field. During compaction and/or growth of the gels in the sediment and sometimes bioturbation, these magnetic particles become locked in position, giving the sediment an overall magnetization parallel to the geomagnetic field at that time. This recording process is particularly efficient in fine-grained homogeneous sediments, such as muds, that are characteristic of so many recent lakes. Particularly advantageous are the sediments deposited in lakes in volcanic craters (maars), because often these craters are unconnected with the local water table and, without inlet or outlet for water (other than rain or evaporation), the sediments have little chance of being disturbed. When these lakes finally dry up and cease to exist, the dry lake sediments still preserve their detailed record of the changes in the ancient geomagnetic field. Verosub (1977) has reviewed aspects of depositional and post-depositional processes in the magnetization of sediments. Barton *et al.* (1980) have carried out some detailed laboratory studies of the acquisition of magnetic remanence in slurries of fine grained organic muds. Henshaw and Merrill (1980) have reviewed the effects chemical changes can have on the remanence in sediments.

Mackereth (1971) first discovered long-period declination oscillations in cores taken of the post-glacial organic sediments deposited at the bottom of Lake Windermere in England. Sampling was carried out using a pneumatically controlled Mackereth corer (Mackereth, 1958). These corers take 6 m long cores and have been extensively used for studies of lake sediments throughout Europe and in Australia. An extended version for 12 m cores has been developed by Barton and Burden (1979) and used in Australia. In North America wet lakes have been cored using conventional gravity corers or Livingstone piston corers, but extensive work has also been carried out on dry lakes.

Creer *et al.* (1972) subsequently confirmed the early work of Mackereth (1971) and found that the inclination variations in Lake Windermere were apparently of irregular period and amplitude. The results from Lake Windermere were later confirmed by further work in north-west England (Thompson, 1975), Scotland (Turner and Thompson, 1979), Finland (Stober and Thompson, 1977), Poland (Creer *et al.*, 1979), Switzerland (Thompson and Kelts, 1974) and from cave deposits in the Lebanon (Creer and Kopper, 1976). The most detailed record for the past 7000 years has been obtained from Loch Lomond (Turner and Thompson, 1979) and this shows that there are identifiable features of variations in inclination as well as in declination. By combining all radiocarbon and palynological age determinations from Britain with the observed magnetic data, Thompson and Turner (1979) showed that the inclination data were repeatable between lakes. They then produced a master curve for Britain for the past 10,000 years in which the main features of declination and inclination were dated. This master curve, applied to a core from Lake Windermere, is illustrated in Fig. 4.10. It was suggested that this master curve would be applicable over the whole of Western Europe for the last 10,000 years.

In North America cores from Lake Michigan and Lake Erie were examined by Creer *et al.* (1976a, b). Both records are superficially similar to those obtained at Lake Windermere with distinct oscillations in declination being observed and less pronounced changes in inclination. Further work in Lake Michigan has been reported by Vitorello and Van der Voo (1977) and Dodson *et al.* (1977). Lund and Banerjee (1979) have summarized these studies including also results from Lake St. Croix in the Minnesota lakes. The reproducibility of magnetic features only a few hundred years in length in parallel cores from these lakes provides clear evidence of the fidelity of the recording process. At present no attempt has been made to produce a master curve in which the pronounced features have been dated as for Great Britain.

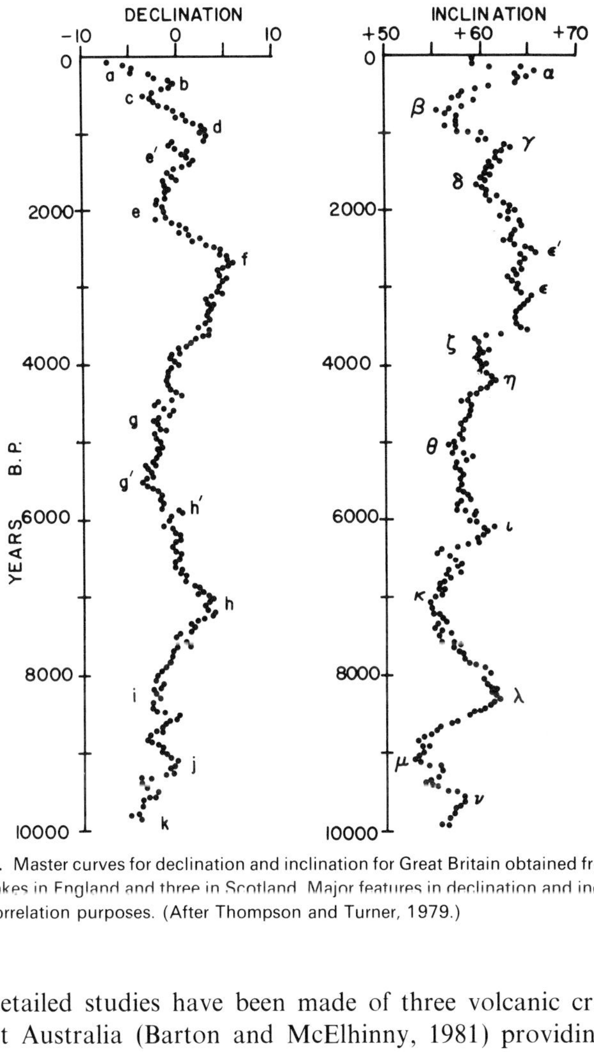

Fig. 4.10. Master curves for declination and inclination for Great Britain obtained from lake sediments from three lakes in England and three in Scotland. Major features in declination and inclination are given letters for correlation purposes. (After Thompson and Turner, 1979.)

Very detailed studies have been made of three volcanic crater lakes in south-east Australia (Barton and McElhinny, 1981) providing records of inclination and declination over the past 10,000 years. Unlike the records in Europe and North America the most pronounced swings are observed in inclination rather than declination. The master curves for south-east Australia are shown in Fig. 4.11. They represent 100-year means obtained by stacking profiles from the three lakes using the most pronounced features and the many radiocarbon dates determined for these lakes.

Dry lake sediments have been most extensively studied in North America, the most recent work being on the non-varved dry lake sediments from the western United States. Detailed secular variation records are available from Mono Lake, California from several studies by Denham and Cox (1971),

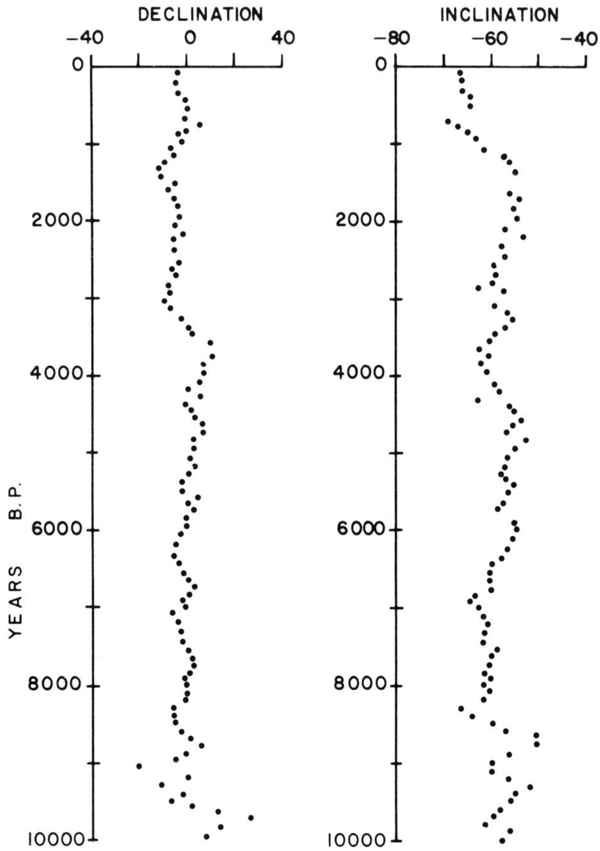

Fig. 4.11. A master curve for south-east Australia gives the declination and inclination variation recorded in lake sediments for the past 10,000 years. Individual points represent 100-year means obtained from stacked profiles from three lakes. (After Barton and McElhinny, 1981.)

Denham (1974) and Liddicoat and Coe (1979). In Mexico, dry lake deposits have also been studied at Tlapacoya by Liddicoat *et al.* (1979). In New England varved sediments have been studied by Verosub (1979).

4.2.2 The Westward Drift and Runcorn's Rule

Runcorn (1959) has shown that a moving magnetic source (such as a dipole) in the earth's core can cause the magnetic field vector at a fixed observatory to rotate clockwise (when viewed along the vector direction) for a westward motion of the source and anticlockwise for an eastward motion. If the total field at the surface is a function of time t and position $\boldsymbol{r}$ of the observatory, then

$$\boldsymbol{H}(\boldsymbol{r}, \boldsymbol{r}'(t)) = \boldsymbol{H}_0(\boldsymbol{r}) + \boldsymbol{H}'(\boldsymbol{r}'(t)) \tag{4.6}$$

where $\boldsymbol{H}_0(\boldsymbol{r})$ is the steady component primarily due to the main axial dipole, and $\boldsymbol{H}'(\boldsymbol{r}'(t))$ is the perturbation field due to a small, moving, non-axial dipole located at $\boldsymbol{r}'$. As the small dipole moves beneath the point of observation, the motion of the total field vector $\boldsymbol{H}$ will generate an elliptical cone and the tip of $\boldsymbol{H}$ will describe an ellipse. The situation is illustrated in Fig. 4.12. The sense of motion along this ellipse is opposite for a westward moving small dipole to an eastward moving one. It is important in visualizing this to recognize that one is only referring to the motion on the plane perpendicular to H. This proposal has become known as *Runcorn's Rule*.

As a result of the perturbation of the surface field by a moving dipole source, a plot of inclination versus declination (so-called Bauer plots) describes an elliptical curve corresponding to the motion above. The axes of the plot are of course oriented corresponding to geographical directions. The presence of several such curves (all clockwise) in geomagnetic records was first noticed by Bauer (see, for example, Bauer, 1895). Skiles (1970) and Dodson (1979) have examined Runcorn's rule in some detail and shown that under certain limited conditions anticlockwise rotation of the vector could still be produced by westward drifting sources.

If the perturbation of the field is caused by a single westward drifting

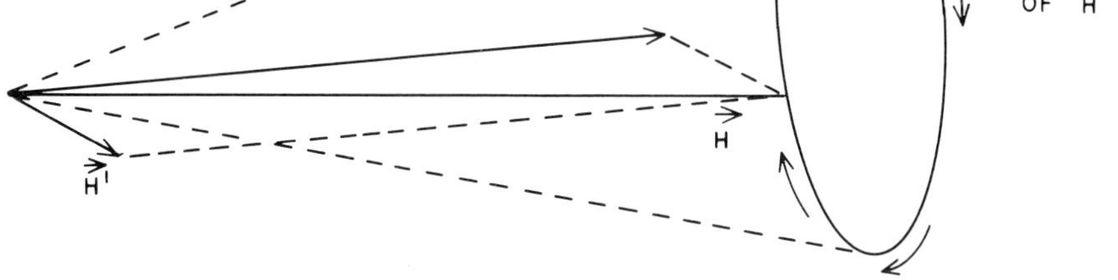

Fig. 4.12. Runcorn's rule can be described in terms of a change in the total instantaneous magnetic field $\boldsymbol{H}$ about its steady value $\boldsymbol{H}_0$. $\boldsymbol{H}$ is the vector sum of $\boldsymbol{H}_0$ and a small perturbation field $\boldsymbol{H}'$. The tip of $\boldsymbol{H}$ describes an ellipse about $\boldsymbol{H}_0$. A clockwise sense of motion around this ellipse, when viewed along the vector direction, indicates westward drift.

harmonic $P_l^m(\cos\theta)$, the perturbing vector traces out an ellipse whose plane is inclined at an angle ψ to the horizontal with:

$$\tan\psi = \frac{-(l+1)P_l^m(\cos\theta)}{\mathrm{d}P_l^m(\cos\theta)/\mathrm{d}\theta} \tag{4.7}$$

The magnetic vector corresponding to this harmonic then traces this ellipse with period $2\pi/m\omega$, where ω is the angular velocity relative to the mantle of the westward moving source in the core. Runcorn (1959) has shown that his clockwise rule will apply provided

$$I_0 - 180° < \psi < I_0 \tag{4.8}$$

where I_0 is the inclination of the axial dipole field vector. This condition is satisfied by all but a few low degree harmonics. If condition (4.8) is not satisfied then the magnetic vector will rotate anticlockwise for a westward drifting dipole source. However, any realistic source field might consist of several harmonics of different degree. Dodson (1979) has examined several such situations and shown that under certain circumstances Runcorn's rule will be violated. The exception will be elaborated on shortly, but Runcorn's rule can still be regarded as applicable in a statistical sense.

Runcorn's clockwise rule has been used as a method of inferring the direction of drift of the geomagnetic field from data at a single observatory. It is particularly applicable to the palaeomagnetic data from lake sediments. Over the past 10,000 years there appears to be a predominant clockwise motion observed in the data so far available, although specific sections of some records show times when anticlockwise motion is seen. The Bauer diagrams corresponding to the younger sections of the master curves from Britain (Fig. 4.10) are shown in Fig. 4.13 (Turner and Thompson, 1979). Although the dominant motion is clockwise there is an extensive anticlockwise portion between 1100 and 600 years B.P. For older records Skiles (1970) observed clockwise motion as dominating results from Japan in the time range around 50,000 years ago. The record from Mono Lake, California at around 25,000 years is considered to be the best existing evidence for anticlockwise motion and eastward drift (Denham, 1974). However, Dodson (1979) points out that this motion could also be produced by two westward drifting dipoles, rather than a single eastward drifting one.

Dodson (1979) has considered the two extreme cases of zero probability of eastward drift and equal probability of westward and eastward drift. In the case of no eastward drift the following should be observed:

(1) Anticlockwise motion would seldom be observed at high latitudes.
(2) Clockwise loops would be more numerous unless there is a bias towards high latitude anomalies and/or towards positive anomalies.

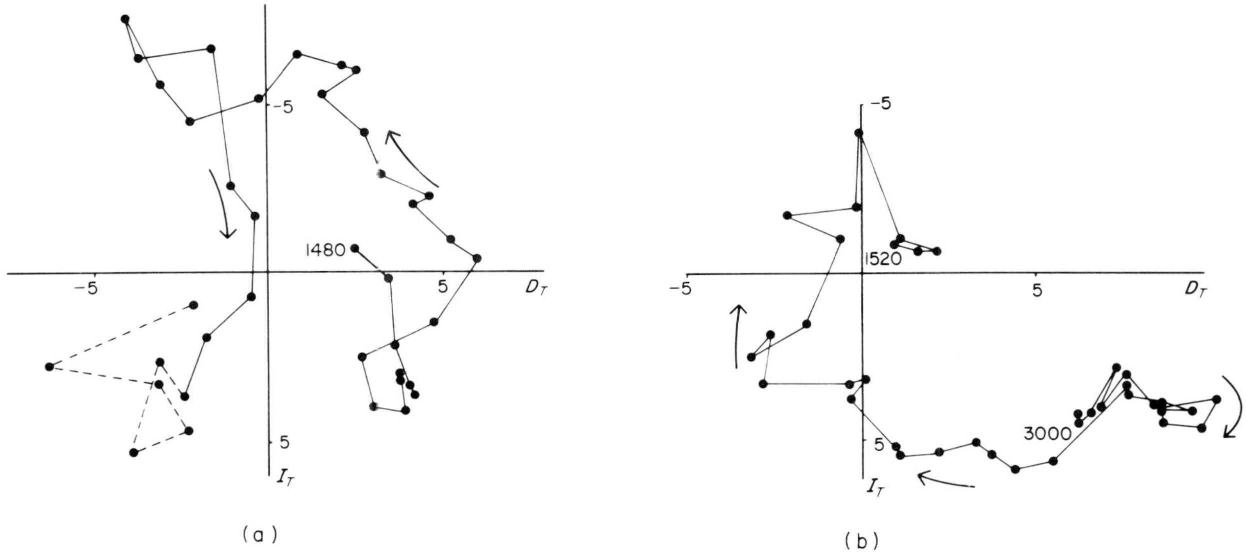

Fig. 4.13. Bauer plots of transformed declination D_T and inclination I_T for two lake cores from Great Britain. Individual points represent 40-year intervals for (a) 1500 years to present and (b) 3000 to 1500 B.P. A general clockwise motion is present and suggests westward drift according to Runcorn's rule. (After Turner and Thompson, 1979.)

(3) If weak anomalies are more common than strong anomalies, clockwise loops would be larger on the average than anticlockwise loops.
(4) Clockwise VGP loops can be symmetric about either the observer's meridian or anti-meridian, but anticlockwise loops can only be symmetric about the meridian.
(5) Anticlockwise motion would always be accompanied by a decrease in the total field intensity, but clockwise motion could be accompanied by either a decrease or an increase in intensity.

If eastward and westward drift occur with equal probabilities, none of the above asymmetries would be expected.

The first possibility cannot yet be tested because there are as yet no continuous records of sufficient resolution from high latitudes. The data from Mono Lake seem consistent with the fifth possibility, but the interpretation of intensity measurements in sediments is a hazardous procedure. The Mono Lake data also appear consistent with the fourth possibility, but neither the British nor the older Japanese data show any conclusive dependence of the orientation of the VGP loops and their direction of precession. At the present time there are too few continuous records of sufficient quality to test the second and third possibilities. On balance the evidence probably favours a dominant westward drift.

4.2.3 Interpretations in Terms of Dipole Sources

Yukutake and Tachinaka (1968) noted that the westward drift of the historical field is most pronounced in the large-scale features of the non-dipole field corresponding to harmonics of degree $l = 2$ to 4. These large-scale features have been modelled by Lowes (1955) and Alldredge and Hurwitz (1964) by radial dipoles situated in the outer part of the core. The dipoles on the model of Alldredge and Hurwitz (1964) are situated at radii $r_0 = 0.28R$ from the earth's centre where R is the earth's radius. Although the dipoles may not be physically realistic, they provide a convenient way to describe the individual major non-dipole features with non-geocentric sources for which only four parameters (dipole moment p, r_0/R, colatitude θ_0 and longitude ϕ_0) are required for each major non-dipole feature. These large anomalies dominate the non-dipole field.

Each of the large non-dipole anomalies may be approximated non-uniquely by an equivalent dipole in the outer core with parameters as described above. The coefficients of the potential of any such dipole (Hurwitz, 1960) given in SI units of tesla are:

$$\left.\begin{aligned} g_l^m &= \frac{\mu_0 p}{4\pi R^3}\left(\frac{r_0}{R}\right)^{l-1} l P_l^m(\cos\theta_0)\cos m\phi_0 \\ h_l^m &= \frac{\mu_0 p}{4\pi R^3}\left(\frac{r_0}{R}\right)^{l-1} l P_l^m(\cos\theta_0)\sin m\phi_0 \end{aligned}\right\} \tag{4.9}$$

The potential of the dipole is thus

$$V = \frac{p}{4\pi R^2}\sum_{l=1}^{\infty}\sum_{m=0}^{l}\left(\frac{r_0}{R}\right)^{l-1} l P_l^m(\cos\theta_0)\{\cos m(\phi-\phi_0)P_l^m(\cos\theta)\} \tag{4.10}$$

A typical large feature of the present non-dipole field as modelled by a dipole at the equator directed radially outwards has $\mu_0 p/4\pi R^3 \approx 10\ \mu\mathrm{T}$ ($p \approx 2.5 \times 10^{22}\ \mathrm{A\ m^2}$) (c.f. Fig. 2.8 and Eqn. (3.39)).

Creer (1977) has used the radial dipole model of Alldredge and Hurwitz (1964) to account for the geomagnetic secular variations observed in Europe and at Lakes Michigan and Erie in North America from lake sediment cores. To explain the present geomagnetic secular variation, Alldredge and Hurwitz (1964) have a nine-dipole model (central dipole plus eight radial dipoles) in which the positive radial dipoles drift predominantly eastward and the negative ones westward. The mean drift velocity for all the dipoles is westwards. Creer (1977) argues that it is difficult to account for the observed variations in the lake sediment data by geomagnetic sources drifting either eastwards or westwards around the geographic axis. The observed pattern should be the same at sites occupying the same latitude but with a time-lag due to the longitude difference. The observed differences do not appear to be as simple as this but include different oscillation periods for changes in declination and in inclination (see also §4.4.3).

Creer (1977) has chosen a model in which the radial dipoles of Alldredge and Hurwitz (1964) remain essentially fixed in position over the time interval under study, but are allowed to oscillate with the same amplitude and each with its own particular frequency. He finds that he can explain the main features observed in the lake sediment data from Europe and the Great Lakes of North America using such a model. The main problem with these models is the conveniently large number of variables that can be adjusted to explain any given observation.

4.3 Geomagnetic Excursions

4.3.1 The Laschamp and Lake Mungo Excursions

Apart from reversals of the geomagnetic field which will be discussed in detail in Chapter 5, wide departures from its normal value have been observed to

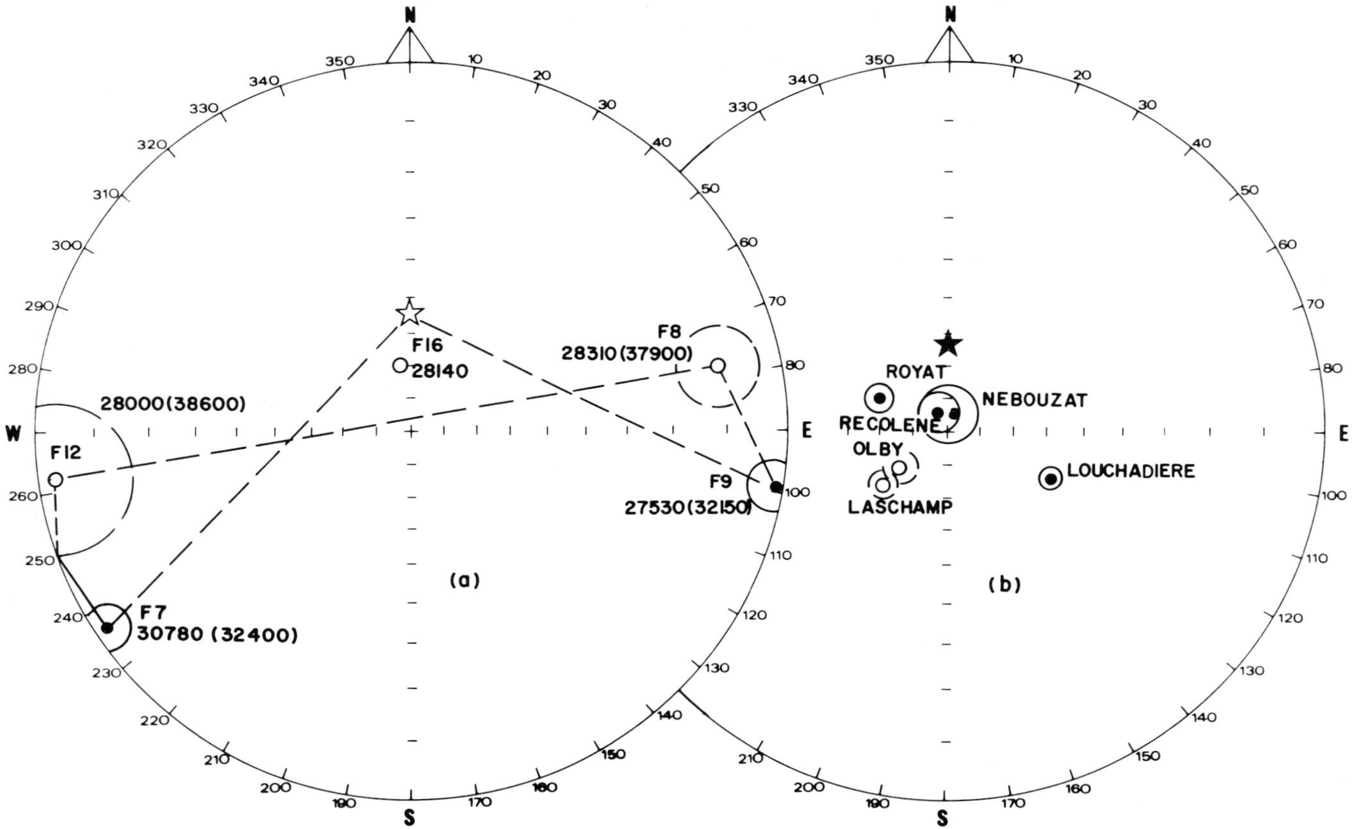

N
W
E
S
F16
28140
F8
28310(37900)
28000(38600)
F12
F9
27530(32150)
F7
30780 (32400)
(a)
ROYAT
NEBOUZAT
RECOLENE
OLBY
LOUCHADIERE
LASCHAMP
(b)

occur in the geomagnetic field direction at a single locality. Such departures, when the field does not appear to change polarity but returns to its previous state, have been termed geomagnetic excursions. *Geomagnetic excursions* are generally defined to have occurred when the VGP calculated from the field direction at the locality departs more than 45° from its time-averaged position for that epoch and is not associated with a polarity transition. Sometimes it is difficult to distinguish whether the latter has occurred, because short polarity intervals ($\simeq 10^5$ years) referred to as *events*, are known to be present in the geomagnetic record (Chapter 5). Examples of excursions have been observed in several studies where a succession of lava flows have recorded the time variations of the magnetic field in some detail (e.g. Doell and Cox, 1972). They are also recorded in sedimentary sequences but the interpretation is often equivocal because their time scale is short and only a narrow band of sediment is often involved. The departures can be argued to be due to some effect of the sedimentation itself rather than the geomagnetic field (Verosub and Banerjee, 1977). This problem is discussed in more detail in §4.3.2.

Of particular interest are the youngest recorded excursions, because they have the potential at least of being identified at several places on the earth's surface and their global morphology investigated. For times much older than say 500,000 years it is very unlikely that correlation would ever be possible with any certainty. For times less than 50,000 years ago, however, the possibility does exist and in this time range two excursions have been recorded as a TRM. They have aroused particular interest because arguments about the viability of the sedimentary recording process do not arise. These excursions occur in lavas of the Chaine des Puys in France, the *Laschamp Excursion* (Bonhommet and Babkine, 1967; Bonhommet and Zahringer, 1969), in prehistoric aboriginal fireplaces in Australia, the *Lake Mungo Excursion* (Barbetti and McElhinny, 1972, 1976) and in welded tuffs in Japan (Tanaka and Tachibana, 1981). Initially the Laschamp Excursion was regarded as a genuine reversal of the earth's magnetic field (Cox, 1969b), but now it is usually referred to as an excursion. The detailed direction changes at Lake Mungo and at Laschamp are illustrated in Fig. 4.14. Heller (1980) argues that the Olby, and also possibly the Laschamp flow, have self-reversing properties, in contrast to earlier rock-magnetic work (Whitney *et*

Fig. 4.14. Equal-angle stereographic plot of VGP for (a) the Lake Mungo Excursion (Barbetti and McElhinny, 1972, 1976) and (b) the Laschamp Excursion (Bonhommet and Babkine, 1967). The star represents the direction of the axial dipole field at each locality. Solid symbols are on the lower hemisphere and upper symbols on the upper hemisphere. Ages B.P. are given for each particular aboriginal fireplace (F) in (a) and the names of the lava flows studied are given in (b). 95% confidence levels are given as circles about the VGP.

al., 1971). Heller's work is preliminary and substantially more rock-magnetic work is required to demonstrate self-reversal, particularly considering some rock-magnetic evidence to the contrary. In the discussion that follows it is assumed that the Laschamp Excursion truly reflects changes in the magnetic field and is not of rock-magnetic origin. At Lake Mungo the detailed radiocarbon dates indicate a time span for the excursion of 2500 to 3000 years (Barbetti and McElhinny, 1976). Two radiocarbon ages date the reversely magnetized welded tuffs in Japan close to 30,000 years (Tanaka and Tachibana, 1981).

The exact date of the Laschamp Excursion has been the subject of much debate. Initially it was placed at around 8730 years from radiocarbon measurements and potassium–argon determinations suggested an older limit of 20,000 years (Bonhommet and Zahringer, 1969), but later analyses suggested ages between 40 and 40,000 years (Hall and York, 1978). The most recent thermoluminescence data give an age of 34,300 ± 3000 years (Guerin and Valladas, 1980) and this is virtually identical to 33,500 ± 4300 years obtained by thermoluminescence for the Lake Mungo Excursion (Huxtable and Aitken, 1977). There is thus a strong indication that these refer to the same geomagnetic event observed on opposite sides of the globe.

The Lake Mungo Excursion is characterized by a very large increase in field strength from between 20 and 30 μT (recorded after the excursion) to between 100 and 200 μT during the excursion (Barbetti and McElhinny, 1976; see also Fig. 4.9 and further discussion in §4.3.3). There are no palaeointensity measurements available for the Laschamp or Olby flows that record the maximum deviation during the Laschamp Excursion. However, the record from Lake Mungo also suggests a second excursion occurred at a radiocarbon age of 26,000 years B.P. (4000 to 5000 years after the first excursion) and this is associated with low palaeointensities of from 10 to 20 μT. The Royat flow (Fig. 4.14) has a thermoluminescence age of 25,800 ± 2200 years (Huxtable *et al.*, 1978) and also gives a low palaeointensity of 15 μT (Barbetti and Flude, 1979a). The unusual direction recorded by the Royat flow suggests this may be a record of the second excursion as seen at Lake Mungo.

4.3.2 Excursions Observed in Lake and Deep-sea Sediments

Verosub and Banerjee (1977) have summarized the various excursions that have been observed in sedimentary sequences covering the last few hundred thousand years. Because there are many ways in which isolated distortions of the palaeomagnetic recording process can arise, proof of the existence of an excursion must depend on consistent results from within a given lake or marine environment as well as from adjacent sedimentary environments. A critical appraisal of the many proposed excursions reveals that there is not yet sufficient evidence to confirm their existence.

Several excursions have been proposed from sedimentary sequences in Sweden by Noel and Tarling (1975), Morner *et al.* (1971) and Morner and Lanser (1974). The time around 12,400 years B.P. is associated with the so-called Gothenburg event and Morner and Lanser (1975) claim this is also recorded in deep-sea core A179-15. Opdyke (1976) regards the data from one of the oldest deep-sea cores as unreliable. Verosub and Banerjee (1977) find the evidence for all these excursions and the Gothenburg event to be singularly unimpressive. The event has to be reconciled with its non-observation anywhere in Europe for the past 14,000 years or from cores in the Black Sea back to 25,000 years (Creer, 1974) or the Aegean Sea back to 27,000 years (Opdyke *et al.*, 1972). Clearly further work needs to be undertaken to discover if these excursions are a sedimentological (climatic?) or rock-magnetic effect or whether they represent real geomagnetic behaviour.

In North America Vitorello and Van der Voo (1977) observed a well-documented excursion in Lake Michigan sediments but no such excursion was recorded by Creer *et al.* (1976a) or Dodson *et al.* (1977) in a parallel study of the same sediments. Creer *et al.* (1976b) observed an excursion called the Eriea excursion, in Lake Erie sediments, but its expression is greatest at lithological boundaries and it is recorded with different characteristics at different parts of the lake. All of the above occur in the time range 8000 to 14,000 years and Verosub and Banerjee (1977) are even less impressed with the North American excursions than they are with the Swedish ones.

At Lake Biwa, Japan excursions have been observed in a deep core at 18,000 and 49,000 years (Yashikawa *et al.*, 1973; Nakajima *et al.*, 1973). The older excursion is only very approximately dated by interpolation between widely separated and uncertain dates. However, the 18,000-year excursion may also have been recorded in sediments for Imuruk Lake, Alaska (Noltimeier and Colinvaux, 1976). The large inclination changes coupled with only small changes in declination are consistent with the observations from Lake Biwa (Verosub and Banerjee, 1977). Also, in the Gulf of Mexico, Clark and Kennett (1973) and Freed and Healy (1974) observed two excursions in at least a dozen deep-sea cores whose ages, based on foraminiferal zonation, lie at 17,000 $\pm$ 1500 and 32,500 $\pm$ 1500 years respectively. The main problem is that the magnetic signature in each core is different, so that the question of sedimentological disturbance arises. Stupavsky *et al.* (1979) also observed an excursion in three out of five core samples from the Meadowfield till, Ontario. Unfortunately the three horizons recording the excursion do not appear to be the same age so that the

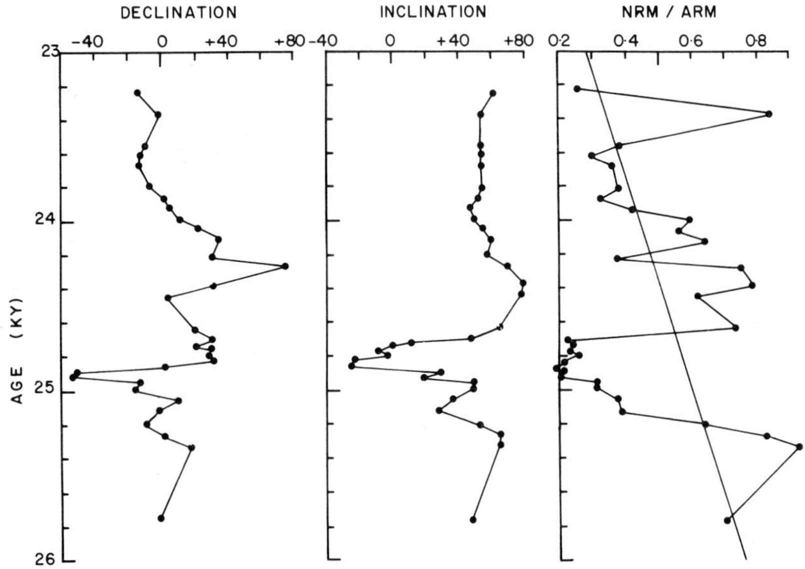

Fig. 4.15. Declination and inclination data plotted against age B.P. in Mono Lake, California. Sedimentary cores suggest that a magnetic excursion occurred near 25,000 years ago (left and middle). Natural Remanent Magnetization (NRM) to Anhysteretic Remanent Magnetization (ARM) ratios are sometimes useful to obtain a first-order estimate of the relative palaeointensity down a core. The relative palaeointensity during the excursion (right side) appears to be much lower than at other times. (After Liddicoat and Coe, 1979.)

interpretation is equivocal. Radiocarbon measurements place the age of the till between 38,000 and 31,500 years B.P. Abrahamsen and Knudsen (1979) have also observed an excursion in a single core at Rubjerg, Denmark with an age between 23,000 and 40,000 years B.P.

The best documented excursion from lake sediments is the Mono Lake Excursion recorded in dry lake sediments by Denham and Cox (1971) and Denham (1974) and subsequently investigated in more detail by Liddicoat and Coe (1979). They found that the record at four different localities indicated a much more elaborate and extended excursion than first documented. The excursion has a duration of about 1200 years beginning at 25,200 years B.P. The major feature is an inclination swing from a normal value around $+50°$ to $-23°$ (Fig. 4.15). NRM/ARM ratios for one locality suggest the field intensity may have fallen well below the dipole field intensity during this excursion. Detailed direction and intensity changes are illustrated in Fig. 4.15.

4.3.3 Models of Geomagnetic Excursions

The evidence from the preceding two sections suggests that a geomagnetic excursion was observed at Lake Mungo, Australia and at Laschamp, France 33,000 years B.P. Less secure evidence suggests it is recorded in the Gulf of Mexico, Canada and Denmark and possibly in Lake Biwa, Japan. This is the only evidence to suggest an event of global proportions occurred during the past 50,000 years. However, the synchroneity of palaeomagnetic excursions needs to be examined very carefully to see if the "reinforcement syndrome" (Watkins, 1972) is operating. The Lake Mungo Excursion is the only excursion for which direct palaeointensity measurements are available at any locality. Unfortunately, some chemical changes occur on heating the Lake Mungo samples that do affect the magnetic properties. Elaborate experiments have been conducted to delineate these chemical changes and it is believed that the palaeointensity indicating an increase in field strength of five times its post-excursion value is probably a valid result for this excursion (Barbetti and McElhinny, 1976). However, because this is the highest palaeointensity obtained anywhere and because of possible adverse effects resulting from incomplete delineation of the actual chemical changes (see review by Merrill, 1975), it is necessary to obtain similar results elsewhere to conclude with confidence that the earth's magnetic field can, at least locally, obtain such large values.

Around 18,000 years B.P. a more localized excursion may be recorded at Lake Biwa, Japan and Lake Imuruk, Alaska. At 25,000 years B.P. an excursion was recorded at Mono Lake, California and possibly at Lake Mungo, Australia. In both cases there is indirect palaeointensity information suggesting a decrease in field strength occurred during the excursion. At Mono Lake the field strength changes suggest the excursion was a manifestation of the non-dipole field (Liddicoat and Coe, 1979).

Various models proposed for geomagnetic excursions all involve extreme situations in which one of the non-dipole sources of the Alldredge and Hurwitz (1964) type changes in strength and/or location to produce the desired effect (Barbetti and McElhinny, 1976; Creer, 1977; Coe, 1977; Liddicoat and Coe, 1979; Harrison and Ramirez, 1975). One might visualize three basic situations to explain excursions:

(i) the dipole field underwent a dramatic change in direction, or
(ii) the main dipole field decreased in strength and the non-dipole field dominated over a large portion of the globe, or
(iii) one of the non-dipole "sources" increased dramatically in strength.

Both the first two cases imply a global phenomenon. The extent of this in the second case is obviously related to how much the dipole field decreases. In case (i), if the large Lake Mungo palaeointensity estimate is substantiated, then it would imply unusually large amounts of energy in the field for a very short time. In addition, one would expect considerably more evidence of such an event considering the wide distribution of marine, and to a lesser extent, lake cores. Therefore, this case seems highly unlikely. Cases (ii) and (iii) really are variations of the situation when the ratio of the non-dipole field mean intensity to dipole field intensity is considerably larger than present, allowing for non-dipole field domination, at least on a local scale. Case (iii) is the most difficult to test because it implies that the excursion may be restricted to a very small region. Indeed in a spherical harmonic description of the field, the field can be restricted mathematically to as small a region as one desires, simply by choosing high enough degree harmonics to describe the restricted feature. This is also very plausible physically. Harrison and Ramirez (1975) have shown that the non-dipole field can be represented by a dipole source in the outer core that produces a localized "reversal" of field direction at a site, while less than 15° (1665 km) away the field still exhibits its original polarity.

At Mono Lake, Liddicoat and Coe (1979) model the excursion with a single eccentric dipole of varying moment displaying a complex pattern of westward, eastward and northward drift. The average moment of this dipole is comparable with the largest calculated for the present non-dipole field and the maximum moment is almost twice as great. This model would predict a continental size effect observed only over a restricted portion of the globe. At Lake Mungo, Coe (1977) has discussed possible source models. To explain the observations a single radial dipole of one to three times the present geocentric dipole moment is required. This would produce extremely high fields over very large areas and significantly high anomalous fields over the entire globe. A combination of two radial dipoles of opposite sign, in their most efficient combination to account for the observations, would produce significant effects as far away as south-east Asia and India, southern Africa and much of Antarctica and the South Pacific Basin. Thus the test as to which model is correct depends on one's view as to the synochroneity of the Lake Mungo and Laschamp Excursions. Of course in terms of spherical harmonics there are an infinite number of sources and variations on the above models that will satisfy the data.

Perhaps the most significant observation is that the geomagnetic dipole moment was abnormally low over the time span 20,000 to 50,000 years ago (§4.1.5). In that case changes in the non-dipole field sources will have much more significant effects on the magnetic field observed at the earth's surface than if it were its normal value. It may be that a requirement for the occurrence of geomagnetic excursions is simply that the dipole field remain abnormally low so that non-dipole effects are more readily seen.

4.4 The Geomagnetic Power Spectrum

4.4.1 Time Series Analysis

A geomagnetic record typically gives some magnetic signal as a function of time, on which is superimposed a background of random noise and unwanted signals. Suppose for the moment the record is continuous and denoted by $f(t)$, where t is time. It is possible to extract information on dominant frequencies present in the record (e.g. associated with westward drift of the non-dipole field). To do this it is convenient to transform from the time domain to the frequency domain:

$$A(\omega) = \frac{1}{\sqrt{2\pi}} \int_{-\infty}^{\infty} f(t)\, \mathrm{e}^{-i\omega t}\, \mathrm{d}t \tag{4.11}$$

where ω is the angular frequency and $A(\omega)$ is the Fourier transform of $f(t)$. The inverse transform is given by:

$$f(t) = \frac{1}{\sqrt{2\pi}} \int_{-\infty}^{\infty} A(\omega)\, \mathrm{e}^{i\omega t}\, \mathrm{d}\omega \tag{4.12}$$

Usually the signals are analysed in digital form and then the analysis is referred to as a *Time Series Analysis*. A digitized signal can be handled analogously by simply using the discrete Fourier transform (e.g. Aki and Richards, 1980).

In practice the geomagnetic record is finite and requires a truncated series representation of $f(t)$:

$$f(t) = \sum_{j=0}^{N-1} A_j\, \mathrm{e}^{i\omega_j t} \tag{4.13}$$

in which $\omega_j = 2\pi j/N$ and A_j are given by

$$A_j = \frac{1}{N} \sum_{j=0}^{N-1} f(t)\, \mathrm{e}^{-i\omega_j t} \tag{4.14}$$

Referring to §2.3.2, A_j are the least squares estimates of the corresponding parameters in the infinite Fourier series. The relative power in a given frequency can be defined by

$$W_j = 2|A_j|^2 \tag{4.15}$$

Generally the power is spread over a continuum of frequencies and $W(\nu)$ is referred to as the *power spectrum* of the process. Here ν, the linear frequency, has been used instead of the angular frequency ($\nu = \omega/2\pi$). The variance of this process is given by:

$$\int_{-\infty}^{\infty} W(\nu)\, d\nu$$

The method described above to obtain the power spectrum is essentially the periodogram method developed in the late nineteenth century (Schuster, 1898). Unfortunately (4.15) provides a poor estimate of the power, in that the variance of the frequency estimate is typically very large with respect to the mean (Childers, 1978).

Subsequently several methods have been introduced to obtain estimates of the power spectrum with smaller variances (e.g. Kanasewich, 1973; Childers, 1978). Moreover the periodogram methods, modified by filtering and tapering, have returned to prominence to some extent due to the greatly improved computational procedures associated with the fast Fourier transform algorithm (Cooley and Tukey, 1965). Perhaps none of the new methods has stirred more debate than the maximum entropy method (MEM) first introduced by Burg (1967, 1972). Only this method will be discussed further, because it is probably one of the most widely used and poorly understood methods in geomagnetism (e.g. Ulrych, 1972; Currie, 1974; Denham, 1975; Phillips and Cox, 1976; Barton and McElhinny, 1982).

Subsequent to its formulation, MEM has been shown to be directly related to least squares error linear prediction and autoregression (Van der Bos, 1971). The discussion here of discrete autoregression follows that given by Barton (1978). This process assumes that $f(t)$, in digital form, is linearly dependent on the previous m values of f plus a Gaussian (white) noise:

$$f(t) = a_1 f(t-1) + a_2 f(t-2) + \ldots + a_m f(t-m) + B(t)$$

where m is referred to as the *order of the autoregressive process*, a_j are constants and $B(t)$ is called the *innovation* of the process. The mean values of $f(t)$ and $B(t)$ are taken to be zero and $B(t)$ is derived from a Gaussian process. The autoregressive process can be viewed as providing a prediction of $f(t)$ based on its previous m values and $B(t)$ as the prediction error. The spectrum of the mth order autoregressive process is known to be (e.g. Ulrych and Bishop, 1975):

$$W(\nu) = 2\beta_m \left| 1 - \sum_{j=1}^{m} a_j e^{i2\pi\nu_j} \right|^{-2}$$

in which β_m is the variance of $B(t)$. $W(\nu)$ is to be evaluated between $\pm\nu_N$, the Nyquist frequency. Basically, the MEM involves fitting an autoregressive model to the data in such a way to maximize the information in the actual geomagnetic record and minimize information about the series outside the observed record. Although this is clearly desirable, the method of doing it is at present very controversial. In particular, one of the fundamental problems in applying the analysis involves selection of the optimum value for the order m of the process. It is known that if m is chosen too low the resulting spectrum is too smoothed, so that information on the higher frequencies is lost. On the other hand, if m is chosen too large, spurious detail is introduced into the spectrum. Swingler (1979, 1980) has applied Burg's MEM algorithm to synthetic data and finds that the spectral peaks are often mislocated and their widths too broad. Thus both the frequencies and their associated power are inaccurately obtained. This indicates that the choice of a particular MEM algorithm must be done with care. At present there is no consensus upon an algorithm for optimum selection of the order of the process. Somewhat different results can thus emerge from the same record if analysed by different workers (e.g. Burg, 1967, 1972; Akaike, 1969a, b, 1970; Smylie *et al.*, 1973; Ulrych and Bishop, 1975; Swingler, 1979, 1980; Barton, 1982, 1983).

4.4.2 Spectrum from Historical Records

There is considerable controversy over the interpretation of the geomagnetic variations with periods from one to 30 years (e.g. analyses of Yukutake, 1965; Bhargava and Yacob, 1969; Currie, 1973, 1975; Rivin, 1974; Alldredge, 1976, 1977; Courtillot and Le Mouël, 1976; Srivastava and Abbas, 1977; Chen, 1981) (see §2.4.2). Part of the reason for this controversy involves the methods used for analysing the power in particular periods and part involves the interpretations. For example, the use of MEM on data from Hong Kong (Chen, 1981) indicates significant peaks in the spectrum of the horizontal component of the field near 20 years, 10.6 years, 7.1 years and 5.8 years, roughly consistent with world-wide peaks found by Currie (1973). The corresponding periods for the vertical component of the magnetic field in Hong Kong are at 20, 6.7 and 5.8 years (Chen, 1981). The vertical component spectrum peaks at 6.7 and 5.8 years can barely be resolved. It could be that this is due to peak splitting resulting from too large a choice for the order of the process (§4.4.1). The strong correlation between the vertical and horizontal components argued by Chen (1981) would then be based on only two periods, one near 20 years and one near 6 years. The analysis of Chen (1981) is probably correct, but it shows how the interpretations and the methods of analyses are interwoven into a complex web that is difficult to untangle without more agreement on what are the proper procedures for time series analyses.

In spite of these problems with analysis, it is possible to combine data at different sites to obtain an estimate of the power spectrum of internal origin (Currie, 1967, 1968) (see Fig. 4.16). The discussion in §2.4.2 regarding whether spectral peaks near 22 years were of internal or external origin (see also Alldredge, 1976, 1977) highlights the controversial nature of the interpretation of this spectrum. Barton (1982, 1983) used a different algorithm

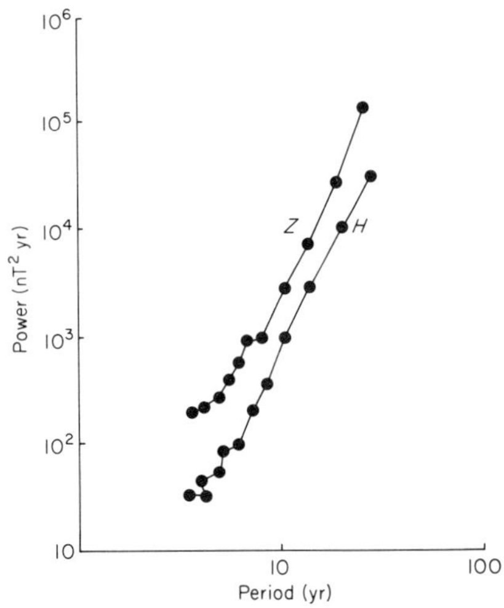

Fig. 4.16. Power in the vertical (Z) and horizontal (H) components of the geomagnetic field are shown as a function of period. (After Currie, 1967, 1968.)

for spectral analysis discussed in §4.4.3 (see Fig. 4.18). His spectral power estimates are significantly higher than those of Currie (1967, 1968) (even if Z and H in Fig. 4.16 are combined). Nevertheless a large increase in power with increase in period is evident in both analyses. This indicates that most of the power in the geomagnetic field of internal origin probably resides in much longer periods than can be deduced from the analysis of historical records.

4.4.3 Spectrum from Lake Sediment Data

Denham (1975) first suggested that the quasi-periodic fluctuations in the geomagnetic field recorded by the magnetization of lake sediments (§4.2) could profitably be analysed by power spectrum techniques. He showed that time series analysis could provide valuable information on the dominant periods present and also be used to obtain the direction of drift of the non-dipole field (Fig. 4.17). However, there are several problems associated with

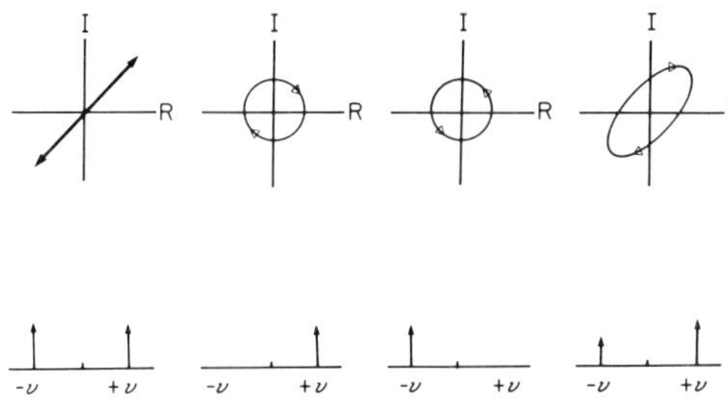

Fig. 4.17. Hypothetical power spectra (bottom) for four classes of simple two-dimensional (complex: *I* = imaginary, *R* = real) fluctuations (top) at arbitrary frequency ν. Planar motion (far left), regardless of angle, yields a symmetric spectrum. Circular motion (centre) produces a one-sided spectrum that can be used to distinguish the sense of rotation, while (far right) elliptical motion yields an asymmetric spectrum weighted to one side or the other depending on the sense of the looping. (After Denham, 1975.)

the time series analysis of palaeomagnetic data which are not encountered with the analysis of historical data:

(i) there might be changes in sedimentation rates that are not properly delineated by radiometric dating (see discussion by Clark and Thompson, 1979);
(ii) there might be dating errors;
(iii) the noise might increase with sediment core depths (time), as suggested to be the case in the Australian lakes studied by Barton and McElhinny (1981); and
(iv) significant changes in apparent periodicities might be present in the geomagnetic spectrum over the longer time periods sampled (non-stationary processes; see Chapters 6 and 10).

There is a basic problem in deciding the most appropriate variables to use for the spectral analysis of palaeomagnetic data. Declination and inclination are the two most common variables measured in sediment cores. However declination and inclination time series can be radically different at a given site. Consider the simplified example of a magnetic field consisting of the sum of an axial geocentric dipole and an equatorial geocentric dipole. A change in the intensity of the axial dipole field affects only the inclination and not the declination, whereas a change in the equatorial dipole direction, constrained to lie in the equatorial plane, affects only the declination. Declination is a poor choice because it becomes highly unstable in the region of the dip poles. Although inclination is better, any spectrum obtained using this as a variable would be difficult to compare with that obtained from the historical record or to analyse for drift direction by Runcorn's rule. Barton (1982) has probably suggested the best compromise to the above problems by summing the contributions to the spectral power at a given frequency from each component:

$$W(\nu) \equiv W_x(\nu) + W_y(\nu) + W_z(\nu)$$

where z is the vertical component and x and y are the orthogonal horizontal components. He bypasses the problem that absolute palaeointensity data cannot be obtained from sedimentary cores, by assigning a constant magnitude to the magnetic vector equal to that due to the present-day dipole field (8×10^{22} A m^2).

In an attempt to obtain a first order estimate of the global power spectrum, Barton (1982, 1983) analysed lake sediments data, deep-sea core data and archaeomagnetic data from ten locations within the latitudinal zones of 30°–55°. Three of these are in the southern hemisphere (all Australia) and the remainder are in the northern hemisphere. Barton also analysed the historical records from 12 magnetic observatories reasonably well distributed over the earth. The time series analysis used was a form of the periodogram method described by Bloomfield (1976), involving detrending, tapering and filtering methods that are significantly different from those used by Currie in his analysis of the historical record (§4.4.2).

Figure 4.18 presents a summary of the power spectrum as a function of

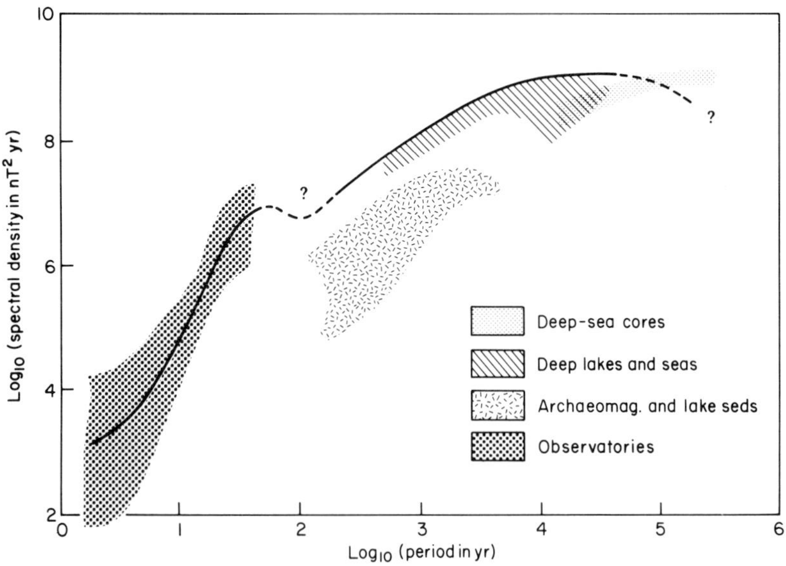

Fig. 4.18. Continuous geomagnetic power spectrum derived from observatory and palaeomagnetic data. (After Barton, 1983.)

period. The part of the spectrum obtained from historical records is similar in shape to that deduced by Currie (Fig. 4.16) using different data and a different method of spectral analysis. At longer periods the power generally increases showing an apparently broad maximum near 10^3–10^5 years. Details in this spectrum can probably be attributed to the use of different types of data and to particular methods of analysis. For example, the dip near 10^2 years coincides with the change from using low period historical data to long period palaeomagnetic data. The broad peak in the power in the range 10^3–10^5 years is probably a manifestation of the westward drift, whose period probably varies with time (Chapter 10). In that case a broad peak is to be expected. Figure 4.18 suggests that most of the power in the geomagnetic spectrum is associated with the westward drift of the geomagnetic field.

Chapter Five

Reversals of the Earth's Magnetic Field

5.1 Evidence for Field Reversal

5.1.1 Self-reversal in Rocks

David (1904) and Brunhes (1906) first observed magnetizations in lava flows that were roughly opposed to that of the present earth's magnetic field. The problem was to decide whether or not it was the earth's magnetic field itself that had changed polarity or whether some self-reversal mechanism existed that produced a thermoremanent magnetization on cooling antiparallel to the applied field. The first theoretical models for self-reversal were produced by Neél (1955) and since then other possible self-reversing mechanisms have been proposed by a number of authors (Verhoogen, 1956; Uyeda, 1958; Ishikawa and Syono, 1963; O'Reilly and Banerjee, 1966; Hoffman, 1975; Stephenson, 1975; Petherbridge, 1977; Tucker and O'Reilly, 1980). Although self-reversal can also occur in other remanences, such as CRM, only that which might occur in TRM will be considered here.

Self-reversal can occur only when there are (or were) two magnetic phases in a rock. One phase becomes magnetized first parallel to the external magnetic field and subsequently the second phase becomes magnetized antiparallel to the first phase. This occurs either because there is a negative exchange interaction acting between the two phases or because the magnetic field of the first phase swamps that of the external magnetic field (*magnetostatic interaction*). An example of the latter is illustrated in Fig. 5.1. Phase A is assumed to have a higher Curie Temperature than phase B and is magnetized parallel to the external field $\boldsymbol{H}$. Phase B becomes magnetized in the total field $\boldsymbol{H} + \boldsymbol{H}_A$, where $\boldsymbol{H}_A$ is the magnetic field due to phase A. In the example in Fig. 5.1, $\boldsymbol{H}_A$ would be opposite to that of $\boldsymbol{H}$. If $H_A > H$ and if the total magnetization of B exceeds that of A at room temperature, then the sample will have *self-reversed*. Unfortunately not all self-reversals are detectable in a laboratory experiment, because chemical changes including ordering, that occur because of the laboratory heating, sometimes remove one of the phases.

Experimentally the best understood and most studied self-reversal mineral is a titanohematite containing roughly 50 mole percent ilmenite and hematite (Uyeda, 1958; Ishikawa and Syono, 1963; Westcott-Lewis and Parry, 1971; Hoffman, 1975). The self-reversal mechanism is very complicated, involving Fe–Ti ordering and possibly exsolution. However, the essence of the mechanism is that a negative exchange interaction occurs between a weakly

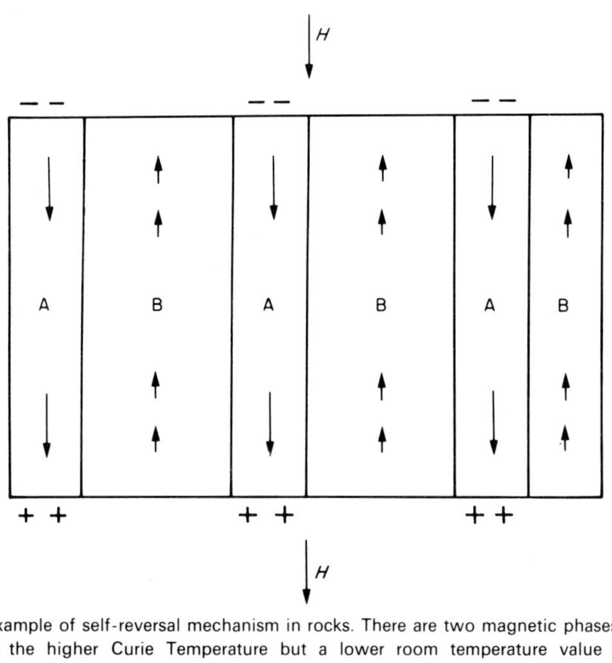

Fig. 5.1. Example of self-reversal mechanism in rocks. There are two magnetic phases A and B for which A has the higher Curie Temperature but a lower room temperature value of saturation magnetization than B. A becomes magnetized first on cooling in a direction parallel to the external field *H*. Inside the grain the magnetic field arising from the magnetization of A external to this phase is antiparallel to *H*. If this internal field is larger than *H* then phase B will acquire a magnetization opposite to *H* on further cooling. The sample will then have a self-reversed magnetization at room temperature.

magnetic phase that has a high Curie Temperature and a much stronger magnetic phase with a lower Curie Temperature.

Although self-reversal mechanisms involve some challenging problems in geophysics, actual examples of self-reversal in nature are rather rare. In addition to the titanohematite given above, only a few pyrrhotites, titanohematite near 90 mole percent hematite, and possibly a few restricted compositions of exsolved titanomagnetites have exhibited self-reversing tendencies (Schult, 1968; Carmichael, 1961; Ozima and Ozima, 1967; Readman and O'Reilly, 1970; Peterson and Bliel, 1973; Ryall and Ade-Hall, 1975; Petherbridge, 1977; Tucker and O'Reilly, 1980). The bulk of evidence as set out below favours the view that the vast majority of observed reversed magnetizations observed in rock arise from changes in the polarity of the earth's magnetic field.

Therefore it was somewhat surprising to find an apparent correlation between oxidation state and polarity, as was alleged to be the case in the late 1960s (e.g. Wilson and Watkins, 1967; Smith 1970b). Although this is sometimes still referred to (e.g. Jacobs, 1981), this "correlation" has been shown to be due to a data artifact (see review by Merrill, 1975). It is very doubtful that many (if any) erroneous magnetic field reversals have been reported because of self-reversals in the rocks being studied.

5.1.2 Baked Contacts

Three lines of evidence each provide compelling arguments that favour the view that most observations of reversed magnetizations in rocks are due to reversals of the earth's magnetic field. These are:

(1) Studies of baked contacts adjacent to intrusive igneous rocks or underlying lava flows should show agreement in the polarity of the magnetization between the igneous rock and the baked rock.
(2) There should be world-wide simultaneous zones of one polarity.
(3) There should be records of the magnetic field observed in rock sequences showing the polarity changing continuously from one state to the other.

Each of the above lines of evidence will be considered separately in subsequent parts of this chapter. The first two lines of evidence rely on simultaneous observations of different rock types. In general a baked contact will have a much different magnetic mineralogy from the igneous rock that provides the heat. Simultaneous zones of one polarity observed world-wide should also involve rocks of widely different magnetic mineralogy.

The evidence from the studies of baked contacts was first compiled by

Wilson (1962b). This was later updated by Irving (1964) and McElhinny (1973). The successive increase in the overwhelming evidence showing the correspondence in polarity between the baked and baking rocks is given in Table 5.1. Conventionally rocks whose magnetization is in the same sense as the present magnetic field are termed *normal* (N) and those in the opposite sense are termed *reversed* (R). Directions intermediate between these (defined as rocks giving virtual geomagnetic poles inclined at an angle greater than 45° to the axis of rotation) are termed *intermediate* (I). Comparisons of polarities for the case of reversed magnetizations were particularly important to confirm the reality of the field reversals and this is seen by the much greater emphasis given to this situation in Table 5.1. The three cases of disagreement reported in the table refer to measurements made before the technique of magnetic cleaning was used in palaeomagnetic studies. The subject has advanced so much in the past decade that baked contact studies are now used not so much as to check on the reality of field reversals, but as a check on the age of the magnetization of intrusive rocks. This is especially true in studies of old Precambrian rocks where the importance of discovering evidence relating to the age of the magnetization being measured is paramount.

A particularly convincing study was made by Wilson (1962a) into the magnetic record in a doubly-baked rock. The situation is illustrated in Fig. 5.2. A lava flow, reversely magnetized, had heated an underlying laterite. The baked laterite had a direction of magnetization the same as that of the lava, reversed. Subsequently both lava and baked laterite were intruded by a dyke whose magnetization was also reversed, but different by about 25° from that of the lava and baked laterite. A study of the zone of second heating of the laterite caused by the dyke showed a gradual change from the dyke direction, adjacent to the dyke, to the lava direction at a distance where the second heating had produced no effect. Thus in the same region both superimposed magnetizations were of reversed polarity. It seems impossible to explain this

TABLE 5.1

Comparison of polarities observed in igneous rocks and their baked contacts. Original analysis by Wilson (1962b), updated by Irving (1964) and McElhinny (1973). N, normal; R, reversed; I, intermediate

Igneous	Baked contact	No. of observations		
		Wilson (1962b)	Irving (1964)	McElhinny (1973)
N	N	14	34	47
R	R	34	49	104
I	I	1	2	3
N	R	3	3	3
R	N	0	0	0

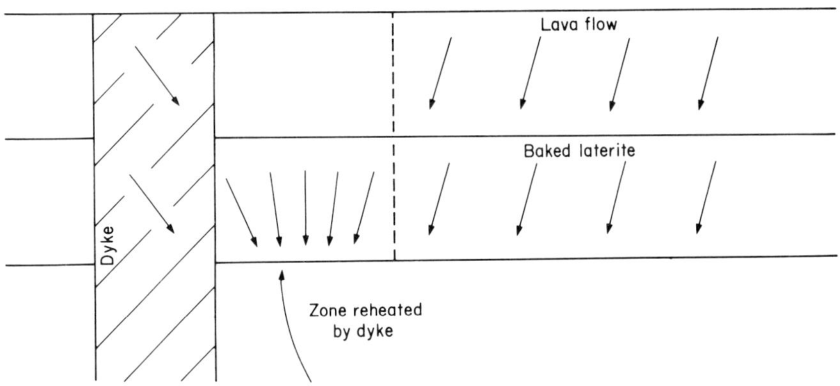

Fig. 5.2. Evidence for field reversal from the magnetic record in a doubly-baked rock (Wilson, 1962a). A lava flow originally heated the laterite horizon. Subsequently a dyke intruded both, reheating both lava flow and laterite in its adjacent region. In this region the laterite was thus first reheated by the lava and then reheated a second time by the dyke.

observation by any other mechanism than the fact that the earth's magnetic field had changed polarity and was reversed during both heating episodes.

5.1.3 Polarity Time-scale

Mercanton (1926) first realized that if rocks containing reversed magnetizations were due to reversals of the earth's magnetic field, this should be registered in rocks world-wide. Rocks of the same age should have the same polarity of magnetization. Matuyama (1929) demonstrated that early Quaternary lavas had reversed polarity whereas younger lavas had normal polarity. Observations by Roche (1951, 1956) of volcanic rocks from the Chaîne des Puys in France led him to conclude that the last reversal of the earth's magnetic field took place in the middle of the Lower Pleistocene. A similar conclusion was reached by Opdyke and Runcorn (1956) on rocks from the United States. These observations suggested an ordered sequence of polarity inversions might exist in the geological record. This conclusion was further emphasized by the evidence that almost all rocks of Permian age were reversely magnetized (Irving and Parry, 1963).

The above observations were based upon the relatively imprecise methods used for dating rocks from the occurrences of fossils. It was only in the early 1960s that developments (by Evernden, McDougall and Dalrymple) in the K–Ar isotopic dating method enabled quite young volcanic rocks to be dated with some precision. The first example of the combined use of magnetic polarity and K–Ar dating was that of Rutten (1959). He concluded that the present normal polarity extended from at least 0.47 million years ago and that an earlier period of normal polarity existed about 2.4 million years ago.

Systematic studies attempting to define a polarity time-scale using joint magnetic polarity and K–Ar age determinations on young lava flows were undertaken both in the USA and Australia. The first time-scale put forward by Cox *et al.* (1963a) suggested a periodicity of magnetic reversals at about one million year intervals. However, as new data quickly appeared in the literature (Cox *et al.*, 1963b, 1964a; McDougall and Tarling, 1963, 1964) it soon became apparent that there was no simple periodicity, the lengths of successive polarity intervals were sometimes long ($\sim$1 Ma) or quite short ($\sim$0.1 Ma). Cox *et al.* (1964b) proposed that within intervals of predominantly one polarity lasting of the order of 1 Ma, there were short intervals of opposite polarity of the order of 0.1 Ma. The longer intervals were termed magnetic polarity epochs and the shorter intervals were called events. The

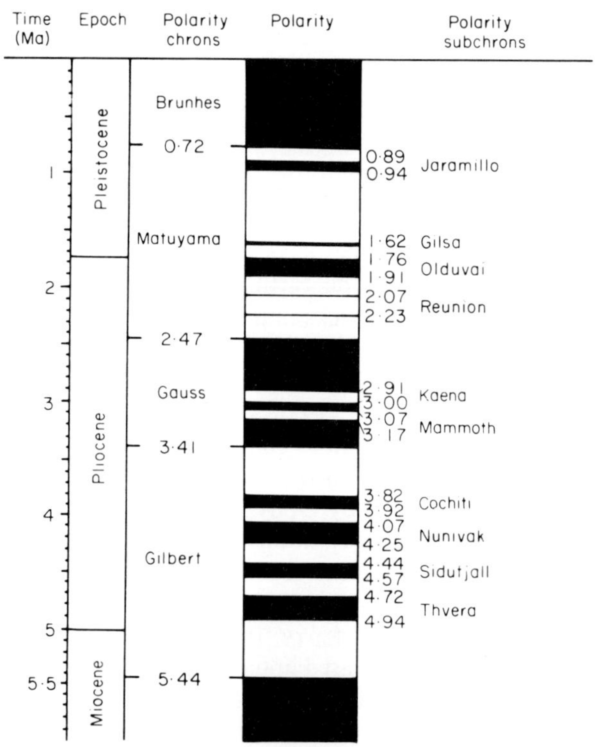

Fig. 5.3. Geomagnetic polarity time-scale for the past 5 Ma based mainly on K–Ar and palaeomagnetic data on igneous rocks (after McDougall, 1979). Black represents normal polarity, white represents reversed polarity.

epochs were named after pioneering scientists in geomagnetism, Brunhes, Matuyama, Gauss and Gilbert, whereas the events were labelled from the location of their discovery. The International Sub-Commission on Stratigraphic Classification has now adopted the terms *chron* and *subchron* to replace the terms epochs and events.

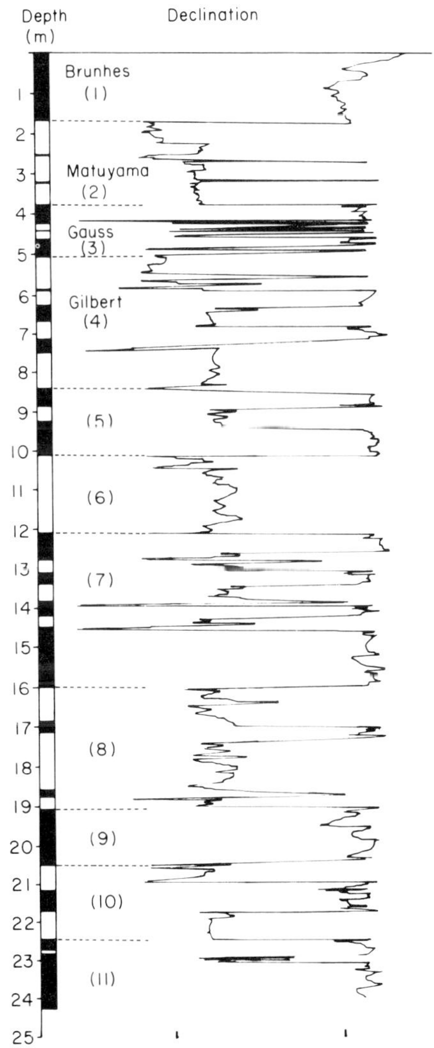

Fig. 5.4. Change in magnetic declination in deep-sea core RC12-65 (Lamont-Doherty Geological Observatory) with respect to the split face of the core as a function of depth (after Opdyke, 1972).

A compilation of the geomagnetic polarity time-scale for the past 5 Ma has been made by McDougall (1979) in which corrections are made to existing scales, such as that of Cox (1969b), for a revision in the ^{40}K decay constant. This time-scale is shown in Fig. 5.3 and is based on the recommended decay constants for ^{40}K (Steiger and Jäger, 1977). Lack of resolution in K–Ar dating makes it very difficult to extend this scale beyond 5 Ma using stratigraphically unrelated samples because the typical precision of ages at 5 Ma is about $\pm 2\%$. Errors thus increase beyond ± 0.1 Ma which is well within the range of duration of many of the shorter polarity intervals shown in Fig. 5.3. However, careful studies of thick volcanic sequences in Iceland by McDougall *et al.* (1976a, b, 1977) have enabled extrapolation of the scale to about 7 Ma.

A continuous record of polarity changes can be obtained by investigating the palaeomagnetic record preserved in deep-sea sediments by taking cores from the ocean bottom. Although the cores are usually not oriented, they are taken nearly vertically into the ocean bottom so that changes in sign of the magnetic inclination measured in the cores or changes of 180° in declination can easily be identified as records of polarity change. Typical oceanic sedimentation rates range from 1 to 10 mm per 1000 years so that the last polarity change corresponding to the Brunhes-Matuyama boundary at 0.72 Ma will occur at depths of 0.72–7.2 m. Likewise the Gauss-Gilbert boundary at 3.41 Ma will occur at depths of 3.41–34.1 m.

The earliest investigations of marine sediment cores soon established the reality of the polarity time-scale observed on land (e.g. Ninkovich *et al.*, 1966; Opdyke and Glass, 1969). Using cores up to 30 m long and with suitable sedimentation rates, the polarity sequences can be extended back to 10 Ma or more (Foster and Opdyke, 1970; Opdyke, 1972; Opdyke *et al.*, 1974). An example of such a sequence is shown in Fig. 5.4. The largest errors in the marine sediment record are associated with dating (fossils) and variations in sedimentation rate. These errors generally increase with the age of the sediment.

5.2 Marine Magnetic Anomalies

5.2.1 Measurement and Calculation

Magnetic anomaly is used here to mean the difference between the observed magnetic field intensity at a location and that expected from some reference field, such as that derived for the last magnetic epoch (e.g. the 1980 mean field). Observed anomalies in total intensity are believed to be due to crustal sources and vary significantly over short distances ranging from a few metres to a few hundred kilometres in the case of some major marine magnetic anomalies. Such short wavelength features will not significantly affect the lowest order spherical harmonics used to describe the field for a given epoch.

The analysis of marine magnetic anomalies has made a major contribution to our understanding of the earth's magnetic field in the past. The intensity of the magnetic field at sea is usually measured by a magnetometer towed behind a ship at the sea surface, although deep-tow magnetometers are sometimes used to measure the intensity close to the ocean bottom (e.g. Spiess and Mudie, 1970; Atwater and Mudie, 1973; MacDonald and Holcombe, 1978). Vine and Matthews (1963) and Morley and Larochelle (1964) independently recognized that the linear marine magnetic anomalies observed at sea (Mason and Raff, 1961) were due to alternating blocks of normally and reversely magnetized volcanic rocks that form the upper part of the oceanic crust. In this case the strong remanent magnetization swamps the induced component that occurs in the presence of the present earth's magnetic field.

Because only the intensity of the magnetic field is measured at sea, some additional assumptions are required to determine the polarity of the magnetization from magnetic anomalies. Calculation of the external magnetic field can be made by first calculating the potential ψ at an external point, $\boldsymbol{r}$, from a magnetized body from

$$\psi(\boldsymbol{r}) = -\int_V \boldsymbol{M} \cdot \nabla \frac{1}{(r - r_0)} \, \mathrm{d}r^3$$

where the integration is carried out over the entire body of volume V, the dimensions of which are assumed to be known. The distribution of the magnetization M throughout the body needs to be assumed. The total magnetic field consists of the geomagnetic field $\boldsymbol{H}_0$ and the "anomalous" magnetic field, $\Delta\boldsymbol{H}$, where

$$\boldsymbol{H} = \boldsymbol{H}_0 + \Delta\boldsymbol{H} = \boldsymbol{H}_0 - \nabla\psi(\boldsymbol{r})$$

Usually the total field magnetometers measure only the field magnitude without sensing its direction.

Some of the more popular methods for determining magnetic anomalies were introduced by Bott (1967) and these have been augmented by Schouten and McCamy (1972), Parker (1973), and Parker and Huestis (1974). When both the shape of the body and the direction of magnetization are known, it is possible to solve the linear problem to determine the magnetic anomaly. More general conditions (e.g. unknown shape) often lead to the necessity of solving non-linear equations.

A development by Bott (1967) illustrates one method for handling the linear problem. Assume the problem to be two-dimensional with the anomaly obtained in the x direction, $A(x)$, and the magnetization is a function of depth only, $M(z)$, and confined between two planes $z = z_1$ and $z = z_2$. The anomaly is measured at the surface $z = 0$. Under these conditions:

$$A(x) = \int_{-\infty}^{\infty} M(z)K[z_1, z_2, \beta, (x - z)]\, dz$$

where β is the angle between M and the external field, here for simplicity assumed to be constant. K is a kernel which is independent of M. K can be rewritten as a function of x minus z, $k(x - z)$ (since these are the only variables present):

$$A(x) = \frac{1}{\sqrt{2\pi}} \int_{-\infty}^{\infty} M(z)k(x - z)\, dz = M * k \tag{5.1}$$

where the asterisk denotes Fourier convolution. Let $A(s)$ be the Fourier transform of $A(x)$, then:

$$A(s) = \frac{1}{\sqrt{2\pi}} \int_{-\infty}^{\infty} A(x)\, e^{ixs}\, dx \tag{5.2}$$

Replace $A(x)$ by a smoothed anomaly $A'(x')$ given by:

$$A'(x') = \frac{1}{\sqrt{2\pi}} \int_{-\infty}^{\infty} A(x)\omega(x' - x)\, dx = A * \omega \tag{5.3}$$

where ω is a weighing factor that will be discussed further shortly. Applying the Fourier convolution theorem to (5.2) and (5.3) one obtains:

$$A(s) = k(s)M(s)$$

$$A'(s) = \omega(s)A(s)$$

or

$$M(s) = \frac{\omega(s)A(s)}{k(s)} \tag{5.4}$$

where ω must be chosen to ensure that M is a Fourier transform. Inverting (5.4) then:

$$M(z) = \frac{1}{2\pi} \int_{-\infty}^{\infty} e^{-izs} \frac{\omega(s)}{k(s)}\, ds \int_{-\infty}^{\infty} e^{+ixs} A(x)\, dx$$

and this allows M to be calculated formally from the anomaly $A(x)$.

Applications of the method and similar methods to data have been illustrated by Bott (1967), Schouten and McCamy (1972) and Parker (1973). Parker and Huestis (1974) pointed out that the linear formulation of the problem made it suitable for the inverse theory of Backus and Gilbert (1967, 1968, 1970) (see §2.3.5). In particular they showed how a magnetic annihilator could be obtained (that included topographical variations in the magnetized material) providing that certain assumptions were met, including that no polarity changes occur as a function of depth.

5.2.2 Sea-floor Spreading

The sea-floor spreading hypothesis was first formulated by Hess (1960, 1962). It is assumed that the mid-ocean ridge system that circulates the globe is the site of the rising limbs of convection cells where hot magma comes right through to the surface. New oceanic crust is formed as the magma cools, and the intrusion of new material was thought to force the cooling crust to move away from the ridge symmetrically on either side. However, it is now generally believed that the crust and upper mantle, referred to as the *Lithosphere*, are under tension at a spreading centre. Thus the lithosphere is pulled apart allowing magma to rise to the surface and the whole oceanic crust is part of a conveyor belt system, coming up at the mid-ocean ridges and going down eventually at the oceanic trenches.

Dramatic support for the sea-floor spreading hypothesis was provided independently by Vine and Matthews (1963) and Morley and Larochelle (1964). Interpreting the linear magnetic anomalies observed at sea as representing alternating blocks of normal and reversely magnetized oceanic crust, they realized that as new crust was formed at the mid-ocean ridge according to the sea-floor spreading hypothesis, it would cool through the Curie Temperature and become magnetized parallel to the earth's magnetic field. If the earth's magnetic field reversed itself periodically, then strips of crust of alternating polarity would be produced parallel to, and in the simplest case, distributed symmetrically about the axis of the ridge. The model is illustrated in Fig. 5.5.

The hypothesis therefore is that the crust near the oceanic ridges acts as a crude tape recorder of the earth's magnetic field in the past, with the polarity and smaller changes in the field being recorded horizontally away from the spreading centre. This hypothesis was one of the most important leading to the plate tectonics model (McKenzie and Parker, 1967; Morgan, 1968; Le Pichon, 1968). The model is now accepted as a first-order explanation of global tectonics and is often described as having caused a revolution in the earth sciences (see, for example, the review by McElhinny, 1973). Magnetic

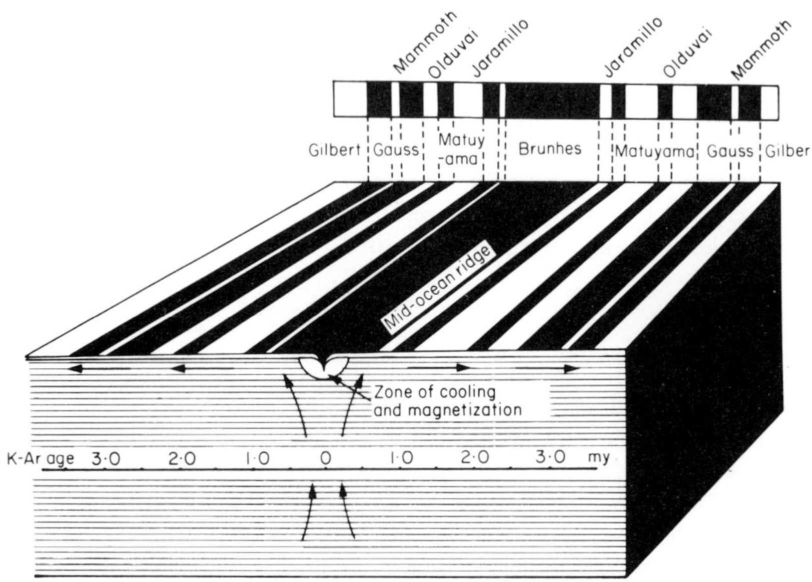

Fig. 5.5. Schematic representation of sea-floor spreading and the formation of linear magnetic anomalies due to reversals of the earth's magnetic field. Normal polarity zones are shaded.

anomalies have also continued significantly to our understanding of the earth's magnetic field as first became apparent in the work of Heirtzler *et al.* (1968) and which will be discussed in more detail later.

The first deep (0.5 km of basement or more) hole drilled into the igneous basement (layer 2 of the oceanic crust) in the Atlantic Ocean yielded polarity changes (Johnson and Merrill, 1978). Such down hole reversals of magnetization were not expected on a model based on injection of material along an infinitely thin spreading centre. Several years earlier Matthews and Bath (1967) and Harrison (1968) suggested that the Vine–Matthew's hypothesis required minor modification to account for certain features of magnetic anomalies. In particular, they suggested that the *average* magnetization changed smoothly between normal and reversed polarity anomalies, but that the detailed magnetization in any crustal section could be very complex, including the recording of reversals in the vertical direction. Subsequently, other models have been suggested which predict complex magnetic vertical structures in the oceanic crust (e.g. Blakely, 1976; Kidd, 1977; Johnson and Merrill, 1978; Schouten and Denham, 1979; also see reviews by Blakely and Cande, 1979; Johnson, 1979). The various factors which must be considered in the construction of such models include (i) the width of the magma injection zone, (ii) the rate of spreading, (iii) the cooling rate, and (iv) the rate and extent of chemical alteration, particularly that associated with hydrothermal circulation. Although no consensus yet exists as to the detailed magnetic structure of the oceanic crust, there is agreement that the magnetic structure is generally more complex than originally envisaged by Vine and Matthews. The thickness of the magnetized layer is generally greater than the 0.5 km thick layer typically used in modelling magnetic anomalies (see reviews by Blakely and Cande, 1979; Johnson, 1979). Schouten and Denham (1979) have shown that complex magnetic structures can give rise to the observed magnetic anomalies. Thus the core of the Vine–Matthew's hypothesis remains intact, even though the magnetic structure is more complicated than originally envisaged.

5.2.3 Aspects of Magnetic Anomaly Interpretation

As suggested above, the inversion of magnetic anomaly data (intensity only) is highly non-unique. Because there is no field outside a uniformly magnetized infinite sheet (an example of a magnetic annihilator—see §2.3.5), it becomes immediately apparent that an infinite set of models for magnetization distribution exist for any anomaly pattern. It becomes necessary to use additional geological and geophysical considerations to constrain the problem. The work of Parker and Huestis (1974) probably does this best at present, but further refinement is desirable. Associated with this problem is the fact that the earth itself effectively acts as a filter, as illustrated below.

Usually the marine data are obtained at the sea surface, although sometimes they are obtained by a deep tow and there is presently a NASA satellite program (MAGSAT) to obtain a complete global picture for the field above the earth's surface. Consider therefore the effects elevation changes have on filtering the magnetic data. The magnetic potential just above the sea-floor can be written as a function of horizontal distance x in a flat earth model in terms of its Fourier components

$$\psi(x, 0) = \sum_{n=1}^{\infty} A_n \cos \frac{n\pi x}{L} + B_n \sin \frac{n\pi x}{L} \tag{5.5}$$

where $2L$ is the length of the traverse in which the measurements were taken. A_n and B_n are constants to be determined (A_0, representing an arbitrary constant, has been dropped). In this situation Laplace's equation holds so that

$$\frac{\partial^2 \psi}{\partial x^2} = -\frac{\partial^2 \psi}{\partial z^2}$$

A solution for $\psi(x, z)$ similar to (5.5) is

$$\psi(x, z) = \sum \left[C_n \cos\left(\frac{n\pi x}{L}\right) e^{P_n z} + D_n \sin\left(\frac{h\pi x}{L}\right) e^{Q_n z} \right]$$

To satisfy the boundary condition $A_n = C_n$ and $B_n = C_n$, it follows that:

$$P_n = Q_n = -\frac{n\pi}{L}$$

This last result is critical, because it shows that each term representing the potential (and also the magnetic field given by the negative gradient of the potential) is damped by the term exp $(-n\pi z/L)$ where n corresponds to the order of the term. Because the higher order terms are associated with shorter wavelength features, this means that the rate of damping increases exponentially with the order n. This is effectively the same effect seen earlier in dealing with spherical harmonic analyses that dealt with global features (Chapter 2). As elevation is increased there is an exponential filtering of the higher wavelengths which smooths the data. This has the converse effect that any minor error in measuring a field (say in a satellite) will be greatly amplified if downward continuation is made. Thus a complete coverage as close to the source as possible would provide the most reliable data in principle (assuming equal errors in the measurements), but at substantially greater financial cost. Also one would not expect to gain much information about crustal magnetic anomalies that have wavelengths shorter than the distance from the anomaly source to the instrument. Therefore, the details of the magnetic field history are not as likely to be resolved from satellite data as they would be from anomaly data obtained at sea.

One method to reduce the noise problem discussed above was introduced by Blakely and Cox (1972) who stacked the data obtained from several

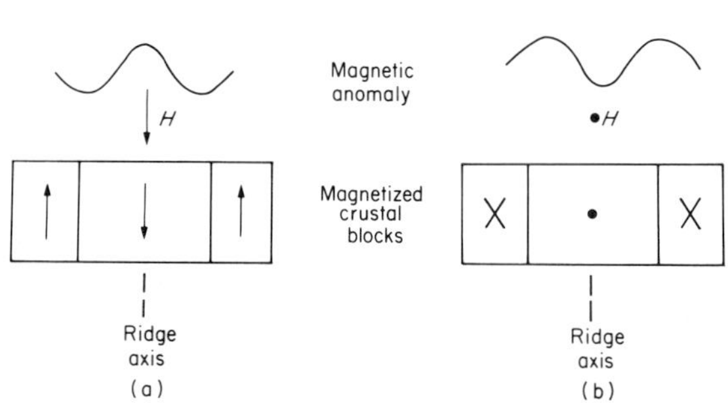

Fig. 5.6. Latitude variation of the shape of the magnetic anomaly due to the magnetized crustal blocks adjacent to the axis of a spreading ridge. The external magnetic field H due to an axial geocentric dipole is shown for the two cases. (a) At the north pole the intensity of the field is increased over the ridge axis, but (b) at the equator it is decreased.

traverses across spreading centres. However, before this was done, a geocentric axial dipole field assumption was utilized and the data were transformed to the geographic pole. To see why this is necessary, consider Fig. 5.6 showing a spreading ridge at the North Pole and an E–W ridge at the equator. Only the first magnetic reversal is shown following the sea-floor spreading hypothesis. During normal polarity time the magnetic anomaly is positive reinforcing the field's intensity at the pole, but it is negative at the equator. This illustrates that the anomalies vary as a function of latitude. In this example the anomalies are symmetric about the spreading axis. Simple stacking by adding the anomaly strengths after adjustment for different spreading rates would result in destructive interference, even though the anomalies reflect identical reversal histories. Blakely and Cox (1972) recognized that if all anomalies were transformed to the pole using the axial geocentric dipole field assumption (§6.1), this problem could be eliminated. Stacking the anomaly data in this way would therefore reduce the noise.

It might be added that the assumption that anomalies are symmetric about a spreading axis if the magnetization is symmetric is not necessarily true nor always observed. The observed symmetry about several spreading centres played an important role in convincing geophysicists that plates were spreading about this axis, but the symmetry is partly an accident of the strike of ridges under investigation. Following the dyke model of Vine and Matthews (1963), consider the magnetization of two dykes symmetric in all respects, including magnetization, that are displaced some distances from the spreading axis. The external field due to sources in the earth's core will usually be different (in intensity and direction) above the two dykes and therefore so will the magnetic anomaly. Again, this represents no particular

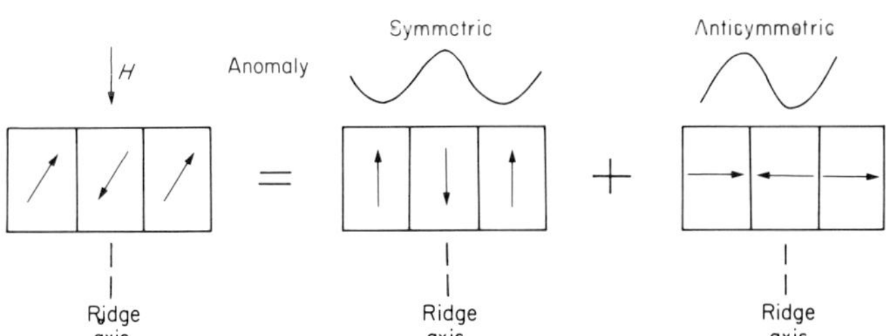

Fig. 5.7. Symmetry of magnetic anomalies about a ridge axis. A symmetric magnetization of the crustal blocks (left) is not parallel to the external field *H*, taken to be vertical. The magnetization can be divided into a component parallel (centre) and perpendicular (right) to the external field. These components produce symmetric and asymmetric anomalies respectively. Normally the magnetization associated with any given anomaly will not be parallel to the present magnetic field in the region because spreading has displaced the lithosphere from its place of origin. Thus there will normally be both a symmetric and asymmetric component. The observed symmetry in many marine magnetic anomalies over oceanic ridges can be traced to the predominant north–south orientation.

problem if one transforms to the pole, where symmetric magnetic anomalies do imply symmetric magnetization and vice versa. This symmetry exists, however, only if the axial geocentric dipole field assumption is correct. This point and some insight into magnetic symmetry is illustrated in Fig. 5.7.

5.2.4 Extension of Sequences to 165 Ma and Older

A series of papers were published in 1968 showing that the same sequence of magnetic anomalies parallel to the ridge crest is present in much of the Pacific, Atlantic and Indian Oceans (Pitman *et al.*, 1968; Dickson *et al.*, 1968; Le Pichon and Heirtzler, 1968; Heirtzler *et al.*, 1968). The most prominent positive magnetic anomaly peaks were numbered from 1 at the mid-ocean ridges to 32, the oldest anomalies examined. Following the lead of Vine and Matthews (1963), Heirtzler *et al.* (1968) extrapolated from the well-dated polarity time-scale for the recent few million years (§5.1.3) by assuming uniform spreading rates in the different oceans. However, this extrapolation led to different estimates of the polarity time-scale for all four regions involved, the South Atlantic, the North and South Pacific, and the South Indian Oceans. Various geological and geophysical criteria were used to decide that the South Atlantic magnetic profile should be selected as a standard. A magnetic reversal chronology extending back to 80 Ma was then obtained by assuming a uniform axial spreading rate at mid-latitudes in the South Atlantic. This spreading rate was estimated to be 1.9 cm year^{-1} from the known polarity sequence for the past few Ma (§5.1.3). This reversal chronology was found subsequently to be consistent with ages obtained by palaeontological dating of basement sediments obtained by the Deep Sea Drilling Project (Maxwell *et al.*, 1970; La Brecque *et al.*, 1977). Even today this South Atlantic time-scale provides a reasonable first-order approximation of the reversal chronology during the Cenozoic. An updated version of these results using the polarity time-scale of Lowrie and Alvarez (1981) is shown in Fig. 5.8.

There was some early scepticism that much improvement could even be made on the above chronology. If the reversal time-scale were extended back beyond a few million years using the methods of §5.1.3, errors in dating would become too large to delineate reversals. A 3% error in radiometric dating of a lava flow with age near 10 Ma is larger than the mean polarity time of the last few million years. Therefore, it is not possible to date a lava flow accurately enough to conclude that it erupted in the same polarity interval, or chron, as another lava flow in a different locality. This problem was surmounted by using long *sequences* of lava on land (McDougall *et al.*, 1976a, b, 1977) and long sediment sections exhibiting several reversals on land and beneath the oceans (Kennett, 1980; Lowrie and Alvarez, 1981). Relative ages could then be obtained by applying the well-known principles of stratigraphy, and this approach is referred to as *magnetostratigraphy*. One or more absolute ages are usually also required to compare a magnetostratigraphic section to a magnetic anomaly sequence.

The above approach has been facilitated by the finding of two very long polarity intervals, one in the Permian (Irving and Parry, 1963) and one in the Cretaceous (Helsley and Steiner, 1969), which could be used as marker horizons in many cases. Various names have been used to describe these long polarity intervals, including magnetic quiet zones (so-named because of their rather featureless signature in the marine magnetic anomaly record), long polarity intervals, the Kiaman Interval (application to the 50 Ma or so interval of reverse polarity in the Permian), and superchron recommended by the International Sub-Commission on Stratigraphic Classification.

Several excellent reviews of the reversal chronology are available (Ness *et al.*, 1980; Lowrie and Alvarez, 1981; Cox, 1982). These time-scales include suggested changes in the potassium–argon decay constants (McDougall, 1979; Mankinen and Dalrymple, 1979) and information from older sections

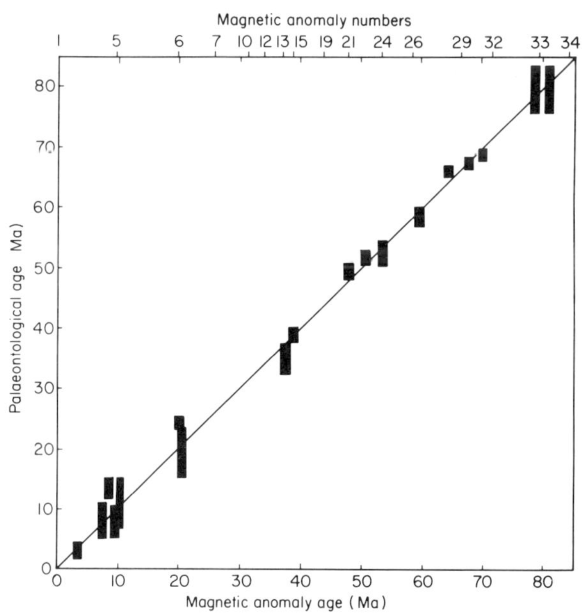

Fig. 5.8. Comparison of the palaeontological ages of basal sediments in deep-sea drill holes with the basement ages predicted from the magnetic anomalies. The 45° line is that expected for perfect agreement. (After Lowrie and Alvarez, 1981.)

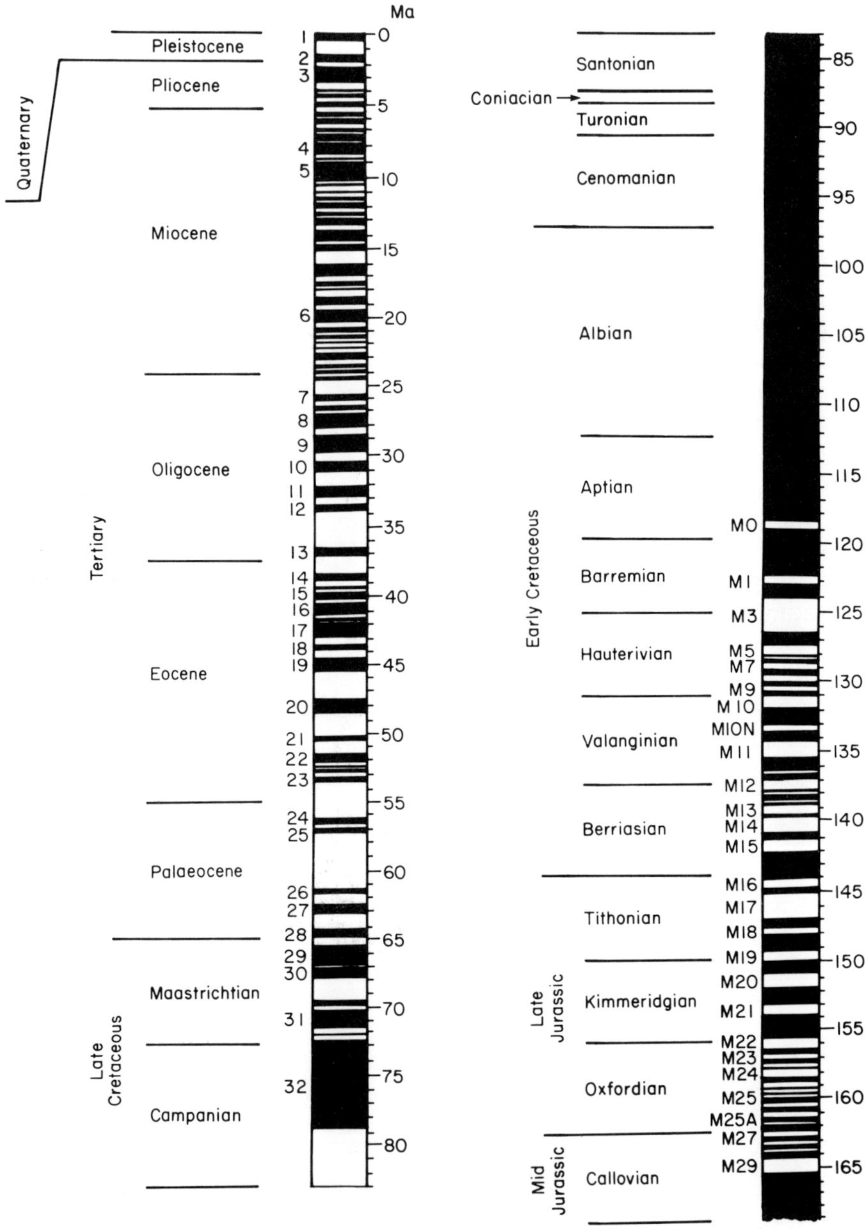

Fig. 5.9. Reversal time-scale for the past 170 Ma (after Cox, 1982).

well-dated by biostratigraphic or radiometric methods (McDougall *et al.*, 1976a, b, 1977; Butler *et al.*, 1977; Lowrie and Alvarez, 1981). Cox's (1982) version of the reversal chronology for the past 170 Ma is shown in Fig. 5.9. Changes are still expected in this chronology, although these will probably come in the form of very minor revisions (McFadden and Merrill, 1984).

5.3 Analysis of Reversal Sequences

5.3.1 Polarity Bias

The reversal chronology for the past 170 Ma is now well determined from analyses of marine magnetic anomalies. However, the reversal record prior to this is absent due to the subduction of older oceanic lithosphere. For older times therefore the reversal chronology has to be based on magnetostratigraphic reconstructions from rock sequences on land.

One of the features of the reversal time-scale for the past 170 Ma (Fig. 5.9) is that the character of the reversal pattern changes markedly with time. The most obvious change occurred 83 Ma ago. For the preceding 35 Ma the field was normal with no intervals of reversed polarity corresponding to the Cretaceous quiet zone in the marine record. Then at 83 Ma the field began to reverse rapidly and symmetrically and this pattern has continued up to the present.

The presence of these long intervals of time in which there were no polarity changes gives rise to the phenomenon of *polarity bias*. During times of *normal polarity bias* the field remains normal for almost all the time, and during times of *reversed polarity bias* it likewise remains reversed. In times of *mixed polarity* the field alternates frequently between the normal and reversed states. The duration of intervals of constant polarity bias seems to range from 20 Ma to about 100 Ma. These intervals are more than an order of magnitude larger than the duration of chrons and subchrons during the Cenozoic, suggesting that the physical origin of polarity bias may be different from that which causes individual reversals. Cox (1982) has used the term *superchron*, the next level above chron in the hierarchy of magnetostratigraphic names, to describe intervals of polarity bias.

Magnetostratigraphic studies of late Palaeozoic sequences in many parts of the world have established the existence of a Permo-Carboniferous reversed polarity superchron (Irving and Parry, 1963). Irving and Pullaiah (1976) in an analysis of global palaeomagnetic data suggest that this superchron occurred between 313 and 227 Ma ago or for 86 Ma. Several very short normal polarity intervals have been described as occurring during this

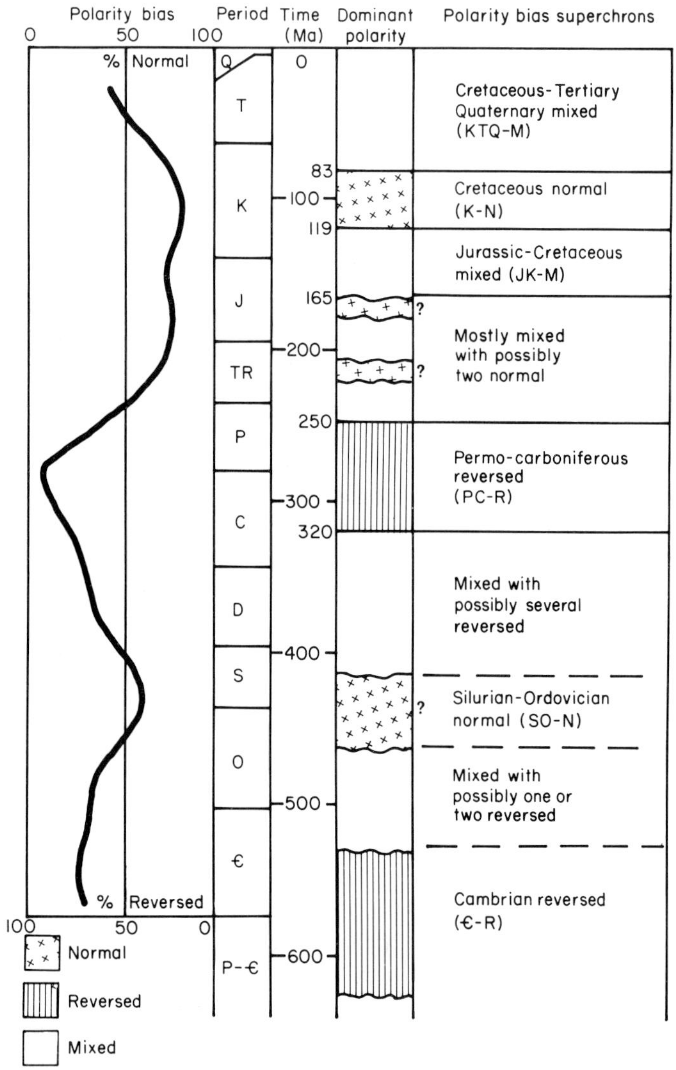

Fig. 5.10. Polarity bias superchrons. The curve on the left shows the percentage of polarities observed in palaeomagnetic results seen through a sliding window of 50 Ma (after Irving and Pullaiah, 1976). The polarity bias superchrons are those defined by Cox (1982) back to 320 Ma. For the remainder of the Palaeozoic they have been inferred from the work of Khramov and Rodionov (1981).

predominantly reversed superchron. Cox (1982) places the boundaries at 320 and 250 Ma as a result of revisions in the geological time-scale, giving a duration of 70 Ma.

Global synthesis of all available palaeomagnetic polarity data can help establish those times when the field exhibited polarity bias. By using a sliding window of 25 or 50 Ma the fraction of samples with normal or reversed polarity can be calculated. This was first carried out by McElhinny (1971) and later updated by Irving and Pullaiah (1976). Figure 5.10 shows this analysis of Irving and Pullaiah (1976) in which a sliding window of 50 Ma has been used. The sub-division of the Phanerozoic into polarity bias superchrons follows that of Cox (1982) back to the late Palaeozoic. For the early Palaeozoic the proposals are speculative and based on extensive studies of sections on the Siberian platform summarized by Khramov and Rodionov (1981).

In Precambrian times the polarity structure is not yet well known. Irving and McGlynn (1976) suggest that during the Proterozoic the field was normal for 77% of the time. However, the definition of normal and reversed polarity for these times depends on precise knowledge of the apparent polar wander path. With gaps in the record for North America upon which this conclusion was based, the certainty with which polarity is defined is much decreased. All that can be said at the present time is that polarity bias appears to have been present during Precambrian times. Polarity bias thus reflects time changes of the order of 10^7 to 10^8 years in the earth's magnetic field. These are the longest time changes that have been discovered to date.

5.3.2 Independence of Polarity Intervals

Suppose t is the lapsed time since the last reversal and $Pr(t)\,dt$ is the probability that a reversal will occur between t and $t + dt$. In a *general renewal process* $Pr(t)$ can increase or decrease with increase in time but the probability cannot depend on the length of prior polarity intervals. If reversals are generated by a general renewal process, the lengths (duration) of polarity intervals will be independent. Such independence was first claimed without convincing proof (Cox, 1968, 1969b; Nagata, 1969) and subsequently challenged (Naidu, 1974). One reason to suspect polarity intervals are not independent is that a reversal cannot occur instantaneously and therefore the magnetic field must have some memory. If the fluid motions powering the earth's magnetic field were to cease, the magnetic field would decay away with a characteristic decay time of 10^4 years (§7.4.3). The fluid motions operating the geodynamo appear moderately complex and therefore this time is probably a reasonable upper limit of the memory time of the core processes. This is more than an order of magnitude smaller than the mean polarity time during the Cenozoic suggesting that the departures from independence might be small.

A second reason for suspecting the independence assumption is that reversal sequences are not stationary as illustrated by Fig. 5.11 (McFadden and Merrill, 1984). The mean polarity interval increases as one goes back in time during the Cenozoic. This means there is some statistical correlation between successive polarity intervals. However, the change is slow relative to the mean polarity interval in the Cenozoic. Again this suggests that departures from independence might be small.

The significance of the above effects can be tested statistically in a manner described by Phillips *et al.* (1975) and Phillips and Cox (1976). Let t_i be the length of time between two successive reversals of the earth's magnetic field. Unbiased estimates $\hat{\rho}_j$ of the correlation coefficients can be obtained from a sequence of polarity intervals with lengths t_i, $i = 1, \ldots, n$ (Phillips *et al.*, 1975):

$$\hat{\rho}_j = \frac{\hat{C}_j}{(C'_{0,j} C''_{0,j})^{1/2}}$$

where

$$\hat{C}_j = \frac{1}{n-j} \sum_{i=1}^{n-j} (t_i - t'_j)(t_{i+j} - t''_j)$$

$$\hat{C}'_{0,j} = \frac{1}{n-j} \sum_{i=1}^{n-j} (t_i - t'_j)^2$$

$$C''_{0,j} = \frac{1}{n-j} \sum_{i=1}^{n-j} (t_{i+j} - t''_j)^2$$

$$t_j = \frac{1}{n-j} \sum_{i=1}^{n-j} t_i$$

$$t''_j = \frac{1}{n-j} \sum_{i=1}^{n-j} t_{i+j}$$

If the observed polarity intervals are independent, then the expected values for the correlation coefficients are all zero for *j* not equal to zero and one for *j* equal to zero. This was found to be the case within 95% confidence limits. Spectral analysis techniques were later used by Phillips and Cox (1976) to show that no significant periodic components in the reversal frequency appeared to be present in the reversal record. Thus to a very good first-order approximation reversals of the earth's magnetic field can be regarded as a purely stochastic process in which the lengths of polarity intervals can be treated as being independent. This is in marked contrast to the Sun, in which reversals appear to occur every 11 years (§12.2).

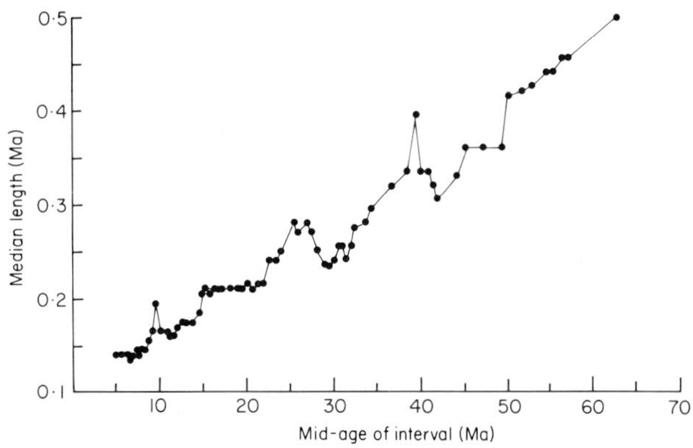

Fig. 5.11. Non-stationarity of geomagnetic reversal frequency. The frequency of reversals appears to have changed continuously since 85 Ma ago (after McFadden and Merrill, 1984).

5.3.3 Statistical Analyses of Polarity Sequences

A common approach to the statistical modelling of reversals of the earth's magnetic field has been to assume that instabilities trigger reversals (Cox, 1968, 1969b, 1981; Nagata, 1969). If these instabilities are randomly distributed in time and are independent, $Pr(t)$ defined in the last section is a constant and the lengths of polarity intervals will be Poisson distributed. As a corollary, the probability density $p(t)$ of the intervals between reversals is given by:

$$p(t) = A\,\mathrm{e}^{-\lambda t} \tag{5.6}$$

where λ is the mean reversal frequency and λ^{-1} is the mean polarity length. Equation 5.6 states that the most likely polarity intervals are the shortest.

Naidu (1971) argued that the lengths of polarity intervals are gamma distributed:

$$p(t) = (\beta\lambda)^{\beta} t^{\beta-1}\,\mathrm{e}^{-\beta\lambda t}[\Gamma(\beta)]^{-1} \tag{5.7}$$

where β is a parameter to be determined. Equation 5.7 reduces to (5.6) in the limit β goes to one. Naidu (1971) used the Heirtzler *et al.* (1968) South Atlantic time-scale (§5.2.4) to find that β was significantly greater than one in the interval between 45 and 75 Ma ago.

Blakely and Cox (1972) suggested that some short events (subchrons) were missing from the Heirtzler *et al.* time-scale and that the Poisson distribution might still be correct. This was consistent with an earlier suggestion by Harrison (1969) that a filtering process was operating that made it difficult or impossible to observe very short polarity subchrons in the marine record.

McFadden (1984) has examined this question rigorously and a summary of his results follows. Suppose polarity intervals (presumably very short polarity subchrons) with duration less than time t_0 are not observed. Further assume that a Poisson distribution describes the reversal frequency in the core, as first suggested by Cox (1968). If u is the probability that an interval is too short to observe, then

$$u = \int_0^{t_0} p(t)\,\mathrm{d}t = 1 - \exp(-\lambda t_0)$$

where (5.6) has been used to carry out the integration. Similarly, the probability that an interval is long enough to be observed is given by $\exp(-\lambda t_0)$. An important point to realize is that if a short interval is lost in the filtering process it must be included in some manner in the preceding and succeeding intervals. The way this is incorporated depends on the number of *contiguous* short intervals lost, as illustrated in Fig. 5.12. McFadden used the above approach to derive a relation between the observed rate of reversals λ_{obs} and the actual rate λ:

$$\lambda_{obs} = \frac{\lambda \exp(-\lambda t_0)}{2 - \exp(-\lambda t_0)}$$

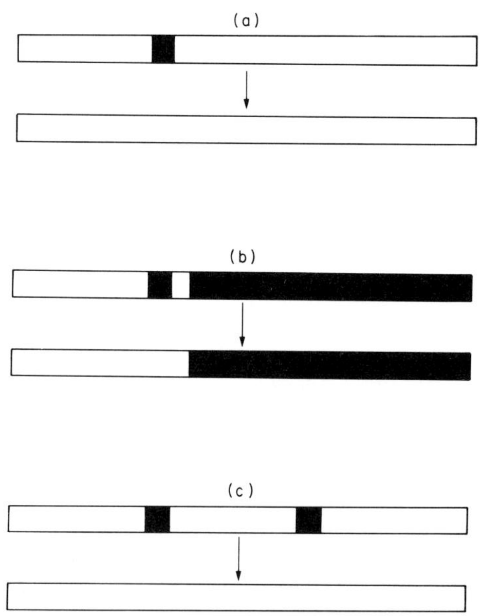

Fig. 5.12. Change in the numbers of polarity intervals caused by preferential filtering of very short intervals. (a) 3 intervals reduced to 1 interval; (b) 4 intervals reduced to 2 intervals; (c) 5 intervals reduced to 1 interval.

Unfortunately for any value of λ_{obs} there are two solutions for λ. For example if λ_{obs} is 2.5 reversals per million years and $t_0 = 0.09$, then both $\lambda = 6.37$ and 11.2 reversals per million years satisfies the above equation.

McFadden (1984) solved this problem by pointing out that if Poisson's distribution is correct and if the lengths of N independent polarity intervals are summed, then the probability $p(z)\,\mathrm{d}z$ that this sum lies between z and $z + \mathrm{d}z$ is:

$$p(z)\,\mathrm{d}z = \frac{1}{\Gamma(N)}\lambda^N z^{(N-1)} \exp(-\lambda z)\,\mathrm{d}z \tag{5.8}$$

where

$$z = \sum_{i=1}^{N} t_i$$

and the mean value of z is

$$\langle z \rangle = \frac{N}{\lambda}$$

Equation 5.8 is the same as (5.7) with $\beta\lambda$ set equal to λ, t to z and β to N. Thus if an interval in the observed sequence is the concatenation of N intervals from the original sequence, it may be considered as a random observation from a gamma distribution with $\beta = N$ and $\lambda_{obs} = \lambda/N$. Note that the λ given in (5.7) should now be considered to be λ_{obs}. β will not generally be equal to one even if $N = 1$, because of the preferential suppression of short intervals. Thus the best fitting value of β is determined by the details of the actual filter. McFadden (1984) shows that the parameter β can be interpreted simply as the ratio of the number of polarity intervals in the original sequence to the number of observed intervals in the filtered sequence. Hence a gamma distribution is to be preferred for analysing the statistical properties of the observed polarity sequence, even though the actual core polarity sequences may be Poisson distributed.

Phillips (1977) used a gamma distribution to describe the reversal frequency since the Late Cretaceous. He found that both the observed normal (N) and reversed (R) polarity sequences separately satisfied a gamma distribution. However there was a discontinuity in the average β (eqn. 5.7) at 45 Ma. Stationarity was observed for the N and R sequences on both sides of this discontinuity. However, the average β for the N and R sequences, β_N and β_R, was 1.19 and 2.28 respectively between 44.8 and 0 Ma. Between 76.3 and 44.8 Ma these values were 1.35 and 2.79 respectively. Hence the shapes of the gamma distribution for the N and R states are different and Phillips shows this difference to be significant at the 95% confidence level. Phillips (1977) interpreted these results as showing differences in stability between polarity states. The result was surprising because the mean length of the N and R intervals were found to be identical. It might have been expected that differences in polarity stability would also lead to differences in mean polarity lengths.

McFadden and Merrill (1984) have examined time-scales published subsequent to Phillips' (1977) analyses (Fig. 5.11). They found that stationarity was now not observed in the revised time-scales, and that there were pronounced changes in the estimated value of β_N and β_R for the different scales published. These differences reflect the instability of β to minor changes in the reversal chronology. They found that the insertion of only two polarity subchrons in the whole Cenozoic time-scale was sufficient to create no significant differences between β_N and β_R over this time interval. Also there was no discontinuity in β near 45 Ma for the most recent time-scale of Cox (1982) (see Fig. 5.9).

The most simple interpretation of the data is that there is no difference in stability between the N and R polarity states. Median polarity lengths decrease continuously throughout the Cenozoic (Fig. 5.11). The longest polarity chron over the last 170 Ma (Fig. 5.9) is the Cretaceous long normal polarity chron. The rate of change of λ with time before and after this long polarity chron is identical, within statistical uncertainties, except for sign (McFadden and Merrill, 1984). Interpretation of these changes in reversal frequency with times of the order of 10^8 years is given in §10.4.

5.4 Behaviour During Polarity Transitions

5.4.1 Records of Polarity Transitions

Polarity transitions are recorded in sedimentary rocks, intrusive igneous rocks and extrusive igneous rocks (lava flows). The best records in sedimentary rocks occur when the remanence is a DRM or post-depositional remanent magnetization (§3.1). In cases of rapid sedimentation, a good continuous record of a transition can be recorded. Unfortunately extracting this record is often difficult. There is no absolute palaeointensity method available for sediments, although normalizing procedures utilizing anhysteretic remanent magnetization or isothermal remanent magnetization may provide valuable information on relative intensity values in some cases. Chemical changes, physical disturbances such as bioturbation, changes in sedimentation rates, and sampling problems are common in sediments and result in distortion of the geomagnetic record in many cases. In spite of these problems, sediment records have proved to be one of the most valuable sources for information on the transitional behaviour of the field (Niitsuma, 1971; Opdyke *et al.*, 1973; Hillhouse and Cox, 1976; Koci and Sibrava, 1976; Freed, 1977; Burakov *et al.*, 1976; Valet and Laj, 1981).

It should be noted that the study of polarity transitions has been restricted mainly to those that have occurred during the last few tens of millions of years. This is fortunate, because generally older rocks are most likely to have acquired secondary components of magnetization than younger ones. A secondary component acquired in the relatively high fields characteristic of non-transition times can produce a large spurious direction when vectorally added to a primary direction acquired in a relatively weak transition field. Rock magnetic noise therefore can be expected to be larger in transition records and more difficult to detect than in non-transition records.

Intrusive igneous rocks usually cool slow enough to provide continuous records of reversal transitions (providing they form during a transition). The magnetic field is recorded as the cooling front sweeps through the intrusive. Because minerals in igneous rocks are seldom, if ever, in chemical equilibrium during cooling and because cooling rates are slow enough in intrusives for chemical changes to occur, chemical alteration of magnetic minerals below their Curie temperatures is probably common. Also estimates of absolute times within a polarity transition are difficult because they must be based on cooling models which are generally poorly determined. Because of these problems only a few successful studies of reversal transitions in intrusive igneous rocks have been completed (Dunn *et al.*, 1971; Dodson *et al.*, 1978).

Terrestrial lava flows have the fewest rock magnetic problems, primarily because they cool very rapidly. Although they likely contain minerals that are in chemical disequilibrium with their environment, alteration often occurs very slowly for kinetic reasons. Although lava flows have desirable rock magnetic properties, unfortunately they only provide spot records of the earth's magnetic field and a continuous record is not available. Hence good records of reversal transitions in lava flows usually require the eruption of a large number of lava flows in rapid succession. In addition, the spacing in time between successive flows erupted during a transition cannot be accurately determined because the errors associated with radiometric ages are typically much greater than the duration of a polarity transition. Nevertheless this source has provided a considerable amount of valuable information on transition directions (Watkins, 1969; Larson *et al.*, 1971; Dagley and Lawley, 1974; Bogue and Coe, 1981, 1982).

Fuller *et al.* (1979) have reviewed all transition records and divide them into two categories. Virtual geomagnetic poles (VGP; see §3.3) were used to describe the paths even though it was recognized that the geocentric axial dipole field assumption was probably not satisfied during a transition.

Alternative methods, such as plotting declination and inclination of directions during a transition, have also been used (Williams and Fuller, 1981a). The category A transitions of Fuller *et al.* (1979) have one or more VGP with latitudes greater than 60° in both the initial and final hemispheres and two or more VGP with shallower latitudes in each hemisphere. Category B transition paths satisfy the weaker requirement that two or more VGP have latitudes less than 60° (no specification of the hemisphere) and one or more high latitude VGP in each hemisphere. Systematic trends in the directional data are discussed in the next section.

Both absolute (only from lavas) and relative palaeointensity data for rocks recording transition directions indicate that the field intensity decreases substantially during a polarity change (Van Zijl *et al.*, 1962; Momose, 1963; Coe, 1967b; Lawley, 1970; Larson *et al.*, 1971; Dagley and Wilson, 1971; Wilson *et al.*, 1972; Opdyke *et al.*, 1973; Hillhouse and Cox, 1976; Dodson *et al.*, 1978; Bogue and Coe, 1981). For the most part these records are from sediments and intrusives and provide only relative intensity information. However, the data are clear and indicate reductions sometimes to only 10% of the usual field intensity outside the transition. There are some notable exceptions, including examples of very strong fields observed right in the middle of transitions (Shaw, 1975, 1977; Prevot, 1977).

Figure 5.13 provides a summary of both absolute and relative palaeointensities relating to polarity transitions for lavas less than about 10 Ma old. Absolute palaeointensities for this time interval have been listed by McFadden and McElhinny (1982) (last 5 Ma only) and Roberts and Shaw (1982). The 288 values include many from the central part of polarity transitions and are characterized by low latitude VGP. In order to allow for the different site locations all intensity values have been converted to virtual dipole moments (VDM; see §3.4.4). Even though the geomagnetic field is probably highly non-dipolar during transitions (§5.4.2), the use of VDM just as with VGP serves to take account of the effects of site location on the surface of the earth. Average VDM for 20° VGP latitude bands between 0° and 180° are shown in Fig. 5.13a. These absolute intensity values show clearly that *on average* the field at the central part of transitions (VGP colatitude 90°) is about 25% its usual value.

Dagley and Wilson (1971) and Wilson *et al.* (1972) have examined the relative intensities of 1314 lavas from Iceland in a similar way. Assuming that the intensity of magnetization of each lava is proportional to the field in which it cooled, they calculated a "pseudo" VDM for each lava and argued that with sufficient numbers the variation in their magnetic properties would average out. The result is shown in Fig. 5.13b in which average VDM for 5° VGP colatitude bands have been used. With more than four times as many lavas as in Fig. 5.13a, the numbers in each band are comparable between the

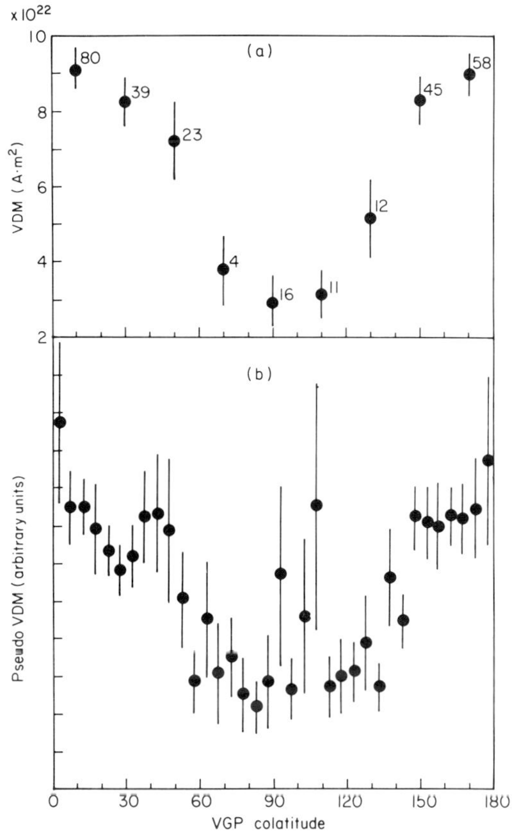

Fig. 5.13. Decrease in field intensity during polarity transitions. (a) Average Virtual dipole moments (VDM) determined from 288 absolute palaeointensities listed by McFadden and McElhinny (1982) and Roberts and Shaw (1982). Averages have been taken over 20° VGP colatitude bands. Standard errors are shown with the number of values in each band. (b) Average intensity of magnetization from 1314 Icelandic lava flows expressed as "pseudo" VDM. Averages have been taken over 5° VGP colatitude bands. Standard errors are shown. (After Wilson *et al.*, 1972.)

two parts of Fig. 5.13. The data again show clearly that the average field at the centre of transitions is about 25% of its usual value. Both sets of data suggest there is a rapid fall in intensity when the VGP colatitude lies between 45° and 135°. Indeed the data of Fig. 5.13b lend some credence to the observations of Shaw (1975, 1977) and Prevot (1977) that high field strengths may occur for very short periods during polarity transitions. However, Kristjansson and McDougall (1982) after making a similar analysis of 2462 lavas from Iceland, suggest that departures from the smooth trend shown in Fig. 5.13b are merely due to lack of data in some of the VGP colatitude bands.

Most early estimates of the length of a polarity transition tended to give times of the order 10^3 to 10^4 years. Probably the best estimates come from oceanic cores with times of 4000 years (Harrison and Somayajulu, 1966), 4700 years (Niitsuma, 1971) and 4600 years (Opdyke *et al.*, 1973). These estimates are for the time of directional changes only. There is some suggestion that there is an intensity decrease associated with the polarity change that occurs prior to the directional changes (Dodson *et al.*, 1978; Hillhouse and Cox, 1976).

5.4.2 Models of Transitions

The earliest attempts to describe polarity transitions involved dipole models in which the dipole field (essentially axial) either decayed to zero and built up in the opposite direction or else rotated through 180° without change in intensity (Fig. 5.14a, b) (Creer and Ispir, 1970). There is now convincing evidence from studies of several transitions that the field is highly non-dipole during a transition (Hillhouse and Cox, 1976; Hoffman, 1977, 1979, 1981a; Hoffman and Fuller, 1978; Fuller *et al.*, 1979; Williams and Fuller, 1981a). Two examples of zonal harmonic non-dipole transition field models are shown in Fig. 5.14c and d. All transition field descriptions are still very uncertain because of lack of data, since there are at present less than 20 category "A" transition records (§5.4.1).

Suppose for a moment that a polarity change occurred by dipole rotation along a great circle as in Fig. 5.14b. A site located on the equator on that circle of longitude would record only changes in inclination. On the other hand, a site on the equator but 90° away from the first site would record only changes in declination. This makes it difficult to compare transitions recorded at different localities. Hoffman (1977) solved this problem by replotting transition VGP paths with respect to a common longitude and noticed an apparent systematic behaviour in the records. The vast majority of the transition data record changes from reverse polarity (R) to normal polarity (N) at sites in the northern hemisphere. Only two out of the fifteen category "A" transition records are $N \rightarrow R$ transitions. For the $R \rightarrow N$ transitions the VGP paths tend to lie in the hemisphere centred about the site meridian. Hoffman (1977) referred to such VGP as *near poles*, and their corresponding VGP paths as *near-sided paths*. *Far poles* and *far-sided paths* then refer to VGP paths that lie in hemispheres opposite to the site meridian.

Hoffman and Fuller (1978) classified transition paths in terms of zonal harmonics. Examples of quadrupole and octupole zonal non-dipole transition field models are shown in Fig. 5.14c and d. The terms quadrupole and octupole refer to the inferred magnetic field configuration at the mid-point of

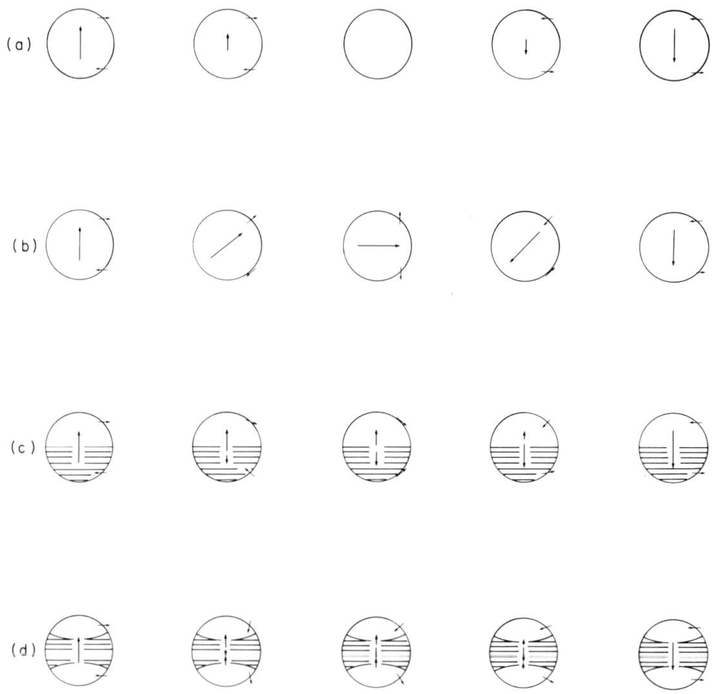

Fig. 5.14. Examples of some models for the transition field for a change from reverse to normal polarity. (a) Decrease in dipole moment which remains axisymmetric during the process. (b) Rotation of the main dipole without change in moment. (c) A quadrupole transition model in which the polarity change is initiated in the southern hemisphere of the core. (d) An octupole transition model in which the polarity change is initiated in the low latitude zone of the core.

the transition when the VGP crosses the equator. In the model of Hoffman and Fuller (1978) quadrupole and octupole transition fields occur depending on which part of the core first changes polarity to initiate the transition. When a polarity change is initiated in one hemisphere of the core a quadrupole transitional field results. When the change is initiated at either high or low latitudes in the core an octupole transitional field results (Fig. 5.15). Whether or not the transition VGP path is near-sided or far-sided depends both on the sense of the transition and the hemisphere of the site. There are eight possible combinations of zonal quadrupole and octupole transition field models and these are all illustrated in Fig. 5.15. Hoffman and Fuller (1978) tentatively conclude that the data are consistent either with a quadrupole model in which the polarity change is initiated in the southern hemisphere (Fig. 5.14c) or an octupole model in which the change is initiated in the low latitude zone (Fig. 5.14d).

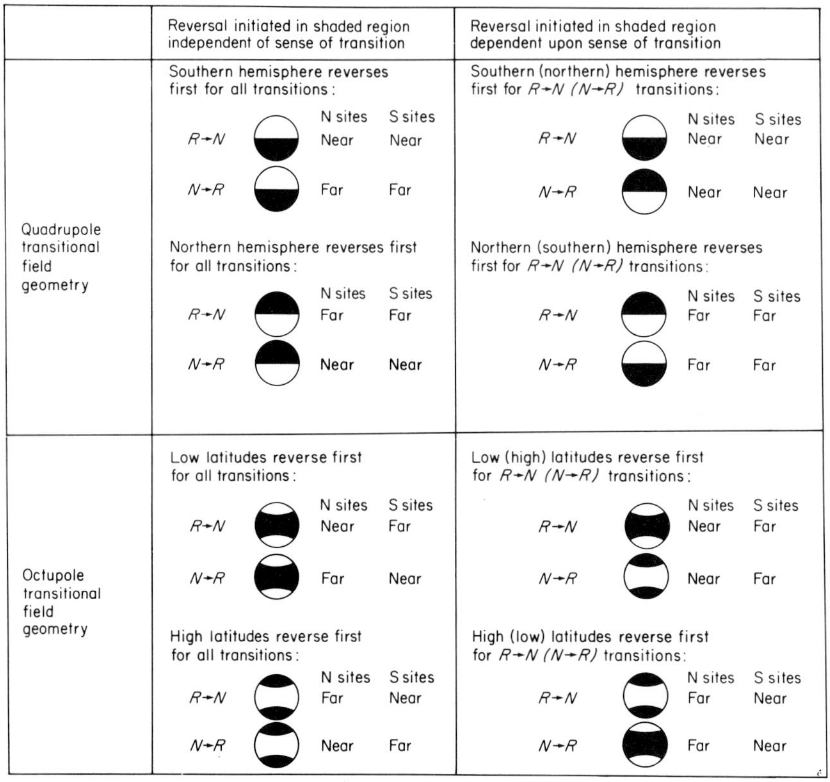

Fig. 5.15. Possible quadrupole and octupole transition field models for polarity changes. (After Hoffman and Fuller, 1978.) For each configuration the VGP path characteristic is indicated for the four possible site location and transition sense combinations. N and S refer to the northern and southern hemispheres respectively. *Near* and *Far* indicate whether the VGP path is near-sided or far-sided. The darkened portion of each figure indicates the region(s) in the core in which the polarity change is initiated.

Williams and Fuller (1981a) have modelled the dynamic properties of transitions by assuming that the dipole field intensity decreases (using a tangent function to represent the time dependence) with part of the dipole energy going into the first three non-dipole zonal harmonics, g_2^0, g_3^0 and g_4^0 (§2.2). They produced a reasonable model for the Matuyama–Brunhes $R \to N$ transition by assuming 75% of the energy in the dipole field was redistributed with the g_2^0, g_3^0 and g_4^0 terms according to the ratio 2:3:5. Hoffman (1981a), however, finds that during the same transition at the time of total axial dipole decay the controlling components are a zonal octupole (g_3^0) and a non-axisymmetric quadrupole (g_2^1, h_2^1). The existence of a significant g_2^0 component could not however be ruled out.

There are large deviations in the observed transition paths relative to the paths predicted by any of the above models. At times there are large changes in declination, not accounted for in pure zonal harmonic descriptions, and even loops in VGP transition paths. Because of the large axial geocentric dipole component in the present instantaneous field and in the time-averaged field (§6.1), transition paths will start out and end up with a high degree of axial symmetry. As a result a bias towards zonal harmonic transition paths will occur.

As more data become available the supposed clear systematics of VGP transition paths have become increasingly cloudy. Hoffman (1977) pointed out the tendency for $R \rightarrow N$ transitions observed at mid-latitudes in the northern hemisphere to have near-sided VGP paths and for $N \rightarrow R$ transitions to have far-sided VGP paths. However, Williams and Fuller (1981b) have now reported a far-sided $R \rightarrow N$ transition path from the Philippines. Bogue and Coe (1982) studied successive $R \rightarrow N$ and $N \rightarrow R$ transitions in lavas from Kauai, Hawaii and found the two VGP paths to be remarkably similar. Valet and Laj (1981) have also studied sequential transitions in Crete in Miocene sediments to find a VGP path that is borderline between a near- and far-sided path for the $R \rightarrow N$ transition and far-sided for the $N \rightarrow R$ transition. The only clear conclusion at the present time is that the field is highly non-dipole during a polarity transition. Systematic directional changes are suggested by the data, but the proof of transition VGP systematics requires careful statistical analyses and more data.

5.4.3 Excursions as Aborted Reversals

Excursions are often interpreted as reversal attempts which have not been successful (Doell and Cox, 1972; Barbetti and McElhinny, 1976; Hoffman, 1981b). Although almost all palaeomagnetists accept the existence of excursions, the actual demonstration of their existence is difficult (§4.3). Hoffman (1981b) has examined four likely (but not proved) excursions recorded in lavas, three from the southern hemisphere and one from the northern hemisphere. VGP paths for these excursions are consistent with those expected if the quadrupole model for reversal (Fig. 5.14c) was initiated but not completed. These data show a consistency between the apparent systematics in VGP transition data and the hypothesis that excursions are aborted reversals.

If excursions are truly aborted reversals, then their existence, extent and duration in the palaeomagnetic record is important to determine as it probably provides valuable insight into the geodynamo. The sun has a general magnetic field that reverses every 11 years (§12.2). Reversals appear to be associated with dynamic processes that are periodic and aborted reversals are not expected in the solar dynamo. On the other hand, reversals in the earth's core appear to be stochastic and consistent with the assumption that there is a small constant probability of reversal in any small increment of time. This implies that a Poisson distribution can describe the reversal process in the core, an implication consistent with the data even though a gamma distribution is preferred to describe the observed pattern of reversal frequency (§5.3.3). Hence, in the case of the earth, aborted reversals appear far more likely than in the sun.

Nagata (1969) has supposed that, when the dipole field increases again after decaying, the probability that its polarity is the same as previously is one half. Thus excursions in the form of aborted reversals should be about as common as reversals in the geomagnetic record. The problem is, however, that unless very detailed records are available the probability of detecting an excursion is much less than the probability of detecting a reversal. Thus in the palaeomagnetic literature it appears that excursions might be fewer and of shorter duration than reversals. Harrison (1980) has examined the occurrence of low-latitude VGP lava flows in nearly a thousand lavas from western and eastern Iceland as recorded by Watkins *et al.* (1977) and Watkins and Walker (1977). The sequence is well dated and could be compared with the known geomagnetic polarity time-scale. The low latitude VGPs were divided into three groups: (i) those that occur during the time of a known transition; (ii) those that clearly occur at times when there are no known transitions; (iii) those that cannot for certain be identified within the first two groups. Harrison (1980) identified 64 lavas in the first two groups having VGP latitudes less than 30°. Of these 27 (42 %) occurred at times when there were no known reversals. If these represent excursions of the geomagnetic field, then their observed rate of occurrence is consistent with the view that excursions are aborted reversals.

Chapter Six

The Time-averaged Palaeomagnetic Field

6.1 Geocentric Axial Dipole Hypothesis

6.1.1 The Past Few Million Years

The interpretation of palaeomagnetic results has always depended upon the fundamental hypothesis that the time-averaged geomagnetic field is that of a geocentric axial dipole. The instantaneous field at the present time deviates substantially from that of a geocentric axial dipole, but because the field undergoes significant long-term secular variation (e.g. the westward drift), the time-averaged field can be expected to be significantly different from the instantaneous field. Hospers (1954), using the newly developed statistical methods of Fisher (1953) (see §3.3.3), was the first to demonstrate that, averaged over several thousands of years, the virtual geomagnetic poles centred on the geographic pole in recent times. Creer *et al.* (1954) first introduced the concept of the apparent polar wander path for the interpretation of palaeomagnetic results from Great Britain. In so doing they explicitly invoked the geocentric axial dipole hypothesis. It is now standard procedure in palaeomagnetism to calculate palaeomagnetic poles (§3.3.2) which, on the basis of the geocentric axial dipole hypothesis, are then assumed equal to the palaeogeographic poles.

It has been customary to illustrate the validity of the geocentric axial dipole hypothesis by plotting all palaeomagnetic poles for the past few million years on the present latitude–longitude grid and showing that they centre about the geographic pole. Successive reviews of Cox and Doell (1960), Irving (1964) and McElhinny (1973) have used this method and indeed this is clearly shown to be the case. Exploration of the oceans has enabled a large number of deep-sea cores to be collected world-wide. Assuming the cores are taken vertically, the magnetic inclination can be determined. Opdyke and Henry (1969) compared these inclinations from different parts of the world with those expected from a geocentric axial dipole and found a very good fit with the model.

Unfortunately the observation that the world-wide palaeomagnetic poles determined for the past few million years centre about the geographic pole does not by itself demonstrate that the time-averaged field is purely a geocentric axial dipole (g_1^0 only). World-wide data from a time-averaged field that consists of a series of zonal harmonics (g_1^0, g_2^0, g_3^0 etc.), when plotted in this way, will always produce a set of palaeomagnetic poles that centre about the geographic pole. Wilson and Ade-Hall (1970) first noted that the palaeomagnetic poles from Europe and Asia for the past few million years all tended to plot too far away from the observation site along the great circle joining the site to the geographic pole. Successive analyses by Wilson (1970, 1971), McElhinny (1973) and Wilson and McElhinny (1974) have confirmed that this occurs on a global scale. This effect is shown in Fig. 6.1 in which global data for the past five million years have been averaged by 45° longitude sectors. The position of the pole is related to the sector in each case, but the average of all eight sector poles still gives the geographic pole within a degree. Note that the reversed data tends to plot further over the pole from the sampling region than the normal data. Wilson (1971) introduced the concept of the common site longitude pole position as a convenient way of analysing the overall far-sided effect. The method is to place all observers at zero longitude by replacing the pole longitude with the common site longitude given by the difference between the pole and site longitudes. The data for the past 5 Ma from 4580 spot readings of the field as used by Lee and McElhinny (1984) have been analysed in this way.

To allow for the widely varying number of measurements in each published result, the analysis gives unit weight to 20 spot readings (sites), this being a typical time-averaged result giving a palaeomagnetic pole. The mean common site longitude poles for the normal and reversed global data are shown in Fig. 6.2. As is apparent in Fig. 6.1, the reversed poles plot more far-sided than the normal ones. At the 95% confidence level the difference between the two means is significant. This point is discussed further in §6.2.3.

The above effects can be imagined to be derived from an axial dipole source that is not geocentric but displaced northwards along the axis of rotation.

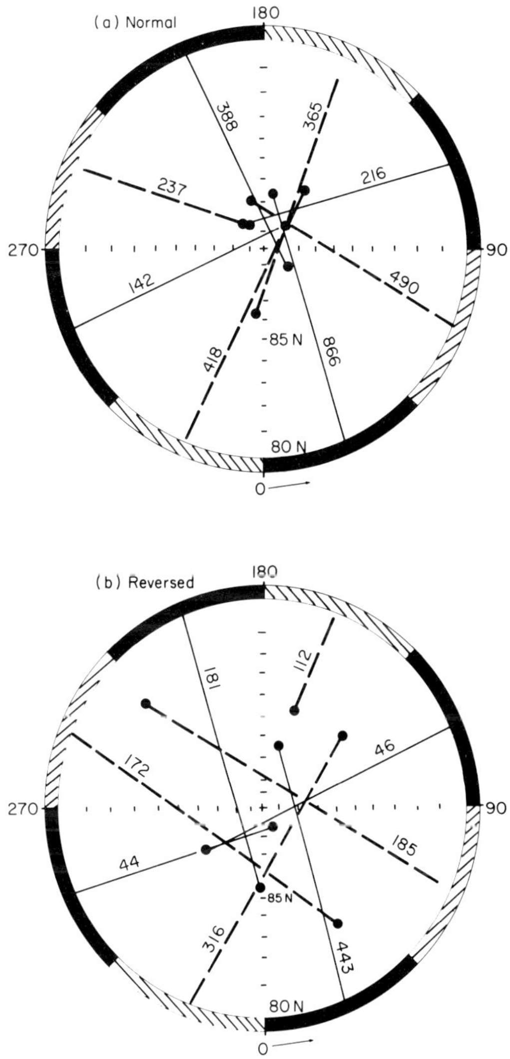

Fig. 6.1. Analysis of global palaeomagnetic results for the past 5 Ma by averaging over 45° longitude sectors. (a) Normal, (b) reversed. The palaeomagnetic poles tend to plot too far and over the geographic pole. The number of spot readings of the field is indicated for each average. Polar stereographic projection at 80°N. (After Lee and McElhinny, 1984.)

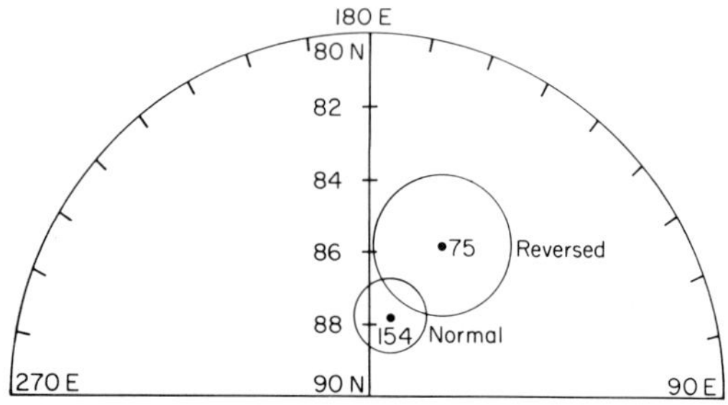

Fig. 6.2. Common site longitude representation of the data in Fig. 1. 95% confidence limits about each mean are calculated by assuming that 20 spot readings of the field are a typical palaeomagnetic result. Thus each spot reading is represented as 0.05 of a unit vector. Polar stereographic projection at 80°N. (After Lee and McElhinny, 1984.)

This offset-dipole model of Wilson (1970) is also equivalent to a geocentric axial dipole (g_1^0) plus a geocentric axial quadrupole (g_2^0) for small dipole offset. The time-averaging of the palaeomagnetic field is performed over intervals of time much longer than the periods of westward drift ($\sim 2 \times 10^3$ years). In that case it might be supposed that the tesseral and sectoral components of the magnetic scalar potential ψ (eqn. 2.13) will be eliminated leaving only the zonal potential. At the surface of the earth, eqn. 2.13 with $a = r$ then reduces to:

$$\psi = \frac{a}{\mu_0} \sum_{l=1}^{\infty} g_l^0 P_l^0(\cos\theta) \tag{6.1}$$

where $P_l^0 = P_l$ reducing to the Legendre polynomials (§2.2.1). Following James and Winch (1967), the potential of a dipole of strength m displaced along the rotation axis at a distance x is:

$$V = \frac{m}{a^3}\cos\theta + \frac{mx}{a^4}(3\cos^2\theta - 1) + \ldots \tag{6.2}$$

If this is related to the expansion of (6.1), namely

$$V = ag_1^0\cos\theta + ag_2^0\left(\frac{3\cos^2\theta - 1}{2}\right) + \ldots \tag{6.3}$$

then

$$g_1^0 = \frac{m}{a^3} \quad \text{and} \quad g_2^0 = \frac{2mx}{a^4}$$

The displacement x of the offset dipole can then be expressed in terms of the zonal coefficients (Wilson, 1970)

$$\frac{x}{a} = \frac{g_2^0}{2g_1^0} \tag{6.4}$$

Although the time-averaged palaeomagnetic field has been that of a geocentric axial dipole to a first order approximation over the past few million years, the above shows there are second order departures from the model. A more detailed examination of these second order terms is given in §6.2.

6.1.2 The Past 600 Ma

Over the past few million years it is reasonable to assume that continental drift has been small and that the relation of present continents to the axis of rotation has remained unchanged. However, for older times the relationship of the continents to the axis of rotation is usually determined from palaeomagnetic data by invoking the geocentric axial dipole hypothesis. So it is possible only to test the dipolar nature of the palaeomagnetic field in the past from palaeomagnetic data alone. Testing the axial nature of the field requires the use of palaeoclimatic information as will be described in §6.1.3.

If palaeomagnetic poles for a given geological epoch are consistent for rocks sampled over a region of continental extent (e.g. the Permian of Europe or North America), this provides compelling evidence that the dipole assumption used in calculating the poles is essentially correct. In determining the apparent polar wander paths for the different continents it is usually observed that the mean poles for successive geological epochs are tightly grouped within each continent. This in itself is strong evidence in favour of the dipole assumption. In very favourable circumstances the palaeomagnetic meridian crosses the length of a continent as for example in Africa during the Mesozoic (McElhinny and Brock, 1975). Extensive palaeomagnetic studies in north-west Africa may be compared with comparable studies in southern Africa. The two regions are separated by an angular great circle distance of 55.4°, almost exactly along the palaeomagnetic meridian. The palaeomagnetic inclinations for the two regions are +32.5° and −55.5° corresponding to palaeomagnetic latitudes of +17.7° and −36.0°. The palaeolatitude difference of 53.7° is not significantly different from the great circle distance thus confirming the validity of the dipole assumption.

Evans (1976) has suggested a most effective test of the dipolar nature of the geomagnetic field throughout the last 600 Ma. For any given magnetic field a definite probability distribution of magnetic inclination $|I|$ exists for measurements made at randomly distributed geographical sites. This is easily obtained by simply estimating the surface area of the globe corresponding to any set of $|I|$ classes. For example the dipole field is horizontal at the equator and has $|I| = 10°$ at latitude 5.0°. At the poles the field is vertical and has $|I| = 80°$ at latitude 70.6°. The surface areas of these two zones imply that if sampling is sufficient and geographically random the $0° \leq |I| \leq 10°$ band would make up 8.8% of results and the $80° \leq |I| \leq 90°$ band would make up only 5.7% of results.

The present uneven distribution of land on the earth's surface and the existence of areas of relatively intense study means that in present-day terms a random geographical sampling of palaeomagnetic data has not been undertaken. However, over the past 600 Ma considerable polar wander and continental drift has taken place and it might be assumed that this has been sufficient to render the palaeomagnetic sampling random in a palaeogeographical sense. Evans (1976) therefore compared the observed frequency distribution of $|I|$ from palaeomagnetic data over the past 600 Ma with that expected for the first four axial multiple fields. The dependence of $|I|$ on colatitude θ for axial multiples can be obtained from the relationship

$$\tan I_l = \frac{-(1 + l)P_l}{(\partial P_l/\partial \theta)} \tag{6.5}$$

where P_l is the Legendre polynomial of degree l.

In Fig. 6.3 the corresponding frequency distributions of $|I|$ for the first four multiples are given and compared with the observed values for 10° intervals of $|I|$. The fit to the dipole curve is very reasonable and certainly favoured over the higher-order multiples. The main discrepancy is at low latitudes and may just reflect sampling inadequacies. The total observations in the range 0–30° is 26.8%, which agrees closely with the expected value of 27.7%.

6.1.3 Palaeoclimatic Evidence

Both palaeomagnetic and palaeoclimatic data provide independent evidence for past latitudes. The factors controlling climate are quite independent of the earth's magnetic field, but depend essentially on the fact that the net solar flux reaching the earth's surface has a maximum at the equator and a minimum at the poles. At present the mean annual equatorial temperature is +25°C and the polar value is about −25°C. The determination of palaeolatitudes from palaeomagnetic data depends on the axial geocentric dipole assumption. Whereas the dipole assumption can be well demonstrated, as in the previous section, to prove the axial assumption requires comparison with some independent measure of past latitudes. For this

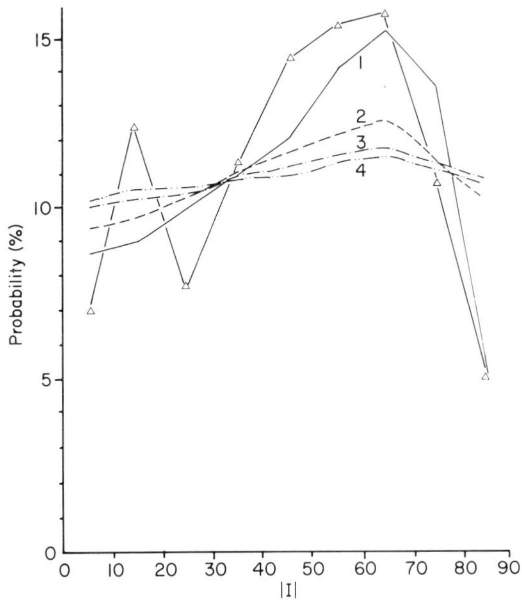

Fig. 6.3. Probability of observing magnetic inclination |*I*| within 10° bands for the first four axial multiple fields (numbered). Triangles give observed values averaged over the past 600 Ma. (After Evans, 1976.)

purpose the comparison between palaeoclimatic and palaeomagnetic data provides the essential test as suggested by Irving (1956).

There are many palaeoclimatic indicators in the geological record. At the present time, coral reefs, evaporites and carbonate deposits, for example, have a density distribution with a maximum at the equator and either a polar minimum or a whole high latitude zone from which they are absent. The comparison between palaeomagnetic latitudes and the occurrences of palaeoclimatic indicators has been made in a number of ways (Irving, 1956; Opdyke and Runcorn, 1960; Blackett, 1961; Runcorn, 1961; Opdyke, 1962; Irving and Gaskell, 1962; Irving and Briden, 1962; Briden and Irving, 1964; Irving and Brown, 1964; Briden, 1968; Stehli, 1968; Drewry *et al.*, 1974; Cook and McElhinny, 1979). The most useful method is to compile palaeolatitude values for a particular occurrence in the form of equal angle or equal area histograms to give the palaeolatitude spectrum of the particular indicator.

Figure 6.4 gives an analysis of the palaeolatitudes of fossil reefs. The distribution of modern coral reefs is symmetrical about the equator (Fig. 6.4a), the maximum frequency being between 10° and 20° latitude and most

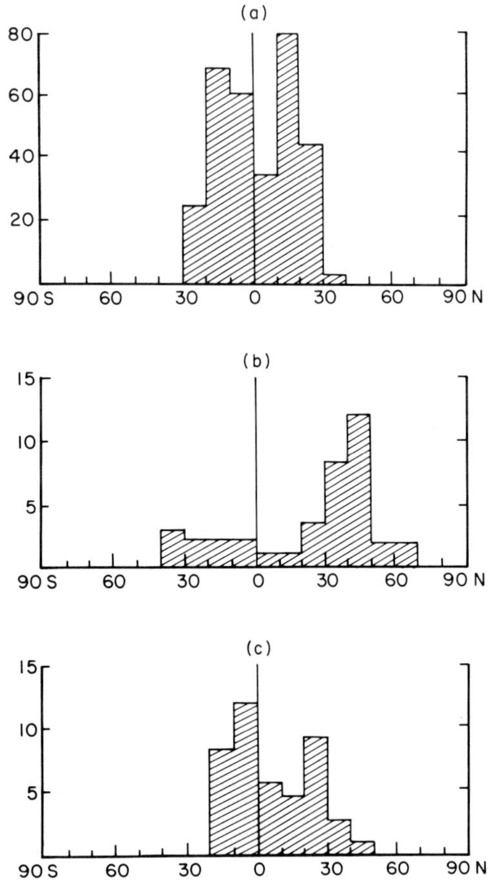

Fig. 6.4. Equal angle latitude histogram for organic reefs. (a) Present latitude of modern reefs; (b) present latitude of fossil reefs; (c) palaeolatitude of fossil reefs. (After Briden and Irving, 1964.)

occurrences lying within 30° of the equator. The present latitude of fossil reefs (Fig. 6.4b) does not show this distribution, but when referred to their palaeolatitudes (Fig. 6.4c) the spectrum is very similar to that of modern reefs, over 95 % of the occurrences falling within 30° of the equator (Briden and Irving, 1964; Briden, 1968). Other palaeoclimatic indicators that have been studied include evaporites (Blackett, 1961; Runcorn, 1961; Irving and Briden, 1962; Opdyke, 1962; Drewry *et al.*, 1974), glacial deposits (Blackett, 1961; Runcorn, 1961; Opdyke, 1962; Drewry *et al.*, 1974), coral deposits (Briden and Irving, 1964; Drewry *et al.*, 1974), oil fields (Irving and Gaskell, 1962), palaeowinds (Opdyke and Runcorn, 1960; Opdyke, 1961), phosphate deposits (Cook and McElhinny, 1979), and the distribution of various fossils

(Runcorn, 1961; Irving and Brown, 1964; Stehli, 1968). There is broad agreement from all these studies that the palaeolatitude distribution of these palaeoclimatic indicators as determined from palaeomagnetism is essentially symmetrical about the equator. This provides the strongest indication that the time-averaged palaeomagnetic field is both axial and dipolar.

6.1.4 Longevity of the field

There are numerous palaeomagnetic results that confirm the existence of the geomagnetic field in Precambrian times. The oldest results are from rocks greater than 2.5 Ga (Archean age). In North America (the Laurentian Shield) the data range back to 2.8 Ga and are summarized by Irving and Naldrett (1977). A particularly good piece of evidence is found where Matachewan dykes (age 2.6 Ga) intrude Archean rocks. A baked contact test confirms the Archean rocks were remagnetized at the time of intrusion. Moving away from the dykes a consistent magnetization is observed that must predate the dykes. Irving and Naldrett (1977) estimate the age of this Archean magnetization at about 2.8 Ga.

In southern Africa the 2.7 Ga Modipa Gabbro is intruded by a 2.3 Ga old granite. At the contact the gabbro has been reheated by the granite providing a baked contact test for the magnetization of the granite (Evans and McElhinny, 1966). The consistent magnetization in the gabbro is thus older than 2.3 Ga and presumably relates to its time of intrusion 2.7 Ga ago. Palaeointensity measurements on this gabbro (McElhinny and Evans, 1968) give a mean VDM about 1.5 times the present dipole moment of the earth.

The oldest magnetization so far observed is from the 3.5 Ga Duffer Formation Volcanics of north-west Australia (McElhinny and Senanayake, 1980). These rocks were folded about 3.0 Ga ago and a positive fold test (§3.3.1) confirms that the magnetization predates the folding. The existence of the geomagnetic field in Archean times is thus well established at least since 3 Ga ago.

6.2 Second-order Terms

6.2.1 The Problems in Time Averaging

It is now well established that, at least for the past 600 Ma, the first order description of the time-averaged palaeomagnetic field is that of a geocentric axial dipole. However, there is good evidence (§6.1.1) that there are important second-order terms. Before attempting to define and then calculate these second-order terms it is necessary to ask the fundamental question as to just what is meant by a time-averaged field and how does one define it. Most debate on the subject of the second-order terms centres around this question.

As a very simple example of the problems involved consider the situation of Fig. 6.5. A travelling wave of the form $A \cos (wt - kx)$ in Fig. 6.5a is modified by the standing amplitude modulation $B(x)$ in Fig. 6.5b resulting in the function of Fig. 6.5c. An observer at P, taking observations at different times, would observe the wave of Fig. 6.5a unmodified. An observer at Q, however, also taking observations at different times, would observe a wave of the same form as the observer at P but with a greatly reduced amplitude. The observer at Q would then claim a far smaller scatter in observations than the observer at P and an analogous situation in the observation of secular variation will produce effects such as the so-called Pacific dipole window of Doell and Cox (1972). Finally, an observer making all observations at the same time but distributed in space would observe the function of Fig. 6.5c. Disagreements between the three observers regarding the form of the function observed are entirely a consequence of the differing restricted perspectives. The only manner in which the true form of the function can be determined is by taking observations distributed in both space and time.

The above illustrates why it is important to define the time-averaged field in a *global* sense, and why the properties of such a field can only be specified through a combination of space-time averaging. Virtually all of the disagreements between results obtained by different authors stem from a lack of appreciation of this time-averaging problem.

6.2.2 Spherical Harmonic Analyses

The representation of the time-averaged palaeomagnetic field in terms of an offset dipole (§6.1.1), although a reasonable first- and second-order model, is obviously not strictly correct and in general it becomes more effective to analyse the data in terms of spherical harmonics. This was first undertaken by Wells (1969, 1973) on late Tertiary and Quaternary data. Benkova *et al.* (1971) first analysed palaeomagnetic declination data, and later Benkova and Cherevko (1972) and Benkova *et al.* (1973) analysed inclination data. Creer *et al.* (1973) and Kono (1973) analysed both declination and inclination data, and subsequently Georgi (1974) and Adam *et al.* (1975) analysed inclination data only so as to be able to use the results from deep-sea cores (Opdyke and Henry, 1969). Different methods of analyses were then used by Cox (1975), Merrill and McElhinny (1977) and Coupland and Van der Voo (1980). The different analytical techniques have produced markedly varying results and

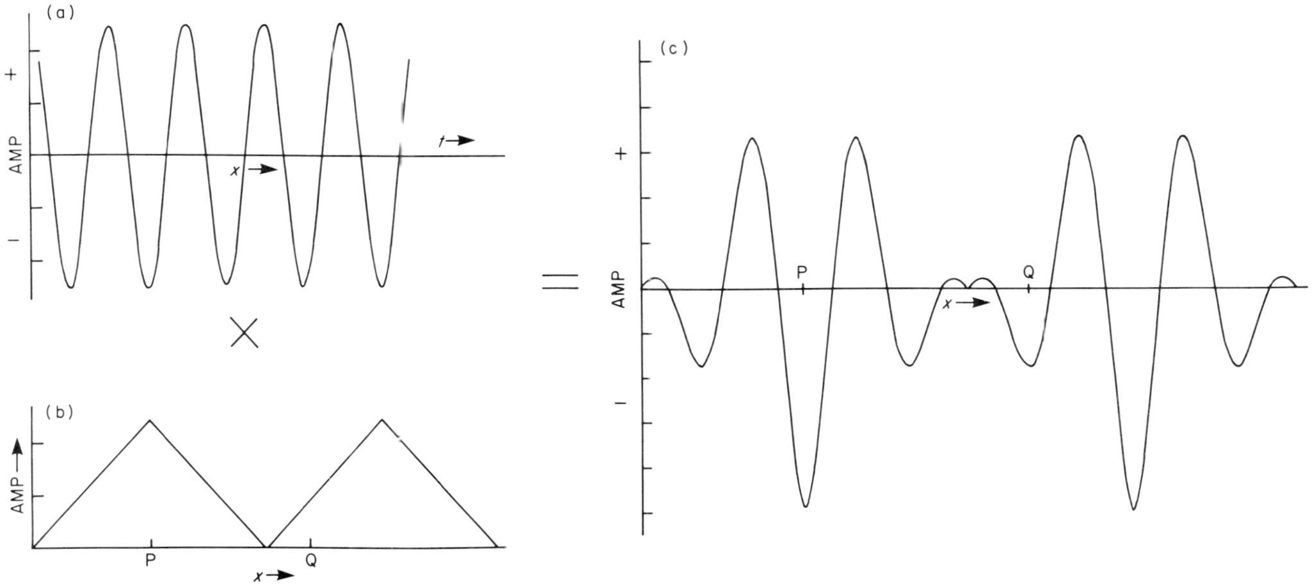

Fig. 6.5. To illustrate the concept of space–time averaging. (a) Travelling wave of form $A \cos (wt - kx)$. (b) Standing amplitude modulation $B(x)$ on (a). (c) Resulting motion. Observers at P and Q see very different amplitudes, but when space-time averaging is performed a consistent result is obtained.

much argument has been generated as to the most appropriate technique to use. As a result the discussion that follows attempts to explain the reason for these differences and produce a technique that is internally self-consistent.

Any spherical harmonic expansion of palaeomagnetic data will be severely truncated because the data are finite. Parseval's theorem (§2.3.2) guarantees that the truncated series of harmonics will be the best that can be done in the circumstances. A better representation of a given data set cannot be obtained by using higher degree harmonics at the expense of lower degree ones. Wells (1973) shows that the accuracy with which the truncated series represents the actual situation cannot be represented statistically unless particular models for the magnetic field are invoked. In particular unevenly distributed data or the use of poor quality data can result in very inaccurate spherical harmonic descriptions. After careful consideration of the errors involved Wells (1973) concluded that only the zonal harmonics (g_1^0, g_2^0, g_3^0) were significant. However, Creer *et al.* (1973) and Georgi (1974) concluded that the non-zonal harmonics (such as g_1^1 and g_2^1) were often comparable in size to the zonal ones. Georgi (1974) concluded it was not possible to say that the zonal terms dominate the time-averaged palaeomagnetic field.

Neither Creer *et al.* (1973) nor Georgi (1974) placed any statistical limits on their determinations of the various spherical harmonic coefficients. Their procedure was to determine regional averages of the palaeomagnetic data at 20 and 32 control points respectively on the surface of the earth. The data distribution is too uneven both in quantity and quality to allow such a procedure and it is not surprising that significant non-zonal harmonics were determined. The analysis of Wells (1973) strongly suggests that only the zonal coefficients are significant in such analyses and that the large and variable non-zonal coefficients determined are a data distribution problem. Merrill and McElhinny (1977), Coupland and Van der Voo (1980) and Lee and McElhinny (1984) used independent methods to come to the same conclusion.

Cox (1975) first realized that the difference in inclination ΔI between the observed field and that due to a geocentric axial dipole is very useful for determining the zonal harmonics. Suppose there is a departure from pure axial symmetry for a given region (the declination D departs significantly from zero). In this case ΔI will be slightly different from the value that is due

Fig. 6.6. Latitude variation of the vertical component Z of the zonal harmonics $-g_2^0$ and $-g_3^0$. When these terms are added to the present geocentric axial dipole field ($-g_1^0$) they produce an inclination anomaly ΔI. The longitude variation of the east component Y produced by the first non-zonal term g_1^1 is shown below. When added to the geocentric axial dipole field, g_1^1 and h_1^1 produce a declination anomaly ΔD which is a function of latitude. This may be normalized to the equator to produce ΔD_0 that has a sinusoidal variation with longitude. (After Merrill and McElhinny, 1977.)

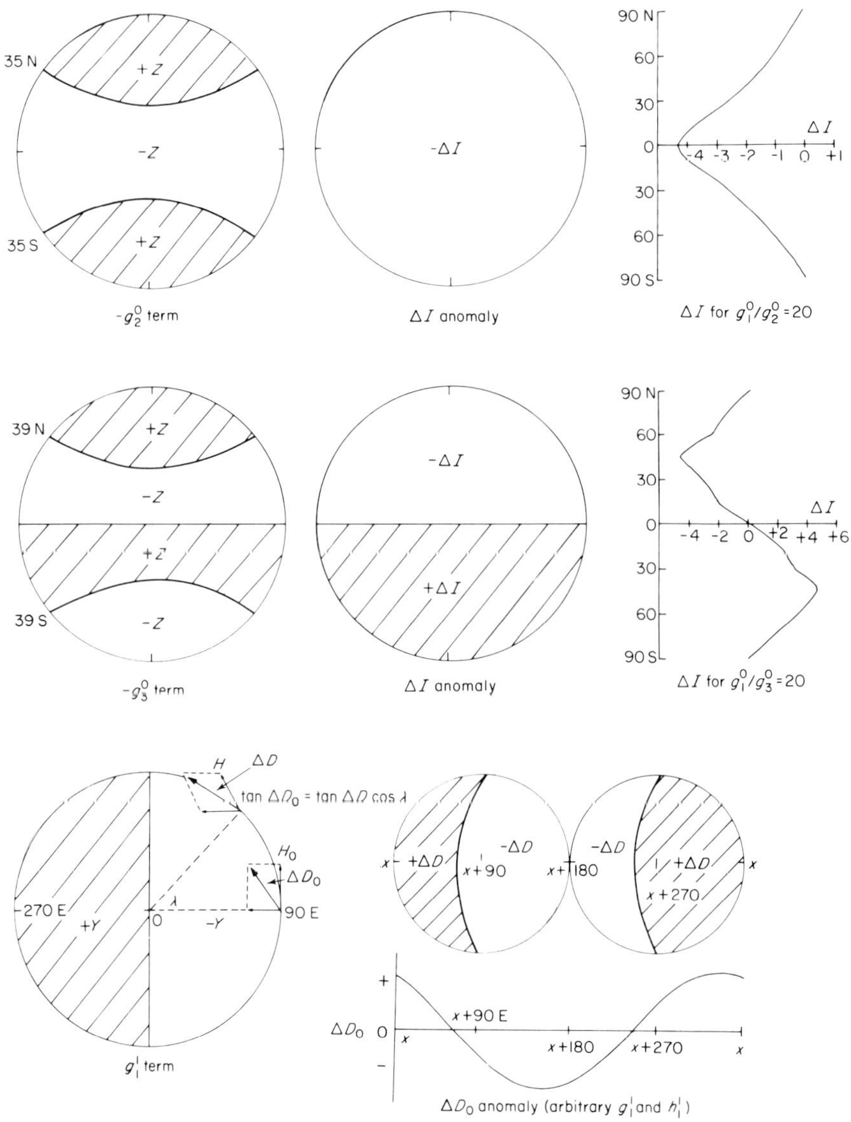

35 N
+Z
-Z
+Z
35 S
$-g_2^0$ term
$-\Delta I$
ΔI anomaly
90 N
60
30
0
30
60
90 S
ΔI
-4 -3 -2 -1 0 +1
ΔI for $g_1^0/g_2^0=20$
39 N
+Z
-Z
+Z
-Z
39 S
$-g_3^0$ term
$-\Delta I$
$+\Delta I$
ΔI anomaly
90 N
60
30
0
30
60
90 S
ΔI
-4 -2 0 +2 +4 +6
ΔI for $g_1^0/g_3^0=20$
H
ΔD
$\tan \Delta D_0 = \tan \Delta D \cos \lambda$
H_0
ΔD_0
270 E
+Y
0
λ
-Y
90 E
g_1^1 term
x
$+\Delta D$
x+90
$-\Delta D$
x+180
$-\Delta D$
$+\Delta D$
x
x+270
+
ΔD_0 0
-
x
x+90 E
x+180
x+270
x
ΔD_0 anomaly (arbitrary g_1^1 and h_1^1)

to a purely zonal harmonic ΔI_z, which is given by

$$\sin \Delta I_z = \sin \Delta I \cos D \tag{6.5}$$

This shows that for departures of D from zero the absolute value of ΔI_z will always be less than the observed ΔI. However, for small values of D this correction will be small. If $\Delta I = 5.0°$, then D can be as large as 10° before ΔI_z differs from ΔI by 0.1°. Thus the observed values of ΔI can be used to obtain estimates of the zonal harmonics with negligible error.

Merrill and McElhinny (1977) and Lee and McElhinny (1984) took this type of analysis one step further. They argued that if the time-averaging of the field has been maximized by using a combination of both time and space averaging (see §6.2.1), then it is to be expected that the non-zonal terms will be eliminated leaving only the zonal potential as in (6.1). Figure 6.6 shows how the ΔI inclination anomaly would look in two simple cases. In the first case the observed field consists solely of an axial geocentric dipole and quadrupole with $g_1^0/g_2^0 = 20$. In the second case the observed field consists solely of an axial geocentric dipole and octupole with $g_1^0/g_3^0 = 20$. The ΔI anomaly is always negative in the first case, and has different signs in each hemisphere in the second case.

In order to check that the assumption of a purely zonal time-averaged field is correct, Merrill and McElhinny (1977) and Lee and McElhinny (1984) then analysed the declination anomaly ΔD (departure of D from zero) to calculate the non-zonal harmonics g_1^1 and h_1^1. The procedure is illustrated in Fig. 6.6c. If the non-zonal terms from a combination of time space averaging can be shown to be small, then the initial assumption of a purely zonal time-averaged field is shown to be a reasonable one. Coupland and Van der Voo (1980) also identified the zonal harmonics as the most important. They then showed that departures from the $g_1^0 + g_2^0 + g_3^0$ best fitting model were essentially random and not due to any significant non-zonal quadrupole terms.

6.2.3 The Past Five Million Years

Most analyses of palaeomagnetic data take each result as published in the literature as a datum. In looking at the time-averaged field in some detail it is necessary to take into account the variability in the amount of data included in each result. For example, Merrill and McElhinny (1977) examined 366 data covering the past 5 Ma, equivalent to about 3500 spot readings of the field over the earth's surface. Lee and McElhinny (1984) then compiled a data base of all individual measurements of the field and came up with 4968 measurements (sites) that were represented by 538 results (364 normal and

174 reversed). The whole data base was then analysed from the original independent measurements avoiding any weighting that could arise from the variable numbers in each result.

For each measurement the inclination anomaly ΔI (departure from the expected geocentric axial dipole field direction) was calculated. These ΔI anomalies were then averaged in time and space by using inclination strips of various widths (depending on the amount of data). Plotting ΔI versus I spreads the data more uniformly, since 0°–10° inclination (0°–5.0° latitude) contains 8.8% of the earth's surface whereas 80°–90° inclination (70.6°–90° latitude) contains 5.7% of the earth's surface. The results of Lee and McElhinny (1984) are illustrated in Fig. 6.7.

For a geocentric axial dipole at colatitude θ,

$$\tan I = 2 \cot \theta \tag{6.6}$$

If g_2^0 and g_3^0 terms are included in the field, then

$$\tan I' = \frac{2\cos\theta + g_2^0/g_1^0(\frac{9}{2}\cos^2\theta - \frac{3}{2}) + g_3^0/g_1^0(10\cos^3\theta - 6\cos\theta)}{\sin\theta + g_2^0/g_1^0(3\cos\theta\sin\theta) + g_3^0/g_1^0(\frac{15}{2}\cos^2\theta\sin\theta - 3\sin\theta)} \tag{6.7}$$

then

$$\Delta I = I' - I \tag{6.8}$$

Using (6.6), (6.7) and (6.8) a model field can be fitted to the data of Fig. 6.7 by least squares to determine the values of g_2^0/g_1^0 and g_3^0/g_1^0. An improved fit can be obtained by including the g_4^0 term in the model for the normal field, but is not significant when included in the model for the reversed field (Table 6.1).

Statistical analysis of the results shows that the normal and reversed values of g_2^0/g_1^0 are significantly different at the 95% confidence level (Lee and McElhinny, 1984). The differences can be modelled by reversing and non-reversing magnetic fields. Analysis of palaeointensities and virtual dipole moments for the past 5 Ma (§6.3.2) shows that g_1^0 maintains its magnitude upon reversing. Thus the g_2^0 term can be imagined to have a reversing (g_{2r}^0) non-reversing (g_{2i}^0) part. The results in Table 6.1 give

$$\frac{g_{2r}^0 - g_{2i}^0}{g_1^0} = 0.034; \qquad \frac{-g_{2r}^0 - g_{2i}^0}{-g_1^0} = 0.069$$

from which $g_{2i}^0/g_{2r}^0 = 0.34$. That is the non-reversing part is roughly one-third of the intensity of the reversing part of g_2^0. The significance of this result is discussed further in §10.2.

The normal and reversed values of g_3^0/g_1^0 are very similar (Table 6.1). This suggests that all the non-dipole terms represented by g_3^0, because of the truncation, change sign when g_1^0 changes sign. Both the g_3^0/g_1^0 terms for the normal and reversed fields are significantly different from zero at the

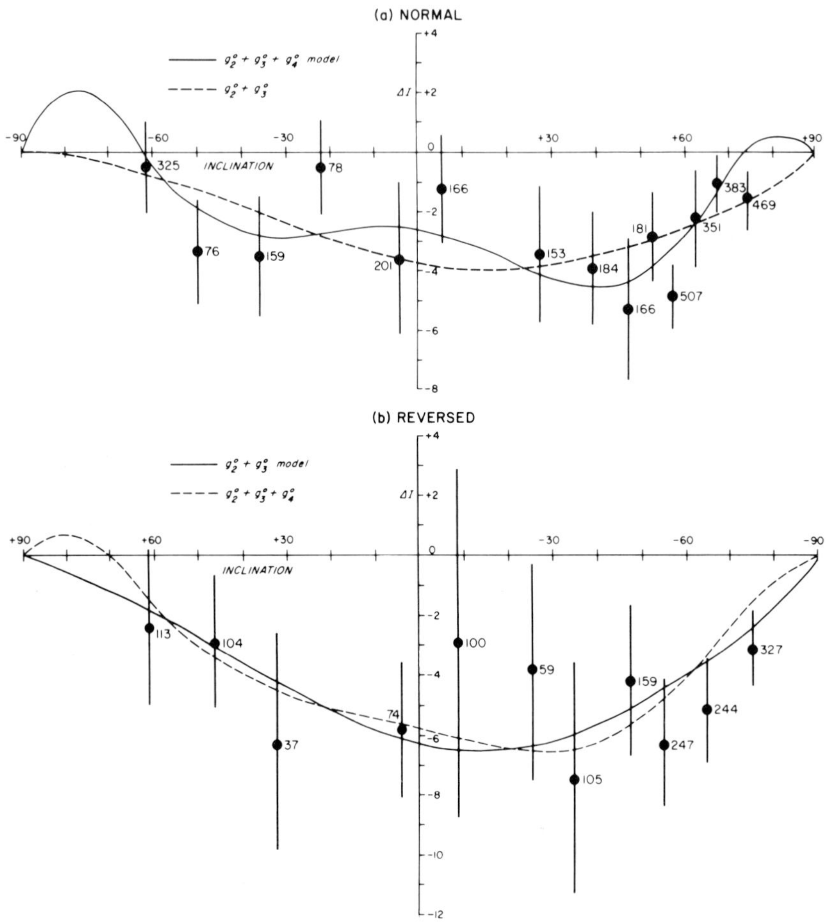

Fig. 6.7. ΔI anomaly averaged over various inclination strips using data for the past 5 Ma. The number of measurements is indicated at each point and 95% confidence limits are shown. Model fits using two or three terms are indicated for (a) normal and (b) reversed data. (After Lee and McElhinny, 1984.)

TABLE 6.1

Coefficients for the time-averaged field over the past 5 Ma normalized to the geocentric axial dipole term g_1^0 for two- and three-term model fits. The significant models are given in bold type. (After Lee and McElhinny, 1984)

Coefficient	Present field	Normal		Reversed	
		$(g_2^0 + g_3^0)$	$(g_2^0 + g_3^0 + g_4^0)$	$(g_2^0 + g_3^0)$	$(g_2^0 + g_3^0 + g_4^0)$
g_2^0/g_1^0	+0.068	+0.040 ± 0.008	+**0.034** ± 0.006	+**0.069** ± 0.016	+0.066 ± 0.010
g_3^0/g_1^0	−0.043	+0.016 ± 0.007	+**0.019** ± 0.010	+**0.019** ± 0.008	+0.021 ± 0.010
g_4^0/g_1^0	−0.031	—	+**0.023** ± 0.012	—	+0.012 ± 0.015
g_1^1/g_1^0	+0.065	−0.012 ± 0.015		+0.020 ± 0.021	
h_1^1/g_1^0	−0.188	+0.012 ± 0.012		+0.017 ± 0.021	

95 % confidence level. However, when the g_4^0 term is included in the model only the normal field has a value significantly different from zero. The best models include $g_2^0 + g_3^0 + g_4^0$ for the normal field and $g_2^0 + g_3^0$ for the reversed field (Table 6.1).

It is necessary to examine the data to see whether this result could have been caused by some artifact of the data. Possible causes are:

(1) Inclination errors in sediments arising from the detrital remanent magnetization (DRM) processes (King and Rees, 1966) or in basalts arising from shape anisotropy (Coe, 1979).
(2) Insufficient magnetic cleaning of secondary components, especially in reversely magnetized rocks.

Both inclination errors in sediments due to DRM processes (King and Rees, 1966) and the effect of shape anisotropy on lava flows (Coe, 1979) follow essentially the same pattern. Magnetic inclinations are deflected towards the horizontal, that is ΔI will be negative for positive I and positive for negative I (Fig. 6.8a). This is the equivalent of the g_3^0 term as illustrated in Fig. 6.6. These effects cannot therefore explain the existence of the predominant g_2^0 term where ΔI is negative in *both* hemispheres (Fig. 6.8a). The hemispherical asymmetry in ΔI due to g_3^0 in Table 6.1 is however consistent in

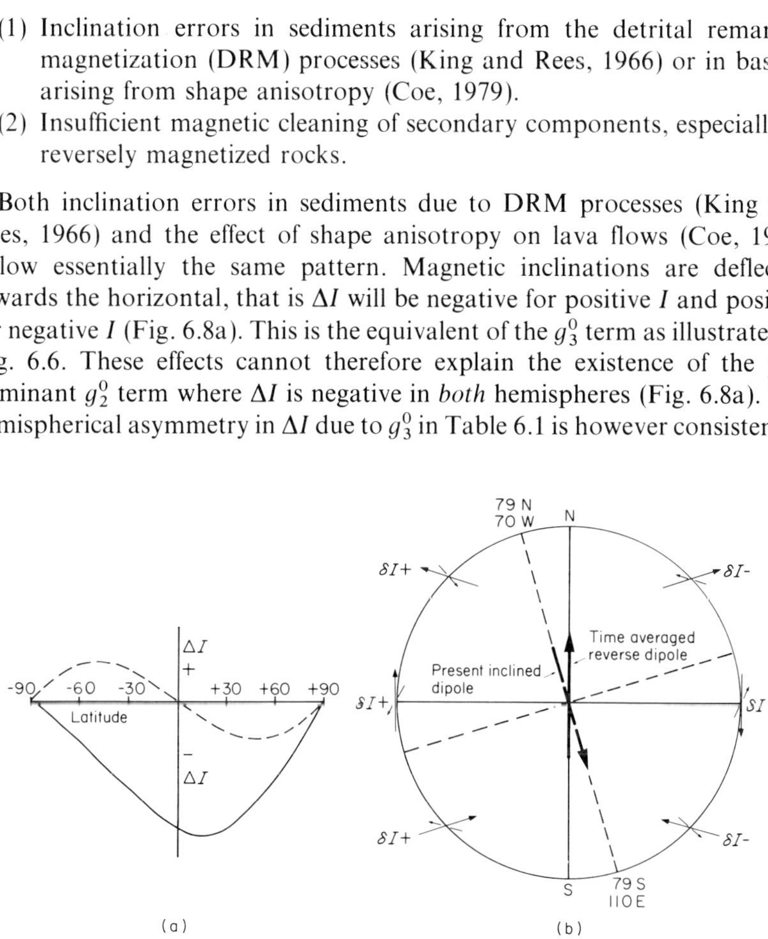

Fig. 6.8. (a) The observed ΔI anomaly cannot be caused by inclination errors in sediments or shape anisotropy in lava flows. ——, observed inclination anomaly ---, effect of inclination errors in sediments and shape anisotropy in lava flows. (b) The significant difference between the ΔI anomaly for the normal and reversed data as expressed in the g_2^0 term cannot be due to incomplete magnetic cleaning of reversely magnetized rocks. Any remaining magnetization due to the present magnetic field will cause an anomaly $+\delta I$ in the eastern hemisphere and $-\delta I$ in the western hemisphere. The method of averaging data over latitude strips cancels out this effect. →, original magnetization direction; - -→ added secondary from present field.

sense and order of magnitude with the shallowing of inclination in subaerial lava flows predicted by Coe (1979). However, the time variation of g_3^0 (§6.2.4) suggests it has changed by nearly an order of magnitude over the past 200 Ma. Such large hemispherical asymmetries are far too great to have arisen either from shape anisotropy in subaerial basalts or from inclination errors in sediments. In any case analyses of inclinations in deep-sea sediments for example suggest that inclination errors are small or non-existent (e.g. Opdyke and Henry, 1969).

Consider the situation of reversely magnetized rocks situated at various points on the earth's surface as in Fig. 6.8b. Imagine a secondary magnetization due to the field of the present (normal) inclined dipole has been added to the magnetization, but was not fully cleaned by the usual demagnetizing procedures (§3.3.1). In the hemisphere towards which the dipole is tilted (160°W to 20°E) any secondary magnetization remaining will cause an anomaly δI that is positive, and in the opposite hemisphere (20°E to 160°W) an anomaly δI that is negative (Fig. 6.8b). The procedure of averaging the observed ΔI anomaly around latitude strips to determine the time-averaged zonal harmonics thus has the effect of randomizing the extra δI from any unremoved secondary magnetization by adding positive values around one hemisphere and negative values around the other. So whereas the scatter in the observed vectors might be increased they will still average to the original time-averaged field value when averaged out over latitude strips around the earth. The increase in scatter, however, probably affects the analysis of palaeosecular variation discussed in §6.4.2. The difference between g_2^0 for the normal and reversed field thus cannot be attributed to an artifact of the data.

To check that the time-averaged field really does consist predominantly of zonal harmonics, the declination anomaly ΔD is used to determine the non-zonal harmonics g_1^1 and h_1^1 (Fig. 6.6) following Merrill and McElhinny (1977). In practice there are larger errors in measuring ΔD than in measuring ΔI. Also the errors in ΔD increase at points closer to the pole. To allow for the fact that ΔD varies with latitude, Merrill and McElhinny (1977) normalized all the values to the equator using the relationship

$$\tan \Delta D_0 = \tan \Delta D \cos \lambda \qquad (6.9)$$

where λ is the latitude and ΔD_0 is the equator-normalized value of ΔD (see Fig. 6.6).

Figure 6.9 shows the analysis of Lee and McElhinny (1984) of ΔD_0 values for the normal and reversed field over the past 5 Ma. In both data sets of Fig. 6.9 there is the suggestion of a sinusoidal variation in ΔD_0 as would be expected from an equatorial dipole defined by g_1^1 and h_1^1. However, the best fit for a sine curve is for a constant base line of 0.3° for the normal field and

1.7° for the reversed field, instead of the expected value of zero. This is essentially the right-handed effect noted by Wilson (1972). In the case of the normal field 0.3° is not significantly different from zero, but for the reversed data 1.7° is just so. This can also be judged from the average common site longitude poles in Fig. 6.2. The average reversed poles lie to the right of the 0°–180° line but the normal poles do not.

Lee and McElhinny (1984) examined the errors associated with determining the first non-zonal harmonics g_1^1 and h_1^1. Their general conclusion is that the determined values are not significantly different from zero (Table 6.1). Comparison with the zonal harmonics deduced for the time-averaged field shows that the non-zonal terms are at best only the same magnitude as the g_4^0 or higher zonal terms. Most of the effects seen in Fig. 6.9 are thus probably due to random errors or possibly the effects of inadequate magnetic cleaning in the case of the reversed data. Coupland and Van der Voo (1980) quite independently reached the same conclusion, concurring with the original conclusion of Wells (1973). Note that in Table 6.1 whereas the time-averaged g_2^0 term is much the same magnitude as the present field value, all other terms are significantly reduced, especially the non-zonal terms. The very signficant g_2^0 term is the most common conclusion reached whatever the method of analysis used.

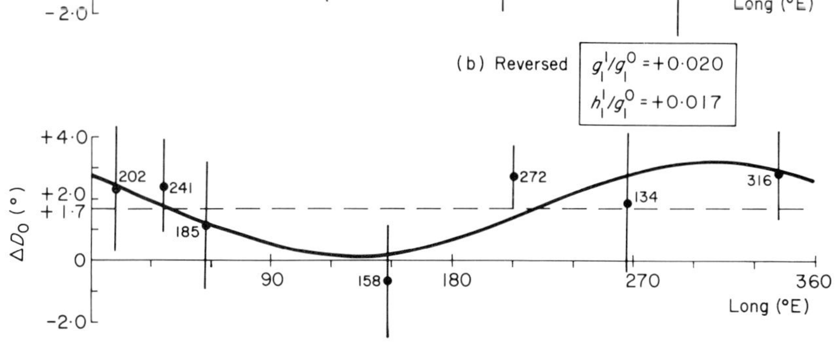

Fig. 6.9. ΔD_0 anomaly averaged by longitude sectors using data for the past 5 Ma. The number of measurements is indicated at each point and 95% confidence limits are shown. The solid line is a model variation using the g_1^1 and h_1^1 terms indicated for (a) the normal data using a base line of 0.3°, and (b) the reversed data using a base line 1.7° as explained in the text. (After Lee and McElhinny, 1984.)

6.2.4 Extension to 200 Ma

It is possible to use the sea-floor spreading magnetic anomalies (§5.2.2) to reconstruct the relative positions of the continents through time back to 200 Ma ago. Most of the world's oceans have now been studied in some detail and the relative rotation parameters are well known. Using this information palaeomagnetic data can be compared with the reconstructed globe and in principle the departures from the geocentric axial dipole model can be determined. Coupland and Van der Voo (1980) first attempted this procedure and later Lee and McElhinny (1984) repeated it.

The continents can be relocated relative to a chosen fixed continent. Coupland and Van der Voo (1980) chose North America, whereas Lee and McElhinny (1984) chose Africa. The problem then is to locate the palaeospin axis. Normally this is done using the palaeomagnetic data itself, but of course these are the data to be used to determine the second-order coefficients. Lee and McElhinny (1984) used the observation of Fig. 6.1 that if the palaeomagnetic poles are averaged round the globe by longitude sectors, then the mean of the sector averages will coincide with the spin axis. In effect this method just serves to cancel out all higher order zonal harmonics. Coupland and Van der Voo (1980) used a more complex minimizing procedure. In both cases it was assumed that the field consisted of zonal harmonics only following the observation that this appears to be the case for the past 5 Ma.

Lee and McElhinny (1984) used the latitude strip ΔI anomaly method of §6.2.2 and 6.2.3. This has the advantage, as illustrated in Fig. 6.8, of helping to cancel out the effects of incomplete magnetic cleaning. Their analysis of data covering the past 5 Ma showed that it was important to analyse as large a data set as possible. The expected scatter in individual values of ΔI over even 10° of inclination is large from palaeosecular variation (§6.4) so that insufficient data leads to spurious results. Lee and McElhinny (1984) analysed all the data for the past 200 Ma compiling an elaborate data set separated into normal and reversed components by recalculation from the original information. Coupland and Van der Voo (1980) used a much more restricted data set and used a sliding window of 20–30 Ma in an attempt to determine the time variation.

The results of Lee and McElhinny (1984) are illustrated in Fig. 6.10. Separate analyses of the normal and reversed data showed that they had the same trends in variation of g_2^0/g_1^0 and g_3^0/g_1^0 with time. The data were insufficient to show any significant differences between the normal and reversed field as was found for the past 5 Ma. In Fig. 6.10 therefore the combined data only are shown. As one goes back in time to 45 Ma ago g_2^0/g_1^0 becomes progressively more negative and g_3^0/g_1^0 progressively more positive.

Coupland and Van der Voo (1980) noted the same effect. However, prior to 50 Ma Coupland and Van der Voo claim that g_3^0/g_1^0 is small and near zero. This is a quite different result to that obtained by Lee and McElhinny (1984) in Fig. 6.10. Interestingly the coefficients shown in Fig. 6.10 are generally much lower than obtained by Coupland and Van der Voo and this suggests there is a random component in their analysis due to insufficient data.

The meaning of the variation in g_2^0 and g_3^0 in Fig. 6.10 is not clear. Certainly there is good evidence for increased hemispherical asymmetry (north–south) in the time-averaged field between 10 and 50 Ma. As mentioned in §6.2.3 this is far too large to be due to any data artifact problems. The data of Fig. 6.10 suggests that g_3^0 is by far the dominant component earlier than 100 Ma. This is quite at variance with Coupland and Van der Voo (1980).

To compare the magnitude of the total non-dipole to that of the dipole

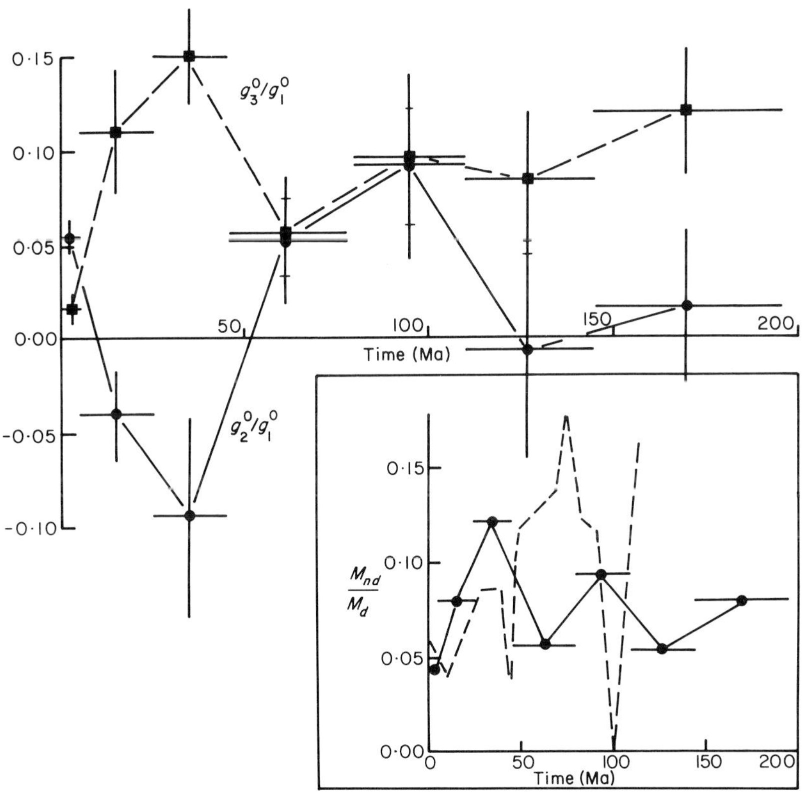

Fig. 6.10. Variation of the zonal harmonics g_2^0 (dots and solid line) and g_3^0 (squares and dashed line) with time. 95% confidence limits for the ratios g_2^0/g_1^0 and g_3^0/g_1^0 are shown, and the time over which the data have been averaged is indicated by the horizontal bar. (After Lee and McElhinny, 1984.) The inset shows the variation in the magnitude of the non-dipole field according to (6.11). Solid line after Lee and McElhinny (1984) from the data above and dashed line after Coupland and Van der Voo (1980).

field for a given field configuration, Coupland and Ven der Voo (1980) devised a single parameter M_{nd}/M_d as follows:

$$\frac{M_{nd}}{M_d} = \left(\frac{\int\int V_{nd}^2 \, dS}{\int\int V_d^2 \, dS}\right)^{1/2} \tag{6.10}$$

where V_{nd} is the geomagnetic potential due to the non-dipole field components, V_d is the potential of the dipole field, and integration is over the earth's surface. If the only important non-dipole coefficients are g_2^0 and g_3^0, then (6.10) becomes

$$\frac{M_{nd}}{M_d} = \left[\frac{3}{5}\left(\frac{g_2^0}{g_1^0}\right)^2 + \frac{3}{7}\left(\frac{g_3^0}{g_1^0}\right)^2\right]^{1/2} \tag{6.11}$$

The factors $\frac{3}{5}$ and $\frac{3}{7}$ arise because the associated Legendre polynomials defined by Schmidt (1935) are only quasi-normalized (see §2.2), and their coefficients must be fully normalized before comparison. The variation of M_{nd}/M_d is shown in the inset of Fig. 6.10. The variation found using a sliding window by Coupland and Van der Voo (1980) is shown for comparison. Lee and McElhinny (1984) analysed data from the Cretaceous quiet interval between 80 and 110 Ma and found no special feature in this interval. Coupland and Van der Voo (1980), however, suggest the non-dipole field reduced to near zero around 100 Ma.

6.3 Variation in the Earth's Dipole Moment

6.3.1 Palaeointensities and Dipole Moments

For the purpose of comparing data from sampling sites at different latitudes it is convenient to calculate the equivalent dipole moment which would have produced the measured intensity at the calculated palaeolatitude of the sample. Smith (1967a) termed such a dipole moment the *Virtual Dipole Moment* (VDM) as was discussed in §3.4.4. This has the advantage that no scatter is introduced in an analysis of VDMs by any wobble of the main dipole because the determined magnetic palaeolatitude is independent of the orientation of the dipole relative to the earth's rotation axis. However, it is not possible to identify that portion of the observed palaeointensity (and therefore of the VDM) which is due to non-dipole components. Hence a *True Dipole Moment* (TDM) cannot be obtained directly from palaeointensity studies. The rapidly varying part of the non-dipole field changes over periods up to about 10^3 years, whereas the dipole field changes over periods of about 10^4 years (Cox and Doell, 1964). These changes can be considered to be superimposed on a slowly varying part. This difference in time constants makes it possible, at least in principle, to separate out the statistical properties of the TDM.

McFadden and McElhinny (1982) have discussed the problem of estimating the mean dipole moment for a given period of time from the observed VDM data. There are three types of means usefully associated with VDM data.

(1) The arithmetic mean of the observed VDMs can be calculated and this is referred to as the *Mean VDM*. The scatter of observed VDMs about the mean is caused by fluctuations in the TDM, the effect of non-dipole components and errors in the palaeointensity determinations.
(2) By making assumptions about the effect of non-dipole components and errors in palaeointensity determinations it is possible to extract the distribution of TDMs from the observed distribution of VDMs. The peak of the determined TDM distribution is then the inherently preferred value of the TDM about which the earth's dipole moment fluctuates. This preferred value of the TDM is then the intensity analogue of the Palaeomagnetic Pole of directional data about which the VGPs fluctuate (§3.3.2) and is referred to as the *Palaeomagnetic Dipole Moment* (PDM) after McFadden and McElhinny (1982).
(3) The third value of interest is the *Mean TDM* which may be different from the PDM depending on the distribution assumed for the TDM. Generally the Mean TDM will be equal to the Mean VDM but the scatter of TDMs will be much less due to the removal of the effects of non-dipole components and errors in the palaeointensity measurements. Because only the distribution of TDMs may be determined from a given set of VDMs, it is a lengthy procedure to determine the scatter of TDMs.

6.3.2 The Past Five Million Years

McFadden and McElhinny (1982) have analysed all palaeointensities determined from rocks that are less than 5 Ma in age. Data from polarity transitions were excluded by using only those data where the latitude of the associated VGP is greater than 45°. Also all data covering the past 50,000 years and already analysed in §4.1.3 and §4.1.5 were excluded so as not to bias the analysis to this period of time. The analysis included 166 VDMs, 92 with normal polarity (72 northern hemisphere and 20 southern hemisphere sites) and 74 with reverse polarity (61 northern hemisphere and 13 southern hemisphere sites).

A histogram showing the distribution of all 166 VDMs is shown in Fig. 6.11 and compared with various models to be discussed below. The distribution of VDMs has four main characteristics that are also evident if the normal and reversed data are analysed separately.

(1) There is a fairly sharp cut off in the observed VDMs at low values. This suggests there is a lower limit to the dipole moment that can be associated with geomagnetic field configurations having VGPs with latitudes greater than 45°.
(2) The data group with maximum frequency about some central mean value.
(3) The data exhibit a skew to the right.
(4) There appears to be some maximum value of the dipole moment that is never exceeded and rarely if ever reached in practice.

Cox (1968) has suggested that the most useful analogy to geomagnetic polarity reversals has been provided by the double-disc dynamo of Rikitake (1958). From this he developed a probabilistic model for reversals in which the dipole moment oscillates sinusoidally about some mean value during stable polarity times. It is not difficult to show (Kono, 1972; McFadden and McElhinny, 1982) that such a model predicts a distribution of dipole moments which groups with minimum frequency about a central mean. This is virtually the opposite to that observed (Fig. 6.11a). The data thus quite clearly do not support a model of an oscillating dipole moment during stable polarity times.

McFadden and McElhinny (1982) have analysed the observed distribution of VDMs statistically. For the combined normal and reversed data of Fig. 6.11, a Gaussian distribution of VDMs can be rejected at the 5% level of significance but a Lognormal distribution cannot be rejected at this level. An empirical model was used to fit the data based upon the recognition that any observed VDM was made up of three basic components. The major component is that due to the TDM; there is a component caused by the non-dipole field which will either increase or decrease the observed VDM; and finally there is the error in the palaeointensity determination.

McElhinny and Senanayake (1982) analysed archaeomagnetic data discussed previously in §4.1. For the past 10,000 years they deduced that the scatter of VDMs due to non-dipole components and errors in the palaeo-

---

Fig. 6.11. Histogram of VDMs determined from rocks less than 5 Ma old from McFadden and McElhinny (1982). (a) Compared with the probability density predicted from Cox's (1968) model of an oscillating dipole moment. (b) Compared with the probability density proposed by McFadden and McElhinny (1982). (c) Cumulative frequency plot of the distribution of VDMs proposed by McFadden and McElhinny (1982) compared with the cumulative observations.

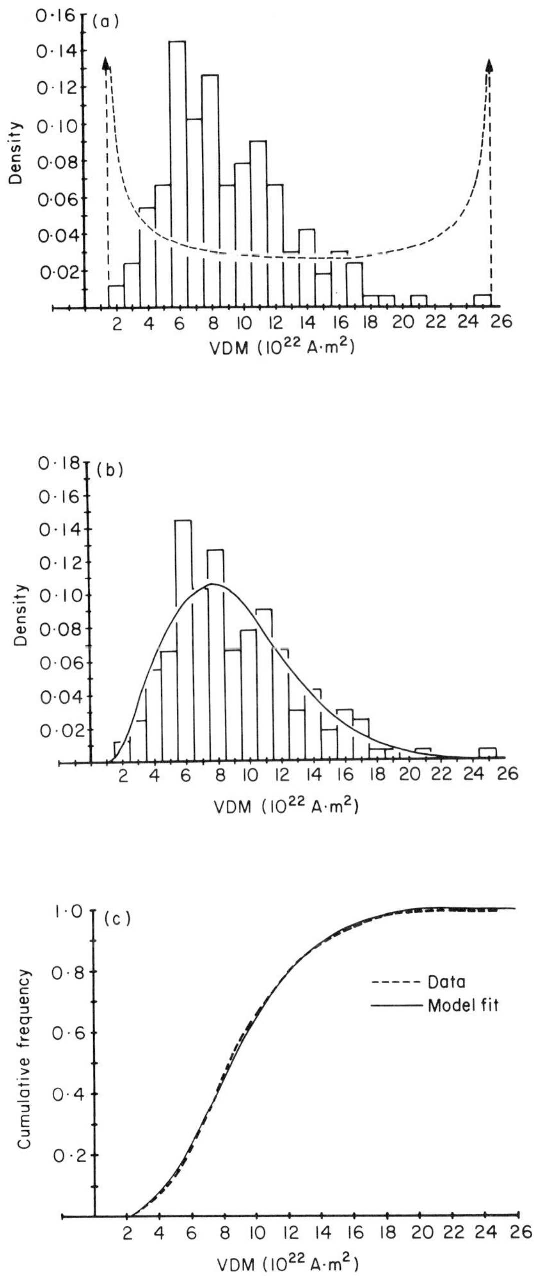
(a)
Density
0·16
0·14
0·12
0·10
0·08
0·06
0·04
0·02
2 4 6 8 10 12 14 16 18 20 22 24 26
VDM (10^{22} A·m^2)
(b)
Density
0·18
0·16
0·14
0·12
0·10
0·08
0·06
0·04
0·02
2 4 6 8 10 12 14 16 18 20 22 24 26
VDM (10^{22} A·m^2)
(c)
Cumulative frequency
1·0
0·8
0·6
0·4
0·2
Data
Model fit
2 4 6 8 10 12 14 16 18 20 22 24 26
VDM (10^{22} A·m^2)

intensity determinations had a standard deviation of 21.1%. Also they noted that in the period 15,000 to 50,000 years B.P., even though the dipole moment was half its mean value for the past 10,000 years, the percentage standard deviation due to the non-dipole components was the same as for the past 10,000 years. Apparently the intensity of non-dipole components can be considered in the first-order approximation to have been linearly proportional to the value of the TDM.

Senanayake *et al.* (1982) have shown that the error in palaeointensity determinations has a standard error of about 10%. Using this value the archaeomagnetic data therefore suggest the standard deviation due to non-dipole components is 18.6%. McElhinny and Senanayake (1982) have analysed the present field by looking at the standard deviation of VDMs calculated from field values at points randomly distributed over the earth's surface. The scatter in VDMs would be due entirely to non-dipole components and equivalent to an analysis of world-wide VDMs as in Fig. 6.11. The average standard deviation derived from 50 samplings each of 100 points on the earth's surface was 17.5%. This lends strong support to the deductions from archaeomagnetic data.

In analysing the VDM distribution for the past 5 Ma, McFadden and McElhinny (1982) therefore proposed an empirical model as follows. The original distribution of TDMs has a Gaussian distribution which is truncated at low values. The observed distribution of VDMs is that of the TDM plus a Gaussian distributed error with standard deviation which is a constant fraction (c) of the TDM. This fraction c is that due to non-dipole components and palaeointensity errors and a value of 21.1% was used as was determined from the archaeomagnetic data (McElhinny and Senanayake, 1982). A maximum likelihood fit of such a model to the VDM data showed that it was an improvement on the Lognormal distribution. The excellent fit obtained (Fig. 6.11c) strongly supports the results from archaeomagnetic data that the non-dipole field intensity can be considered to remain in the same proportion to the dipole field intensity throughout the past few million years, excluding times of polarity transitions or excursions.

TABLE 6.2
VDM scatter—dipole and non-dipole contribution

Time span	Average dipole moment (10^{22} A m^2)	Standard deviation (%)			
		Dipole	Non-dipole	Expt.	Dip + ND
Present field	7.91	—	17.5	—	—
0–10,000 years	8.75	18.0	18.6	10	(26.9)
15,000–50,000 years	4.44	—	—	10	24.5
0–5 Ma (PDM)	8.67	41.9	18.6	10	(45.8)

The value at which the truncated Gaussian distribution of TDMs peaks is the Palaeomagnetic Dipole Moment (PDM). Over the past 5 Ma the PDM is $8.67 \pm 0.65 \times 10^{22}$ A m^2 with 95% confidence. The truncation point is estimated as 2.75×10^{22} A m^2 and exceeds 2.05×10^{22} A m^2 with 95% confidence. This apparently represents the minimum value of the TDM that can exist during stable polarity (non-transition) times. The estimated standard deviation of the untruncated distribution is $3.63 \pm 0.75 \times 10^{22}$ A m^2 with 95% confidence. This variation (41.9% of the PDM) relates solely to the variation in the dipole moment itself. Separate analyses of the normal and reversed VDMs suggested there were no significant differences in the PDM and other parameters.

A comparison of the parameters of the earth's dipole moment for the past 5 Ma with those for the past 50,000 years is given in Table 6.2. The fluctuations in the dipole moment, as represented by the standard deviation of TDMs, are very much greater over the 5 Ma time-scale than those observed over 10^4 years. This suggests that the fluctuations in the earth's dipole moment occur over much longer time-scales of 10^5 or 10^6 years.

6.3.3 Variation with Geological Time

The problem of determining palaeointensities in the geological past becomes increasingly difficult the older the rocks studied. The presence of secondary components and the decay of the original magnetization all serve to complicate the problem. So whereas there are 166 determinations for the past 5 Ma there are only a further 102 determinations for the whole of geological time prior to about 10 Ma ago and back to 2.7 Ga ago. Also some of these determinations have been made using very dubious techniques and no consistency checks (see §3.4). If dipole fluctuations with standard deviations of 40% or more, as observed for the past 5 Ma, have persisted for geological time, there are hardly sufficient numbers of determinations to calculate an average dipole moment for any particular geological epoch.

A few attempts to determine old palaeointensities have been made by Briden (1966), Carmichael (1967), Smith (1967a), McElhinny and Evans (1968), Kobayashi (1968), Schwarz and Symons (1969), Bergh (1970) and Senanayake and McElhinny (1983). The oldest determination is for some rocks from southern Africa which are 2.7 Ga old (McElhinny and Evans, 1968). A palaeointensity about 1.5 times the present field value was obtained. A summary of these results averaged over 100 Ma intervals back to 500 Ma and over 200 Ma intervals for Precambrian times is given in Fig. 6.12. Between 2.7 and 0.7 Ga the mean values fluctuate about the present day value of the earth's dipole moment. The 54 results for this entire 2.0 Ga interval

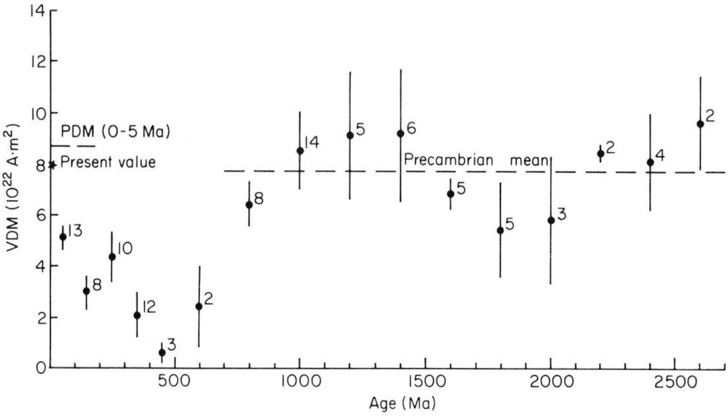

Fig. 6.12. Variation of the mean VDM with time. 100 Ma averages are used from 0 to 700 Ma, and 200 Ma averages from 700 to 2700 Ma. The number of measurements are shown next to each mean and standard errors are indicated.

give a mean VDM of 7.7×10^{22} A m^2 with standard deviation of 4.1 $\times 10^{22}$ A m^2 (53 %). Thus as far as can be deduced from the few available results it appears that the earth's dipole moment averaged over Precambrian times was similar to its present value and also the PDM for the past 5 Ma. Within the uncertainties the standard deviation of dipole fluctuations (48 %, after allowing for the non-dipole field and errors) is also similar to that observed for the past 5 Ma (42 %).

However, the results of Fig. 6.12 suggest the dipole moment was abnormally low during most of the Phanerozoic, especially between 700 and 300 Ma with an apparent minimum around 500 Ma. Indeed all the 100 Ma means for the past 600 Ma are lower than any of the Precambrian values. This is a surprising result that must be viewed with some caution as it may be a chance occurrence related to the particular rocks which have been studied and the methods used. Clearly a lot more work is needed before this result is either confirmed or shown to be a data artifact.

6.4 Palaeosecular Variation

6.4.1 Models of PSV

The angular dispersion of the geomagnetic field over long periods of time represents the aspect of geomagnetic secular variation most accessible to palaeomagnetic measurements. This is because most palaeomagnetic measurements are a set of random observations of the geomagnetic field over a given time. The angular dispersion of palaeomagnetic measurements may be calculated in two ways, either by finding the dispersion of palaeomagnetic field directions or else by finding the dispersion of the corresponding virtual geomagnetic poles (VGP). The variations of angular dispersion as a function of latitude calculated in these two ways are quite different. The dispersion of field directions decreases with increasing latitudes whereas the dispersion of VGP increases with increasing latitude (see, for example, Cox, 1970; McElhinny and Merrill, 1975). Such studies of the angular dispersion of the geomagnetic field are generally referred to as *palaeosecular variation*.

Suppose a set of palaeomagnetic measurements consists of N palaeodirections of the geomagnetic field. The set of field directions maps into a set of VGP, and conversely. Let δ_i be the angle between the ith field direction and the mean direction, and let Δ_i be the corresponding angle between the ith pole and the mean pole. The best estimate of the angular variance of field directions is

$$s^2 = \frac{1}{N-1}\sum_{i=1}^{N}\delta_i^2 \tag{6.12}$$

and that for the VGP is

$$S^2 = \frac{1}{N-1}\sum_{i=1}^{N}\Delta_i^2 \tag{6.13}$$

Most analyses of palaeomagnetic data use the statistical method of Fisher (1953) (see §3.3.3). If k is the best estimate of the precision of field directions, and K the best estimate of the precision of the VGP, then the best estimate of the angular variance of field directions is

$$s^2 = 6561/k \tag{6.14}$$

and that of the VGP is:

$$S^2 = 6561/K \tag{6.15}$$

However, it should be noted that the Fisher (1953) distribution function has azimuthal symmetry about the mean. Field directions with azimuthal symmetry do not in general map into VGP that possess azimuthal symmetry, and conversely. A full discussion of this basic problem and its relation to the analysis of angular dispersions is given by Cox (1970) and McElhinny and Merrill (1975).

It is convenient to separate contributions to angular disperion of the geomagnetic field into three factors:

(i) variation in the intensity and direction of the non-dipole field;
(ii) variation of the geomagnetic dipole moment with time;

(iii) changes in the orientation of the geomagnetic dipole such that on average the dipole axis coincides with the axis of rotation (dipole wobble).

Various models have been proposed to describe the latitude variation of the angular dispersion of the geomagnetic field. These have been referred to as model A (Irving and Ward, 1964), model B (Creer *et al.*, 1959), model C (Cox, 1962), model D (Cox, 1970), model E (Baag and Helsley, 1974), model M (McElhinny and Merrill, 1975), a refinement of model M due to Harrison (1980), and model F (McFadden and McElhinny, 1984).

Model A due to Irving and Ward (1964) considers only variations in the non-dipole field. A geocentric axial dipole field of fixed dipole moment is perturbed by a single randomly directed vector $\boldsymbol{f}$ of constant magnitude. The model predicts an angular dispersion of field directions of the form.

$$s_A = s_N(1 + 3\sin^2\lambda)^{1/2} \tag{6.16}$$

where $s_N = 46.8\ (f/Fo)$ and (f/Fo) is the ratio of the perturbing non-dipole field to the equatorial strength of the dipole field. s_N is thus the equatorial value of the angular dispersion of field directions. Model B due to Creer (1962) and Creer *et al.* (1959) postulates a wobble of the main dipole with VGP which follow Fisher's (1953) distribution. The angular dispersion S_D of VGP due to dipole wobble would be latitude invariant, but expressed as the angular dispersion of field directions the model predicts

$$s_B = S_D\left\{\frac{5 - 3\sin^2\lambda}{2(1 + 3\sin^2\lambda)}\right\}^{1/2} \tag{6.17}$$

Models C, D, E and M are physically more plausible because they include contributions from both the dipole and non-dipole fields. All of these models suffer from several assumptions that are now known to be incorrect. In all cases it was assumed that time variations in the earth's dipole moment are sinusoidal. This would produce a distribution of VDMs totally at variance with that observed (Fig. 6.11a). Models C and D assumed a non-dipole field with the same average intensity at all latitudes. Baag and Hesley (1974) in model E assumed a linear correlation between the dipole and non-dipole angular dispersions that could have different values at different latitudes. There was no observational or physical basis for this supposition. McElhinny and Merrill (1975) in model M assumed the non-dipole field arose from two sources, one with latitude invariance, the other correlated with the dipole field. An error in their analysis has subsequently shown this model is not plausible (Harrison, 1980). However, the attempt by Harrison (1980) to correct this used a refinement which is equally implausible. In any case both McElhinny and Merrill (1975) and Harrison (1980) failed to take into account the imprecise nature of the relationships between the angular variances of VGP and field directions resulting from the mapping functions as pointed out by Cox (1970).

Model F by McFadden and McElhinny (1984) is the most recent model and takes into account the problems of previous models noted above. In this model it is assumed that the non-dipole field originates from some sort of interaction with the poloidal field. On such a model the VGP would be expected to be distributed with spherical symmetry about the geographic pole. Cox (1970) has previously noted that the angular dispersion of field directions of the present field is not symmetric but that of the corresponding VGP is nearly so. The model also implies that the time-averaged energy density of the non-dipole field at any given latitude λ will be linearly related to the energy density of the dipole field at that latitude. That is:

$$\langle f^2 \rangle = k^2 F^2 \tag{6.18}$$

where f is the intensity of the non-dipole field, F the intensity of the dipole field and k is a constant. Analyses of VDMs over the past 50,000 years by McElhinny and Senanayake (1982) (see §4.1) and over the past 5 Ma by McFadden and McElhinny (1982) (see §6.3.2) are entirely consistent with the above in that they show a correlation between the intensity of the non-dipole and the dipole field.

Imagine the distribution of VGPs arising from the non-dipole field to be a set of "non-dipole" VGPs of constant length but uniformly distributed in direction, superimposed upon a "dipole" VGP of constant direction. Then from (6.18) as the non-dipole field changes in proportion to the dipole field, there will be a set of VGPs with spherical symmetry about the geographic pole. This set of VGPs can then be mapped back to the direction and intensity observed at latitude λ. If $\boldsymbol{J}$ is the vector denoting this direction and intensity, then the tip of $\boldsymbol{J}$ will be on a closed surface which in general is not a sphere but an ellipsoid. The relative lengths of the axes x, y and z of this ellipsoid will change with latitude. Thus the angular dispersion of directions will change with latitude even though the energy density of the non-dipole field remains a constant proportion of the energy density of the dipole field. This is an important factor which was ignored in all previous models.

From the above the non-dipole field vector $\boldsymbol{f}$ is simply given by:

$$\boldsymbol{f} = \boldsymbol{J} - \boldsymbol{F} \tag{6.19}$$

Cox (1964, 1970) has already shown that if f is small compared with F and the surface is a sphere ($x = y = z$), then:

$$\delta_d^2 = \frac{2}{3}\frac{f^2}{F^2}$$

where δ_d^2 is the angular variance of directions (δ_d in radians). In the present situation McFadden and McElhinny (1984) have shown that an extremely close fit is provided by the function

$$\delta_d^2(\lambda) = \frac{2}{3}\frac{[1 + a\exp(-b\lambda^2)]}{(1 + 3\sin^2\lambda)^{1/2}}\left(\frac{f^2}{F^2}\right)$$

$$= G(\lambda)\left(\frac{f^2}{F^2}\right) \tag{6.20}$$

where a and b are also functions of $\delta_d(0)$ but over the range of interest they do not vary rapidly.

Let the density distribution of F be $P(F)$ and the density distribution of f given F be $p(f\,|\,F)$. At a given latitude the time-averaged angular variance of *directions* will be given by:

$$\sigma_d^2(\lambda) = \int_F\int_f \delta_d^2(\lambda)p(f\,|\,F)P(F)\,\mathrm{d}f\,\mathrm{d}F$$

Substituting from (6.20) and (6.18) gives

$$\sigma_d^2(\lambda) = k^2G(\lambda) \tag{6.21}$$

Thus $\sigma_d^2(\lambda)$ is independent of the distribution of either f or F! This now has to be transformed to the angular variance of the VGP. This transformation is a function of latitude and of $\sigma_d^2(\lambda)$, the latter having been ignored by previous workers. The angular variance $\sigma_p^2(\lambda)$ of VGP is then given by:

$$\sigma_p^2(\lambda) = \sigma_d^2(\lambda)H[\lambda, \sigma_d^2(\lambda)]$$

$$= k^2G(\lambda)H[\lambda, \sigma_d^2(\lambda)] \tag{6.22}$$

where $H[\lambda, \sigma_d^2(\lambda)]$ is the transformation function.

If $\sigma_d^2(\lambda)$ is very small then $H[\lambda, \sigma_d^2(\lambda)]$ is a function of λ only, and Cox (1970) has given this transform as:

$$H(\lambda) = \frac{2(1 + 3\sin^2\lambda)^2}{5 + 3\sin^2\lambda} \tag{6.23}$$

However, $\sigma_d^2(\lambda)$ is not small enough and Cox (1970) obtained the correct transform by numerical analysis and gave it in tabular form.

If S_N^2 is the angular variance of VGP at the equator due to variations in the non-dipole field, then

$$S_N^2 = \sigma_p^2(0) = kG(0)H[0, \sigma_d^2(0)]$$

and

$$\sigma_p^2(\lambda) = S_N^2 \frac{G(\lambda)H[\lambda, \sigma_d^2(\lambda)]}{G(0)H[0, \sigma_d^2(0)]}$$
$$= S_N^2 W^2 \tag{6.24}$$

If S_D^2 is the angular variance of VGP due to the dipole wobble, then the overall angular variance S^2 is given by:

$$S^2 = S_D^2 + S_N^2 W^2 \tag{6.25}$$

where

$$W^2 = \frac{[1 + a\exp(-b\lambda^2)]}{(1+a)(1 + 3\sin^2\lambda)^{1/2}} \frac{H[\lambda, \sigma_d^2(\lambda)]}{H[0, \sigma_d^2(0)]} \tag{6.26}$$

Tables of appropriate values of a, b and $H[\lambda, \sigma_d^2(\lambda)]$ are given by McFadden and McElhinny (1984).

The latitude dependence of W in (6.26) is illustrated in Fig. 6.13. For comparison the latitude variation of the angular dispersion of the non-dipole

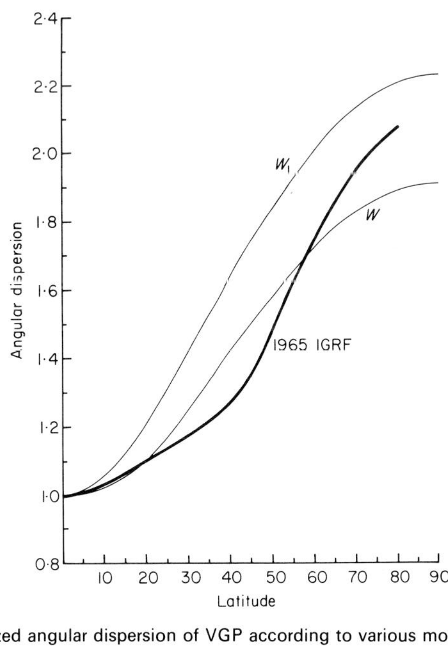

Fig. 6.13. Normalized angular dispersion of VGP according to various models. W is the non-dipole term from model F of McFadden and McElhinny (1984). W_1 is the variation that would be predicted if the various transformation functions are excluded from W. The average curve for the 1965 non-dipole IGRF is shown for comparison after McElhinny and Merrill (1975). (After McFadden and McElhinny, 1984.)

part of the 1965 IGRF is also shown from McElhinny and Merrill (1975). Two forms of W are shown in Fig. 6.13, the generalized form from (6.26) and a simplified form W_1 in which the exponential part of $G(\lambda)$ is omitted ($a = 0$) and the simplified form of $H(\lambda)$ when $\sigma_d^2 = 0$ (6.23) is used. In that case W^2 simplifies to:

$$W_1^2 = \frac{5(1 + 3\sin^2\lambda)^{3/2}}{5 + 3\sin^2\lambda} \tag{6.27}$$

The difference between W and W_1 in Fig. 6.13 is significant, and shows the importance of including the full transformation functions omitted by previous workers (Cox, 1970; McElhinny and Merrill, 1975; Harrison, 1980). The variation in W is a better approximation to that of the 1965 non-dipole field, although it should be realized that the present non-dipole field may not be fully representative of the time-averaged non-dipole field.

6.4.2 The Past Five Million Years

McElhinny and Merrill (1975) have discussed the problems of analysing palaeomagnetic data to obtain estimates of the variation of angular dispersion with latitude. The problem relates to how one defines a time-averaged field and the basic arguments have already been set out in §6.2.1 and illustrated in Fig. 6.5. The properties of the time-averaged field can only be specified through a combination of space-time averaging. The field needs to be defined in a global sense. Thus observations at any one locality can produce spurious effects such as the so-called Pacific dipole window (Doell and Cox, 1972).

Only data from lava flows are presently adequate to study the palaeosecular variation. Data from individual flows are spot readings of the geomagnetic field at one locality. The angular dispersion of VGP from a sequence of lava flows can be measured with respect to the present geographic pole using (6.13) when data for the past 5 Ma are being considered. It is assumed that for the past 5 Ma that the present axis of rotation has remained fixed with respect to all landmasses (continental drift has been neglected). However, several samples are usually collected from each lava and it is necessary to make a correction for the within-lava angular dispersion S_W to determine the angular dispersion of the geomagnetic field S_F (the between lava angular dispersion). If S_T is the total angular dispersion, then:

$$S_F^2 = S_T^2 - \frac{S_W^2}{\bar{n}} \tag{6.28}$$

where $\bar{n}$ is the average number of samples measured in each lava flow. In order to make sure that the correction term for the within-site dispersion is small, it has been traditional to make very detailed measurements for each lava flow (e.g. Doell and Cox, 1965; Doell, 1970, 1972; Watkins, 1973). However, S_W is normally considerably smaller than S_F and in these circumstances it is possible to use results where only three or four samples per flow have been studied.

Lee and McElhinny (1984) have made a very careful selection of all available palaeomagnetic data taking into account the criteria set out above. To make sure that all data from polarity transitions or geomagnetic excursions were excluded, all VGP with co-latitudes greater than 40° were excluded. The resulting data set consisted of 2382 lava flows (1620 normal and 762 reversed) covering the past 5 Ma. This was a much more rigorously selected set than the 2372 flows analysed by McElhinny and Merrill (1975). Following the procedure set out in §6.2.3, the data were analysed in latitude bands globally, combining both northern and southern hemisphere results following McElhinny and Merrill (1975).

Spherical harmonic analysis of the time-averaged palaeomagnetic field over the past 5 Ma has been discussed in §6.2.3. The problem of the effect of incomplete magnetic cleaning of reversely magnetized rocks could be overcome through the space-time averaging procedure around latitude strips. It was pointed out however that the effect of incomplete magnetic cleaning was to increase the overall scatter in reversely magnetized rocks. This is readily seen in Fig. 6.14. A normally magnetized component added to reverse magnetization *increases* the scatter of observed directions, whereas when added to the normal magnetization *decreases* the scatter of observed directions. Thus the combination of all normal and reversed data tends to cancel out this effect, and probably represents the best estimate of the angular dispersion of the palaeomagnetic field.

The data compiled by Lee and McElhinny (1984) are plotted in Fig. 6.15. The 95% confidence limits to the data combined from each latitude strip are obtained following Cox (1969a, 1977). A least-squares weighted fit of the PSV model F of McFadden and McElhinny (1984) is shown. The fit is excellent giving $S_D = 7.2°$ with 95% confidence that $5.3° \leq S_D \leq 8.7°$, and $S_N = 10.6°$ with 95% confidence that $9.8° \leq S_N \leq 11.4°$. The angular dispersion of VGP at the equator is 12.8° (Fig. 6.15) and the corresponding angular dispersion of directions, $\sigma_d(0)$, is 15.5°. From (6.21) this gives $k^2 = 0.06$. Thus the energy in the non-dipole field is about 6% of the energy in the dipole field.

Separate analyses of the normal and reversed polarity data give $S_N = 10.0°$ and 11.2° respectively and $S_D = 7.5°$ and 7.4° respectively. These values are not significantly different, but the higher value of S_N for the reversed data is consistent with the increasing scatter that might be expected from incomplete magnetic cleaning.

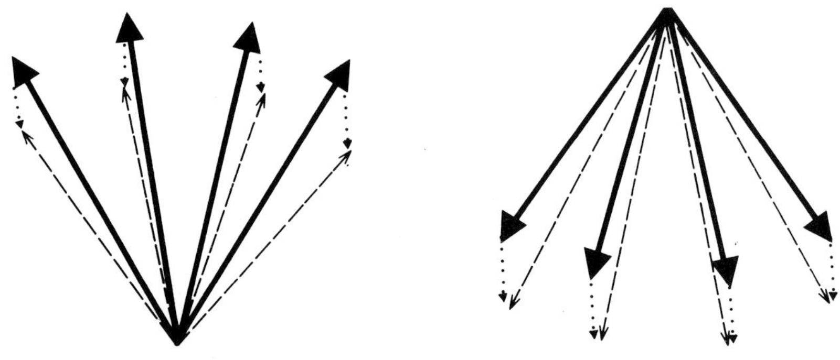

Fig. 6.14. The effect of incomplete magnetic cleaning on the scatter of palaeomagnetic directions. Secondary components in the direction of the present geomagnetic field, (a) increase the scatter in reversely magnetized rocks, and (b) reduce the scatter in normally magnetized rocks.

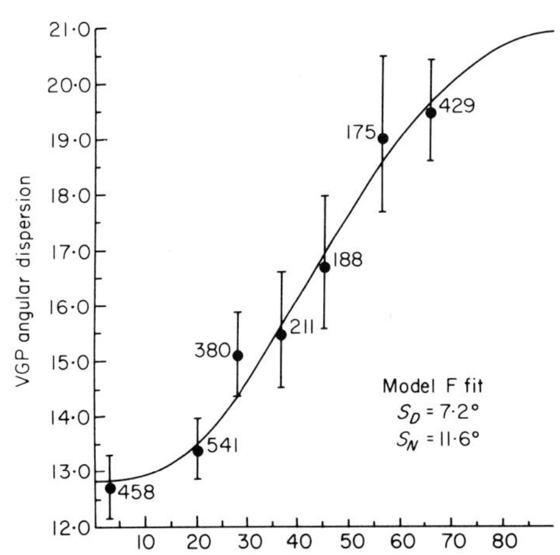

Fig. 6.15. Angular dispersion of VGP from lavas less than 5 Ma averaged over successive latitude strips. Northern and southern hemisphere data are combined and number of measurements used for calculating the PSV dispersion is indicated. 95% confidence limits are shown following Cox (1969a, 1977). Data of Lee and McElhinny (1984). The least squares fit for model F determined by McFadden and McElhinny (1984) is shown by the smooth curve. The values determined for S_D and S_N are not independent of each other. The 95% confidence limits given in the text indicate that variations in values that might provide an acceptable fit.

6.4.3 Predictions from the Analysis of Dipole Moments

The analysis of VDMs over the past 5 Ma (§6.3.2) showed that the standard deviation of VDMs caused by the non-dipole field was linearly related to the dipole field, F. If σ_V is the standard deviation of VDMs, then

$$\sigma_V^2 = c^2 F^2 \tag{6.29}$$

McFadden and McElhinny (1982) showed that the constant $c = 0.186$ determined from archaeomagnetic data by McElhinny and Senanayake (1982) provided a very good fit to the distribution of VDMs. It is possible to use this result derived from the VDM analysis to predict the angular dispersion of directions.

Following McFadden and McElhinny (1984) it is necessary to assume some distribution for $p(f|F)$. Cox (1970) modelled the non-dipole field as a set of vectors of varying length and random orientation. Variations in the magnitude of $\boldsymbol{f}$ can be described by the density function (Cox, 1964),

$$p(f|F) = \frac{4f^2}{\sqrt{\pi f_R^3}} \exp\left(-f^2/f_R^2\right) \tag{6.30}$$

where f_R is the most probable length of $\boldsymbol{f}$. In the present model f_R is linearly related to F, so that

$$f_R = qF \tag{6.31}$$

McFadden and McElhinny (1984) have shown that, given the distribution of directions implied by their PSV model and the distribution of f given by (6.30), for q in the range 0–0.6, the value of c is given to within 0.1 % by

$$c^2 = 0.575q^2 - 0.128q^4$$

For $c = 0.186$, $q = 0.247$. With this particular distribution $\langle f^2 \rangle = \frac{3}{2} f_R^2$ and from (6.21) the prediction is:

$$\sigma_d^2(0) = 15.4°$$

From the analysis of PSV over the past 5 Ma (§6.4.2) the determined value is 15.5°. This remarkable agreement demonstrates the consistency between the analysis of dipole moments over the past 5 Ma and the corresponding analysis of PSV.

6.4.4 Variation with Geological Time

Brock (1971) first attempted to see if time variations in PSV could be seen in palaeomagnetic data. He analysed 83 selected palaeomagnetic results extending back to 2.7 Ga. Because of the nature of the data he had to analyse the angular dispersion of directions about the calculated mean for each result rather than about any known mean direction. The angular variance of such directions s^2 is then as given by (6.14). Brock then used Cox's (1970) model D in the form

$$s^2 = s_N^2 W_N^2 + S_D^2 W_D^2 \tag{6.32}$$

to describe the latitude variation of the angular dispersion. s_N and S_D are the angular dispersions due to the non-dipole field and dipole wobble respectively and W_N and W_D are transformation functions that are functions of latitude. The first term is the equivalent of model A of Irving and Ward (1964) given by (6.16).

A least squares fit to the palaeomagnetic data showed that $s_N = 18.6°$ and $S_D = 2.0°$ suggesting that dipole wobble must have been very low in the past. For separate Tertiary and pre-Tertiary analyses, Brock found $s_N = 20.1°$, $S_D = 1.5°$ and $S_N = 17.2°$, $S_D = 1.8°$ respectively. The more precise PSV model F makes it clear that dipole wobble, over the past 5 Ma, has certainly not been insignificant. Thus model A cannot be a useful model for PSV even though there is an apparently good fit to the observational data for field directions. The problem is that the latitude variation of the angular dispersion of field directions closely resembles the *form* of (6.16) for model A, but it is incorrect then to presume that model A is reasonable. This is especially true since it is known that the non-dipole field cannot be represented by a randomly oriented vector of fixed length independent of latitude.

Making use of Brock's (1971) findings, Irving and Pullaiah (1976) proposed that model A given by the first term in (6.32) and by (6.16) is a good approximation to PSV in the past because dipole wobble can effectively be ignored. Using 188 selected palaeomagnetic results from the literature covering the past 350 Ma, Irving and Pullaiah (1976) then used a 50 Ma sliding window to determine the variation of s_N with time. Their results are illustrated in Fig. 6.16. For the past 5 Ma $s_N = 18.4°$ and there appears to have been a general increase in the magnitude of PSV with time since 350 Ma ago. Of particular interest are the low values of s_N for the Cretaceous quiet interval (110–80 Ma), where $s_N = 14.7°$ and for the Permo-Carboniferous quiet interval (307–237 Ma) where $s_N = 15.6°$. The conclusion drawn is that the non-dipole to dipole field ratio is reduced during times of low or zero reversal frequency. The correct conclusions to be drawn from the analysis of Irving and Pullaiah (1976) is that PSV as represented by the sum of dipole wobble and the non-dipole terms was low during the Mesozoic and late Palaeozoic and especially low during the Cretaceous and Permo-Carboniferous quiet intervals.

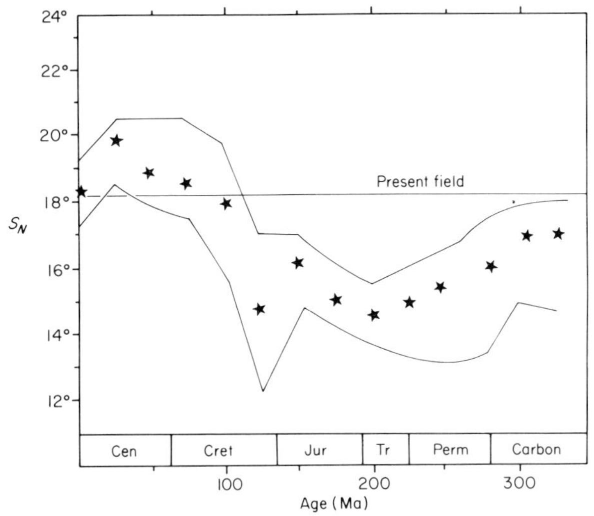

Fig. 6.16. Palaeosecular variation versus time using 50 Ma overlapping averages with standard errors. (After Irving and Pullaiah, 1976.) The angular dispersion of field directions at the equator, s_N from the variation predicted by model A (eqn. 6.16) is used as the measure of PSV.

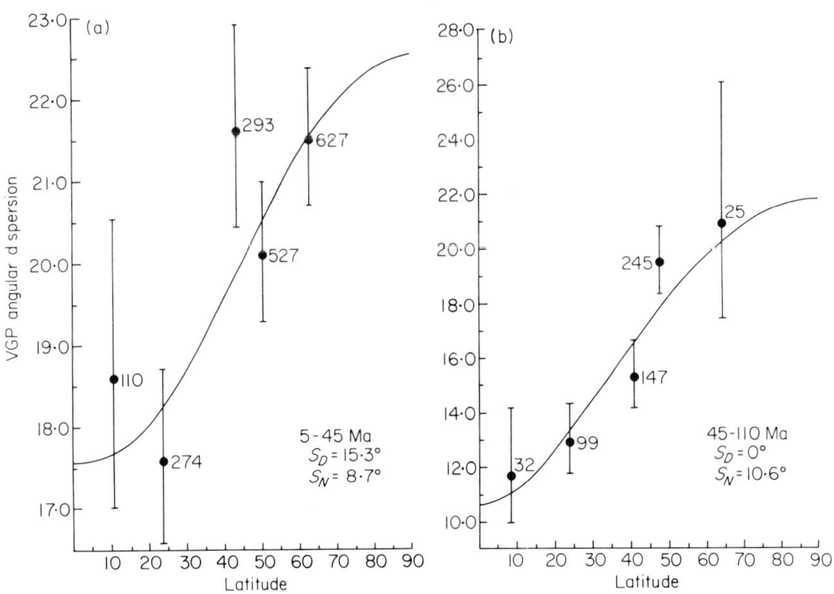

Fig. 6.17. Angular dispersion of VGP from lavas older than 5 Ma determined as described in Fig. 6.15 from Lee and McElhinny (1984). The least squares fit for model F determined by McFadden and McElhinny (1984) is shown by the smooth curve (a) 5–45 Ma interval; (b) 45–110 Ma interval.

Lee and McElhinny (1984) have extended their analysis of the variation with time of the second-order terms in the time-averaged palaeomagnetic field to include PSV. As described in §6.2.4 the best estimate of the position of the axis of rotation after reconstructing continents in the past enabled the zonal terms g_2^0 and g_3^0 to be determined for the intervals 5–22.5 Ma, 22.5–45 Ma, 45–80 Ma, 80–110 Ma, 110–145 Ma and 145–195 Ma. The VGP scatter from studies of lava flow sequences have been determined for each of these time intervals. Unfortunately there is insufficient data to analyse each of the six intervals, so the data have been combined into two broad time-intervals, 5–45 Ma (1831 flows) and 45–110 Ma (550 flows). These correspond to time intervals when the reversal frequency was relatively high (5–45 Ma) and relatively low (45–110 Ma). The latter includes the Cretaceous quiet zone (80–110 Ma).

McFadden and McElhinny (1984) fitted model F to these data and the results are illustrated in Fig. 6.17. PSV was apparently much higher in the interval 5–45 Ma compared with 45–110 Ma. This confirms the general result of Irving and Pullaiah (1976) illustrated in Fig. 6.16. However, the model fits indicate $S_N = 8.7°$ and $S_D = 15.3°$ for the 5–45 Ma interval and $S_N = 10.6°$ and $S_D = 0°$ for the 45–110 Ma interval. The model therefore suggests that the most significant change in PSV is in the magnitude of dipole wobble and that this is the factor which is most related to changes in reversal frequency. Many more data will have to be analysed to confirm this surprising result. In conclusion, it needs to be stressed that there is no simple relationship between the time-averaged non-dipole zonal harmonics g_2^0 and g_3^0 (§6.2) and the angular dispersion of VGP due to the non-dipole field, S_N. This is because they are measuring quite different aspects of the non-dipole field.

Chapter Seven

Origin of the Earth's Magnetic Field 1: Introduction and Physical Insight

7.1 Properties of the Earth's Interior

The properties of the earth's interior, particularly the core, are important to the geomagnetist for evaluating theories dealing with the earth's magnetic field. Geophysicists have done a remarkable job in obtaining information about the earth's interior, considering the difficulties involved. However, often the accuracies quoted for some of the important physical parameters are far higher than warranted. Reviews of some of the properties of the earth's core have been given by Jacobs (1975), Stacey (1977a, b), Brett (1976), Ringwood (1979), Ahrens (1979) and Stevenson (1981).

The inversion of seismic data to obtain the compressional and shear-wave velocity distribution throughout the earth's interior has probably provided the best information on the structure and composition of the core. There are two equations for the three unknown parameters incompressibility, rigidity and density. Shear waves are not transmitted through the outer core, one indication that it is liquid (that is, has zero rigidity). The inner core of the earth appears solid. However, the dimensions and nature of the transition from the outer liquid core to the solid inner core are still hotly debated. Estimates for the density distribution in the earth can be obtained from spheroidal eigenvibration data, or from the body wave data and some equation of state. A parameterized earth model (PREM, Preliminary Reference Earth Model) developed by Dziewonski and Anderson (1981) based on astronomical geodetic data, free oscillation and long-period surface wave data, and body wave data is a good first order estimation of the earth's seismic structure. Table 7.1 gives some of the core parameters taken from this model. These parameters are quite adequate for dynamo theories, with an

TABLE 7.1
Some parameters for the earth's core from the PREM model. (Dziewonski and Anderson, 1981)

Location	Radius	Density (10^3 kg m^{-3})	Pressure (10^{11} Pa)	Gravity (m s^{-2})	Incompressibility (10^{11} Pa)	Rigidity (10^{-11} Pa)
Earth's centre	0	13.0885 $-8.8381\,x^2$	3.6385	0	14.253	1.761
Inner core–	1221.5	12.7636	3.288	4.40	13.434	1.567
outer core "boundary"	1221.5	12.5815 $-1.2638\,x$ $-3.642\,x^2$ $-5.5281\,x^3$	3.288	4.40	13.047	0
Outer core–	3480.0	9,9040	1.3575	10.68	6.441	0
mantle boundary	3480.0	7.9565	1.3575	10.68	6.556	2.938
Top D″	3630.0	—	1.2697	10.48	6.412	2.899

$x \equiv r/a$ where r = radius and a = 6371 km.

exception that the magnitude of the sharp change in density between the inner and outer core is still not well known. The magnitude of this change plays a critical role in determining the relative magnitude of the gravitational energy source in the core (§7.5.4). The properties of the transition zone between the inner and outer core, including its size, are highly model dependent and controversial.

Estimates of the composition of the core are derived from seismic, shockwave, geochemical and geomagnetic data. The latter apply because most geophysicists believe some dynamo mechanism is operating in the earth's core to produce the magnetic field. If so, this necessitates a moderately high electric conductivity in the core. There is general agreement that the core has Fe as its major constituent (e.g. Birch, 1964, 1972; Ringwood, 1977, 1979; Brett, 1973, 1976; Jeanloz, 1979; Ahrens, 1979). Ahrens (1979) summarizes the shock-wave data and concludes that Fe must be alloyed with some lighter element. Sulphur (Hall and Murthy, 1971) and oxygen (Ringwood, 1977, 1979; McCammon *et al.*, 1983) are the leading candidates although C, N, Mg and Si have also been considered. It appears that roughly 6–10% of these lighter elements are required in viable core models to bring the shock-wave estimates in agreement with the seismic data given in Table 7.1 (see Jeanloz, 1979; Ahrens, 1979). However, there are large uncertainties here. For example, McQueen and Marsh (1966) have shown that if a few percent Ni is present, as is often concluded from geochemical studies, the density would be increased without affecting the incompressibility much. This would be hard to detect, particularly given some of the resolution problems associated with shock-wave data. High pressure in shock experiments are highly irreversible; shock pressures rapidly increase on the order of a nanosecond and are maintained at a given point in a sample for only a microsecond or so. Usually, the temperature at these high pressures must be estimated from theory that depends exponentially on the thermodynamic Gruneisen parameter. Small errors in these estimates can lead to serious errors in temperature (Stewart, 1973). Also, the actual temperatures in the core are not well known. Nevertheless, the conclusion that the core consists mainly of iron with small amounts of lighter elements present such as oxygen or sulphur is reasonable, albeit non-unique.

Probably the most important core parameters to the geomagnetist are the conductivity, temperature and viscosity, all of which are imprecisely known. Most recent estimates for the core–mantle temperature have been obtained from predictions of the melting temperature of iron as a function of pressure (Brown and McQueen, 1980). Because the earth's mantle and inner core are believed to be essentially solid throughout, the melting temperature of iron is regarded as an upper limit to the actual temperature in these localities. This approach is often extended to argue that the estimated melting temperature of iron at the core–mantle and the inner–outer core interface gives the actual temperature there, based on the assumption that such boundaries are simple phase transitions. Indeed, one set of temperature estimates made by Higgins and Kennedy (1971; see also Kennedy and Higgins, 1973), based on a semi-empirical melting law by Kraut and Kennedy (1966), greatly stimulated geomagnetic work. They suggested that most of the outer core was stratified and that widespread thermal convection did not occur in the core. Several dynamo theories that did not involve convection, or involved convection in a very restricted zone of the core, were attempted based on this suggestion. Subsequently, it has been demonstrated that the temperatures predicted by Higgins and Kennedy (1971) are wrong. Several criticisms have been made of the Kraut–Kennedy melting "law" and its use in the earth, but the neglect of solid phase changes of iron, as pointed out by Birch (1972), is now known experimentally to be a valid criticism and to have a significant effect on the temperature estimates based on the melting point of iron (Liu and Bassett, 1975).

There are several alternative methods for estimating the melting temperature of iron that are superior to that of Higgins and Kennedy (e.g. see Brown and McQueen, 1980). However, even the best theories of melting are presently somewhat inadequate at atmospheric pressure and great uncertainties are introduced when they are used to predict melting at core pressures. This is illustrated by the observation of Leppaluto (1972) and Hamano (1974) that the experimental results at 10^9 to 10^{10} Pa can be fitted equally well to radically different melting models. Yet, when these models are used to extrapolate to conditions at the outer and inner core boundaries, differences in the predicted temperatures typically vary by 1 to 3 thousand degrees, as indicated by the extremes quoted in Table 7.2. In addition, Stacey (1972, 1977a) has pointed out that even if one could predict the melting temperature of iron correctly, the temperature estimates would probably still be in error primarily because the core does not consist of pure iron. A possible example of this is given by Tolland (1974) and Ringwood (1977) who respectively suggest that eutectic points may exist in the iron–sulphur and iron–oxide systems at those pressures. McCammon *et al.* (1983) have shown that the phase diagram of Fe–FeO is quantitatively similar to the system Fe–FeS with both displaying a eutectic temperature far below the melting point of pure iron. Hereafter, the outer core is assumed to be in convection and its

TABLE 7.2
Summary of some of the physical properties of the earth's core

Mean radius of the earth	6371 km
Depth to outer core	2891 km (Dziewonski and Anderson, 1981)
Depth to inner core	5149 km (Dziewonski and Anderson, 1981)
Possible transition width between the inner and outer core	Uncertain, but may be as large as a few hundred kilometres
Composition	Primarily Fe, possibly a few per cent Ni, and 6–10% of lighter elements, such as O, S, Mg or Si (Brett, 1976; Ringwood, 1977; Ahrens, 1979)
Conductivity (σ)	Around 10^6 Sm^{-1} (Gardiner and Stacey, 1971; Johnson and Strens, 1973; Keeler *et al.*, 1978)
Temperature gradient in outer core	Probably close to an adiabatic temperature gradient near 0.4 km^{-1}
Thermal diffusivity (k)	$\approx 2 \times 10^{-3}$ m^3 s^{-1} (Stacey, 1977a,b)
Range of temperature estimates for the core–mantle boundary	2300 K $< T \leq$ 5000 K (Verhoogen, 1961; Leppaluto, 1972). Probably near 3100 K (Brown and McQueen, 1980)
Limits of temperature estimates for inner core boundary	3000 K $< T <$ 8000 K (Tolland, 1974; Leppaluto, 1972). Probably near 4100 K (Brown and McQueen, 1980)
Limits on viscosity ($\rho\nu$) estimates	10^{-6} m^2 s^{-1} (Gans, 1972) to 10^5 m^2 s^{-1} (Suzuki and Sato, 1970)
Mean density, ρ (outer core)	1.1×10^{-4} kg m^{-3} (Dziewonski and Anderson, 1981)
Pressure core–mantle boundary	1.4×10^{11} Pa (Dziewonski and Anderson, 1981)
Pressure earth's centre	3.6×10^{11} Pa (Dziewonski and Anderson, 1981)

temperature gradient an adiabatic one. Although not necessary to dynamo theory (e.g. Gubbins *et al.*, 1982), this is the *simplest assumption* consistent with all the geophysical and geochemical data.

The usual value for conductivity assumed in the core is around 5×10^5 S m^{-1} (Gardiner and Stacey, 1971; Stacey, 1977a). It is possible to measure conductivity in shock-wave experiments (Keeler *et al.*, 1978). Conductivities of $Fe_{1-x}Si_x$, where x varies from 0.0 to 0.34, have been measured to pressures of 1.8×10^{11} Pa. The conductivity of Fe decreases with increase in pressure, while that of $Fe_{0.66}Si_{0.34}$ increases. A value near 10^6 S m^{-1} is approached nearly asymptotically by both Fe and Fe–Si alloy at 1.8×10^{11} Pa (core pressures). This value appears to be the present best estimate of the outer core electrical conductivity. It is probably good within a half order of magnitude or so.

Watanabe (1977) and Gubbins (1977) have summarized the estimates for viscosity to show that it is one of the poorest known parameters for the outer core. Using shear wave attenuation data Suzuki and Sato (1970) have estimated the viscosity to be near 10^5 m^2 s^{-1}. This can be contrasted with the 10^{-5} to 10^{-6} m^2 s^{-1} estimate obtained by Gans (1972) using a theoretical analysis to extrapolate laboratory measurements of viscosity. On the other hand Hamano (1975; as referenced by Watanabe, 1977) has argued from shock wave data that the viscosity must exceed 1 m^2 s^{-1}. However, estimates of the effective viscosity of the earth's core from shear and shock wave data can be carried out only if one makes (troublesome) assumptions about the core's rheology; such assumptions are required to obtain relationships between damping and viscosity. Thus, the lower estimates for viscosity in Table 7.2 appear far more reasonable, although it is clear that the viscosity of the core is very poorly known, probably not within one order of magnitude.

Table 7.2 gives the values of various parameters applicable to the earth's core; these values can be supplemented by those given in Table 7.1.

7.2 Some Non-dynamo Hypotheses

7.2.1 Permanent Magnetization

Usually the concept that permanent magnetization in the earth's deep interior contributes significantly to the origin of the earth's magnetic field is dismissed on the basis of high temperatures in the core. Although this is a good argument against permanent magnetization, it is not a conclusive argument. At one atmospheric pressure the highest Curie temperatures for rock forming minerals are all well below 1000°C (e.g. 580°C for magnetite,

696°C for hematite and 770°C for pure iron). Measurements of the change of Curie temperature with pressure are not extensive, but show that it is highly variable. In some substances the Curie temperature increases with pressure and in others it decreases. Predictions from theory of the effects pressure has on Curie temperature are not good, primarily because of difficulties in predicting how the overlap of the electron orbitals of the various magnetic minerals change with pressure. In addition, reliable empirical predictions are not likely to be forthcoming, simply because the present measurements are at such low pressures that extrapolations to core pressures are not justified. Therefore, it might be argued that material in the deep interior, such as the inner core, could be above some critical pressure at which it is permanently magnetized. However, such a contrived argument runs into serious problems with respect to reversals of the earth's magnetic field. It is difficult to conceive of a mechanism that uses permanent magnetization to explain these reversals and the other observed properties of the secular variation. Therefore, the presence of permanent magnetization in the deep interior appears highly contrived and improbable.

7.2.2 Thermoelectric Effects

The rotation of a sphere of radius a, containing an electric charge Q, uniformly distributed on its surface gives rise to an external magnetic field, $\boldsymbol{H}$, which has components:

$$H_r = \left[\frac{Q\omega a^2}{3}\right]\frac{2\cos\theta}{4\pi r^3} \qquad (7.1)$$

$$r \geq a$$

$$H_\theta = \left[\frac{Q\omega a^2}{3}\right]\frac{\sin\theta}{4\pi r^3} \qquad (7.2)$$

where ω is the angular frequency, r is the distance from the centre of the sphere to the point of observation and θ is the angle between the axis of rotation of the sphere and the line between the sphere's centre and the observation point. Equations 7.1 and 7.2 show that the magnetic field outside a rotating sphere with surface charge Q is equivalent to a magnetic dipole of strength $Q\omega a^2/3$ located at the centre of the sphere.

Charge separation can arise because of the thermoelectric effect. Insight into the origin of the thermoelectric effect can be obtained by considering an electrically conducting rod that has a thermal gradient superimposed on it. Electrons at the hot end of the rod will have higher mean thermal energy than electrons at the colder end. Therefore, the hotter electrons will have higher mean velocities and initially more of the electrons will travel from the hotter end to the colder end than vice versa. Eventually equilibrium is reached when the difference in electric field energy produced by the charge separation balances the difference in thermal energy.

The magnitude of such an effect for the earth's conducting core can be estimated. The increase in temperature with depth in the core implies that there will be some charge separation. The mean radius of the positive charge is smaller than that for the negative charge. Thus there will be an external magnetic field produced on rotation (eqns 7.1 and 7.2). The external field is zero because of charge balance.

If A is the Gibbs Free Energy, T the temperature, e the charge of the electron and ϕ the electric potential, then for equilibrium conditions and neglecting pressure effects, the following equality must hold:

$$\frac{\mathrm{d}A}{\mathrm{d}T} = e\frac{\mathrm{d}\phi}{\mathrm{d}T} \tag{7.3}$$

From elementary solid state theory it can be shown that

$$\frac{\mathrm{d}A}{\mathrm{d}T} = -\gamma T \tag{7.4}$$

where γ is approximately 4×10^{-11} J Mol^{-1} K^{-2}. Combining (7.3) with (7.4) then:

$$\frac{\mathrm{d}\phi}{\mathrm{d}r}\frac{\mathrm{d}r}{\mathrm{d}T} = \frac{1}{e}\gamma T \tag{7.5}$$

For an order of magnitude calculation, assume a linear temperature gradient through the core:

$$T = T_1 + \beta(R_1 - r)$$

where R_1 = radius of the sphere with positive charge (which may be taken as zero, if desired); T_1 = temperature at R_1; β = a constant that depends on the assumed temperature gradient; r = distance from the earth's centre. Substitution into (7.5) gives:

$$\frac{\mathrm{d}\phi}{\mathrm{d}r} = \frac{1}{e}\gamma\beta[T_1 + \beta R_1 - \beta r]$$

This may be substituted into Poisson's equation:

$$\nabla^2\phi = \frac{1}{r^2}\frac{\partial}{\partial r}\left(r^2\frac{\partial\phi}{\partial r}\right) = -\frac{\rho}{\varepsilon}$$

where ρ is the surface charge density and ε is the dielectric constant. The result of this substitution gives:

$$-\frac{1}{e}\gamma\beta^2 + \frac{2}{r}\frac{1}{e}\gamma\beta[T_1 + \beta R_1 - \beta r] = -\frac{\rho}{\varepsilon} \tag{7.6}$$

Equation 7.6 indicates that the charge density distribution may be written simply as:

$$\rho = C_1 - \frac{C_2}{r}$$

where C_1 and C_2 are constants given by (7.6). This result with (7.1) and (7.2) can be used to obtain magnetic field estimates. Providing one restricts the choice of parameters to "half-way" reasonable values for the earth, the results of such calculations yield values far too small for the earth's magnetic field (typical result: 10^{-16} T at the earth's surface).

Inglis (1955) was the first to consider such effects, although he used a slightly different calculation procedure. He also showed that pressure gradients were more important than temperature gradients and even expanded calculations to include the Hall effect. All estimates yielded magnetic fields that are several orders of magnitude too small to be the cause of the earth's field. Perhaps this result could have been anticipated since magnetic fields resulting from temperature and pressure gradients in a rotating conductor would not exhibit such properties as reversals of polarity. The most important aspect of the work of Inglis (1955) is that his estimates of the fields are so small that it may be reasonable to dismiss such effects as even playing a small role in generating the earth's magnetic field. Calculations by Merrill *et al.* (1979) indicate that although it is conceivable that thermoelectric currents may alter the magnetic field slightly, this is only possible for certain choices of critical core parameters that seem improbable. However, contrary to what is often assumed, it is *not* clear that such effects can be neglected in all astronomical bodies, because the effect increases as the square of the radius (7.1 and 7.2). Indeed, Stevenson (1974) has argued that significant thermoelectric currents might exist in Jupiter.

7.2.3 Other Models

Several other suggestions have been made for the origin of the earth's magnetic field. These include:

(i) residual electric currents set up in the earth's interior early in its history (Lamb, 1883);
(ii) gyromagnetic effect (Barnett, 1933);
(iii) Hall effect (Vestine, 1954);
(iv) galvanomagnetic effect (Stevenson, 1974);
(v) differential rotation effect (Inglis, 1955);
(vi) electromagnetic induction by magnetic storms (Chatterjee, 1956);
(vii) Nernst-Ettinghauser effect (Hibberd, 1979).

Most of these mechanisms have been reviewed by Rikitake (1966) and Stevenson (1974). It is doubtful that any of these by themselves are responsible for the earth's magnetic field, although some may contribute to it.

7.3 The Dynamo Problem

7.3.1 Disc Dynamo

A disc dynamo (or homopolar or Hertzenberg dynamo) is illustrated in Fig. 7.1a. It was this type of dynamo that was suggested by Larmor (1919a, b) to explain the origin of the earth's magnetic field. A torque, usually assumed constant, must be applied to rotate the disc with angular velocity ω in the presence of a field $\boldsymbol{H}$ (see Fig. 7.1a). A Lorentz force will result in an electric current, I, out from the centre of the disc that can be tapped by a wire that has a "brush" contact with the disc. This wire is wound around the axis of the disc in such a way as to reinforce the initial field.

Only an outline of the disc dynamo theory will be given here. Details are

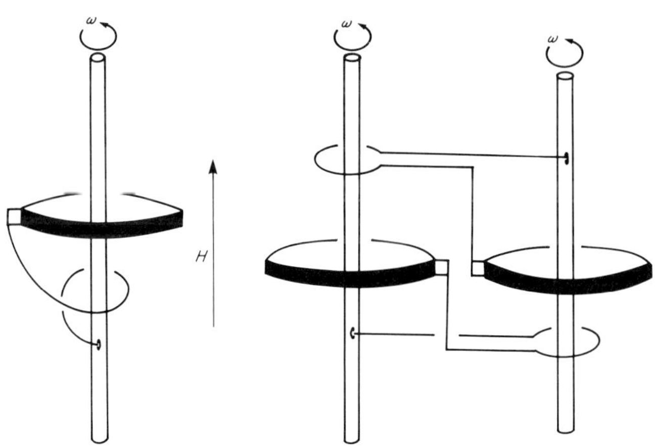

Fig. 7.1. The disc dynamo. (a) Single disc dynamo. A torque ω is applied to rotate a conducting disc in a magnetic field $\boldsymbol{H}$ aligned along the axis of the disc. An electric current, induced in the rotating disc, flows outward to the edge of the disc where it is tapped by a brush attached to a wire. The wire is wound back around the axis of the disc in such a way as to reinforce the initial field. (b) Double disc dynamo. The current induced in one disc is circulated back around the axis of the other. The winding of the wires around the rods are such that reversals of polarity can occur. (Modified after Rikitake, 1966.)

given in Rikitake (1966). The appropriate equations governing the system are:

$$L\frac{dI}{dt} + RI = M\omega I \quad \text{[electric part]} \tag{7.7}$$

$$C\frac{d\omega}{dt} = \Upsilon \quad \text{[mechanical part]} \tag{7.8}$$

where L and M are derived from proper combinations of the self and mutual inductances of the system, t is time, R is the electric resistance of the circuit, C is the moment of inertia of the disc and Υ is the torque. The induced e.m.f. for the disc is $M\omega I$ and the power is $M\omega I^2$. Because the power must be equal to $\Upsilon\omega$ in equilibrium, (7.8) becomes:

$$C\frac{d\omega}{dt} = MI^2 \tag{7.9}$$

Equations 7.7 and 7.8 can be solved simultaneously for I and ω (Rikitake, 1966). The magnitude of the resulting magnetic field can easily be solved once I is obtained for the particular geometry chosen.

Calculations indicate that a single disc dynamo will not give rise to reversals unless the rotation direction is reversed by applying an external torque of the opposite sign. The two-coupled disc dynamo shown in Fig. 7.1b does not suffer from this difficulty. Constant torques applied to the discs result in highly varying fields, including reversals (Rikitake, 1966). However, it is not clear what are the analogues of the disc wire and brushes in the earth's core. Although such dynamo models are interesting and they have played a very important role in stimulating other dynamo research, their analogy with the earth's interior is insufficient to warrant further discussion here.

7.3.2 Magnetohydrodynamics and Plasma Physics

A critical assumption in dynamo theory is that the electric displacement current density, $\partial D/\partial t$, can be neglected. This assumption is often referred to as the magnetohydromagnetic assumption. This compares with plasma physics where $\partial D/\partial t$ cannot be neglected. Valuable insight justifying this assumption can be obtained by considering the microscopic derivation of Ohm's law in a conducting fluid. Let v_d be the drift velocity of electrons, μ_e the electron mass, and τ a constant, which will later be shown to be a relaxation time. Let e be the charge of the electron and E the electric field.

The equation of motion for an electron in the presence of the force eE is:

$$\mu_e\left(\frac{dv_d}{dt} + \frac{v_d}{\tau}\right) = eE$$

The second term on the left side is a friction term originating from collisions with other electrons. The solution of this equation is easily found to be:

$$v = v_0 e^{-t/\tau} + \frac{e\tau E}{\mu_e}$$

Note that as $t \to \infty$ (steady state) one obtains:

$$v_d = \frac{e\tau E}{\mu_e}$$

If N is the electron density, then the steady state current density is (Ohm's Law):

$$J = Nev_d = \frac{Ne^2\tau E}{\mu_e} \equiv \sigma E$$

Setting $f = 1/t$, then f can be interpreted as representing some mean collision frequency.

Suppose an electric field of frequency ν is applied, then three cases arise:

(i) $\nu \gg f$—this is called the plasma domain. Here electrons (and ions) move independently. In effect the high frequency of the applied electric field causes charge separation and $\partial D/\partial t$ cannot be neglected.
(ii) $\nu \ll f$—this is called the magnetohydrodynamic (MHD) domain. There is no charge separation. Particles of opposite sign drift in the opposite directions according to Ohm's law (steady state). In this case, the magnetohydromagnetic assumption, $\partial D/\partial t = 0$, is justified.
(iii) $\nu \simeq f$—this is a complicated situation that will not be considered here.

Pressures exceeding 10^{12} Pa would be required to strip electrons from their nuclei to produce plasma conditions in the earth's core. Thus it appears justified to use the MHD approximation there, that:

$$\frac{\partial \boldsymbol{D}}{\partial t} = \boldsymbol{0}$$

and

$$\nabla \times \boldsymbol{H} = \boldsymbol{J}$$

7.3.3 The Earth Dynamo Problem

The magnitude of the problem involved in obtaining a solution to the earth's dynamo problem can be appreciated by listing the equations that must be solved simultaneously. These are given in Table 7.3. The magnetohydrodynamic assumption, that $\partial \mathbf{D}/\partial t = \mathbf{0}$, is always made and the fluid is usually assumed incompressible, i.e. $\nabla \cdot \mathbf{v} = 0$. The validity of the last assumption can be questioned (Verhoogen, 1980), but at the present level of theoretical development it seems justified. There is roughly a 25% change in density across the entire outer core (Table 7.1). Buoyancy, which requires compressibility, still drives convection but this is incorporated into the theory through the Boussinesq approximation: the fluid is assumed incompressible except for the thermal expansion (Chandresekhar, 1961). Equations 7.10 through 7.19 must be solved simultaneously along with the appropriate boundary and

TABLE 7.3
Equations for the dynamo problem

Maxwell's equations:

$$\nabla \times \mathbf{H} = \mathbf{J} + \frac{\partial \mathbf{D}}{\partial t} \quad (7.10)$$

$$\nabla \times \mathbf{E} = -\frac{\partial \mathbf{B}}{\partial t} \quad (7.11)$$

$$\nabla \cdot \mathbf{B} = 0 \quad (7.12)$$

$$\nabla \cdot \mathbf{D} = \rho_e \quad (7.13)$$

Ohm's law:

$$\mathbf{J} = \sigma \mathbf{E} + \sigma \mathbf{v} \times \mathbf{B} \quad (7.14)$$

Navier-Stokes equation:

$$\rho\left(\frac{\partial}{\partial t} + \mathbf{v} \cdot \nabla\right)\mathbf{v} + 2\rho(\Omega \times \mathbf{v}) = -\nabla P + \eta \nabla^2 \mathbf{v} + \tfrac{1}{3}\eta \nabla(\nabla \cdot \mathbf{v}) - \rho \nabla \phi_g + \mathbf{J} \times \mathbf{B} \quad (7.15)$$

Continuity equation:

$$\nabla \cdot (\rho \mathbf{v}) + \frac{\partial \rho}{\partial t} = 0 \quad (7.16)$$

Poisson's equation:

$$\nabla^2 \phi_g = -4\pi G \rho \quad (7.17)$$

Generalized heat equation:

$$\frac{\partial T}{\partial t} = k\nabla^2 T + (\nabla k' \cdot \nabla T) - \mathbf{v} \cdot \nabla T + \varepsilon \quad (7.18)$$

Equation of state:

$$\rho = \text{Function}\ (P, T, H) \quad (7.19)$$

Notation:

$\mathbf{H} \equiv$ magnetic field	$\rho \equiv$ material density
$\mathbf{B} \equiv$ magnetic induction	$\sigma \equiv$ conductivity
$\mathbf{J} \equiv$ electric current density	$T \equiv$ temperature
$\mathbf{E} \equiv$ electric field	$P \equiv$ pressure
$\mathbf{D} \equiv$ electric displacement vector	$G \equiv$ gravitational constant
$\mathbf{v} \equiv$ velocity	$\phi_g \equiv$ gravitational potential
$\eta \equiv$ viscosity	$\varepsilon \equiv$ heat source term
$\rho_e \equiv$ electric charge density	$k' \equiv$ thermal diffusivity
$\Omega \equiv$ angular velocity of rotation	

initial conditions. As students of mathematics quickly learn, in addition to differences in the actual solutions, even the method used to solve a particular partial differential equation depends on the boundary and initial conditions. The problem represented by (7.10) to (7.19) with the appropriate boundary and initial conditions is referred to as the *earth dynamo problem* and is clearly very formidable. As might be expected from the complexity of the problem, additional simplifications are made to solve different dynamo problems. However, even then the solution is usually not obtained without a considerable amount of mathematical analysis and reliance on the computer.

7.4 The Magnetic Induction Equation

7.4.1 Introduction

One of the most important equations in dynamo theory, the magnetic induction equation, can be derived from combining (7.10) with (7.14) to obtain:

$$\nabla \times \boldsymbol{H} = \sigma \boldsymbol{E} + \sigma(\boldsymbol{v} \times \boldsymbol{B})$$

Then taking the curl of both sides of this equation and using (7.11), (7.12) and $\boldsymbol{B} = \mu_0 \boldsymbol{H}$, one obtains the *magnetic induction equation*:

$$\frac{\partial \boldsymbol{H}}{\partial t} = \frac{1}{\sigma \mu_0} \nabla^2 \boldsymbol{H} + \nabla \times (\boldsymbol{v} \times \boldsymbol{H}) \tag{7.20}$$

Note that this equation also holds if $\boldsymbol{B}$ is substituted for $\boldsymbol{H}$, since $\boldsymbol{B} = \mu \boldsymbol{H}$ and μ, the magnetic permeability, is a constant in the earth's core (μ_0). When $\boldsymbol{v} = 0$, this equation reduces to the vector diffusion equation for $\boldsymbol{H}$. In the absence of a velocity field $\boldsymbol{v}$, a given magnetic field will decay according to the first term on the right-hand side. The last term in this equation gives the interaction of the velocity field with the magnetic field. This interaction can cause breakdown of the magnetic field or build-up of the magnetic field, depending on the nature of the velocity field.

The use of the magnetic induction equation bypasses the need for explicitly considering the electrical currents in the core. Note, however, that the currents can always be explicitly obtained (if desired) when the magnetic field is known by taking the curl of that field. Equation 7.20 has never been solved in closed form, necessitating the need for other approaches, such as the use of expansions and numerical calculations.

In *kinematic dynamo models*, $\boldsymbol{v}$ is specified in some reasonable way (this usually means that $\boldsymbol{v}$ has no sources or sinks and is continuously differentiable), along with some initial magnetic field, $\boldsymbol{H}_0$. The problem then is whether this $\boldsymbol{v}$ field can support an $\boldsymbol{H}$ field that does not decay to zero as time goes to infinity. A subset of the kinematic dynamo problems involves the search for *steady state dynamo* solutions, $\partial \boldsymbol{H}/\partial t = 0$. The kinematic dynamo problem does not require that the $\boldsymbol{v}$ field satisfy the Navier-Stokes equation (7.15). Therefore, there is no feedback from the magnetic field to the velocity field. This means that a magnetic field which approaches infinity, as time approaches infinity, is an acceptable solution to a kinematic dynamo problem.

The problem of simultaneously solving (7.15) and (7.20) represents the *hydrodynamic dynamo* problem. This problem is sometimes solved by assuming that the body force, proportional to $\boldsymbol{v} \times \boldsymbol{B}$ occurring in the Navier-Stokes equation (7.15), is a perturbation. One can then solve the Navier-Stokes equation for some assumed geometry, boundary conditions and initial conditions by neglecting the $\boldsymbol{v} \times \boldsymbol{B}$ term. The solution for the velocity field, $\boldsymbol{v}$, is substituted into the magnetic induction equation (7.20), which in turn is solved for some assumed boundary and initial conditions pertaining to the magnetic field. This last solution provides an estimate for $\boldsymbol{B}$. It is possible to return to the Navier-Stokes equation and use this $\boldsymbol{B}$ to include the magnetic body force term and to solve for a new $\boldsymbol{v}$. Such an iteration procedure can be continued *ad infinitum* or until computer time becomes too expensive. For this to be a valid procedure, it must be demonstrated that the iteration procedure leads to some convergence of $\boldsymbol{v}$ and $\boldsymbol{B}$.

The above approach assumes small Lorentz forces, but if the magnetic forces are assumed to be very strong, this leads to the *magnetohydromagnetic instability models*, pioneered by Braginskiy (1964a, b, 1967). One result of these strong magnetic forces is to make the core fluid more stratified. Braginskiy assumed that the magnetic (Lorentz) force (M) is as important as the Archimedean (buoyancy) force (A) driving the instability. Because the Coriolis force (C) plays an important role in determining the course of the instabilities, Braginskiy named his waves *M.A.C. waves*. Important reviews and further development of this problem have been given by Gubbins (1974) and Soward (1975). These dynamo models will not be discussed further although an introduction to magnetohydromagnetic waves is given in Chapter 9.

Recently, a considerable amount of interest has been generated over the so-called *turbulent dynamo models* or *mean-field electromagnetic models*, first suggested by Moffat (1961, 1970; also see 1978). These will be introduced in §7.5.2 and discussed in §8.3.

7.4.2 Physical Insight

To gain insight into the meaning of (7.20), the role of the two terms on the right-hand side of the equation is considered separately. The first term on the right-hand side is zero for the hypothetical case of infinite conductivity. In this case it can be shown that no induced e.m.f. occurs in a perfect conductor moving in a magnetic field. This theorem is referred to as the *frozen-in-field theorem*. It provides important insight into how a moving conductor can build up or break down a magnetic field. In the case of infinite conductivity, (7.20) reduces to:

$$\frac{\partial \boldsymbol{H}}{\partial t} = \nabla \times (\boldsymbol{v} \times \boldsymbol{H})$$

Consider any area, s, bounded by a line L in a fluid (see Fig. 7.2.). Let $\boldsymbol{n}$ be the normal to s. Then:

$$\int_s \frac{\partial \boldsymbol{H}}{\partial t} \cdot \boldsymbol{n}\, \mathrm{d}s = \int_s \nabla \times (\boldsymbol{v} \times \boldsymbol{H}) \cdot \boldsymbol{n}\, \mathrm{d}s = \int_L (\boldsymbol{v} \times \boldsymbol{H}) \cdot \mathrm{d}\boldsymbol{L} = -\int_L \boldsymbol{H} \cdot (\boldsymbol{v} \times \mathrm{d}\boldsymbol{L})$$

Now $\boldsymbol{v} \times \mathrm{d}\boldsymbol{L}$ is the increment of area perpendicular to $\boldsymbol{L}$ that is swept out in time $\mathrm{d}t$. Therefore,

$$\int_s \frac{\partial \boldsymbol{H}}{\partial t} \cdot \boldsymbol{n}\, \mathrm{d}s + \int_L \boldsymbol{H} \cdot (\boldsymbol{v} \times \mathrm{d}\boldsymbol{L}) = \frac{\mathrm{d}}{\mathrm{d}t} \int_s \boldsymbol{H} \cdot \boldsymbol{n}\, \mathrm{d}s = 0$$

Since the magnetic flux, ϕ, is defined as

$$\phi = \int_s \boldsymbol{B} \cdot \boldsymbol{n}\, \mathrm{d}s = \mu_0 \int_s \boldsymbol{H} \cdot \boldsymbol{n}\, \mathrm{d}s$$

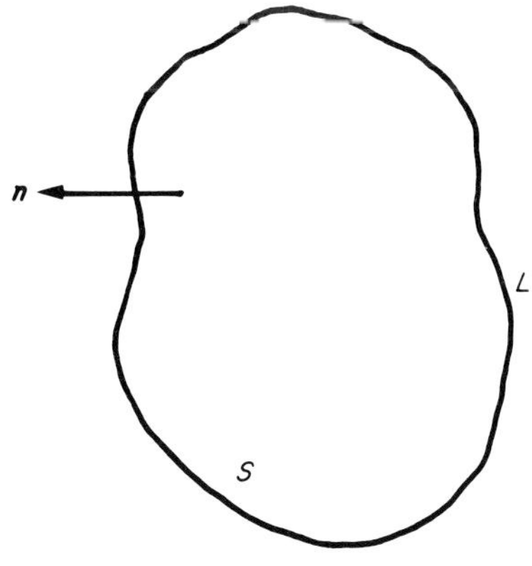

Fig. 7.2. The geometry used in the proof of the frozen-in-field theorem involves any surface with area S, bounded by L. The unit normal to the surface is n.

and the induced e.m.f. is equal to $-\mathrm{d}\phi/\mathrm{d}t$, the above equations imply:

$$\frac{\mathrm{d}\phi}{\mathrm{d}t} = 0$$

This means there is no e.m.f. induced in the perfect conductor and therefore no changes in the magnetic field can ever occur inside a perfect conductor. It is because of this that the magnetic field is sometimes described as being "frozen into" the conductor. Figure 7.3 depicts what will happen when a conductor is moved from a field free space into a magnetic field. Because $\mathrm{d}\phi/\mathrm{d}t = 0$, no field lines can pass through the conductor. Therefore, the field lines will be compressed in front of the moving conductor, and thus the field will be magnified in front of the conductor. Physically, electric current eddies are created on the infinitely-small thin surface of the conductor (by Lenz's Law) resulting in magnetic flux being excluded from inside the conductor.

If only the first term on the right-hand side of (7.20) is present, the magnetic induction equation is reduced to a vector diffusion equation. This explains why $(\sigma\mu_0)^{-1} \equiv k_m$ is often referred to as the *magnetic diffusivity* (or magnetic viscosity) for reasons to be given shortly. This could arise if the velocity field were zero, and in this case the magnetic field would decay with time. For the earth's core one anticipates that both terms on the right-hand side of the magnetic induction equation are usually present. This means there will usually be both diffusion of the magnetic field and build up (or breakdown) of the field due to the interaction of $\boldsymbol{v}$ with $\boldsymbol{B}$. The ratio of the second term to the first term on the right-hand side of (7.20) gives an

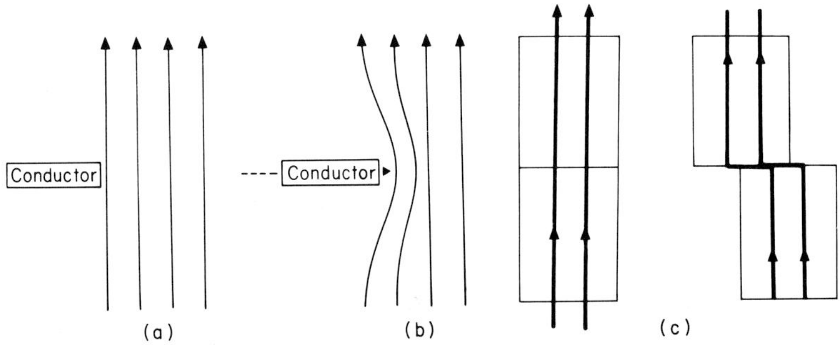

Fig. 7.3. To illustrate the frozen-in-field theorem. (a) and (b) show the effects of moving a perfect electrical conductor from a field-free region into a magnetic field indicated by vertical lines. The magnetic field lines are compressed in front of the conductor and no magnetic field lines ever pass through it as in (b). If the conductivity is finite, the magnetic field can diffuse back into the conductor. In (c) suppose there are two blocks of finite conductivity and initially the magnetic field has diffused back into the blocks. If the lower block is instantaneously moved to the right, it will carry the magnetic field lines with it as shown. Note that the magnetic field lines are concentrated in the *shear zone* at the boundary between the two blocks.

estimate of the rate of build-up of the field to decay. A dimensionless number R_m characterizes that estimate:

$$\frac{\nabla \times (\boldsymbol{v} \times \boldsymbol{H})}{k_m \nabla^2 \boldsymbol{H}} \sim \frac{vL}{k_m} \equiv R_m$$

where L is some appropriate length dimension. The number R_m is called the *magnetic Reynolds number* (by analogy with the Reynolds number in fluid mechanics discussed later). It can be seen that $R_m > 1$ is a necessary, but not a sufficient condition, for a dynamo to be self-sustaining. For small magnetic Reynolds numbers the decay term will dominate and no self-sustaining dynamo is possible. The question of what to use for v and L is difficult because more often than one length scale is present. For example, a treatment given by Parker (1979), applicable to those dynamo models in which convection and non-uniform rotation of the core are assumed, requires that a so-called *dynamo number*, N, be of the order of unity. N is the product of two magnetic Reynolds numbers associated with the non-uniform rotation and convective parts of the earth's core and is given by $\gamma \Gamma R^3 k_m^2$. Here R is the radius of the core, and Γ and γ are parameters related respectively to the magnitude of the $\boldsymbol{v} \times \boldsymbol{B}$ terms associated with the non-uniformly rotating part of the core and the convecting part of the core.

An equation similar to (7.20) can be derived for fluid vorticity by taking the curl of the Navier-Stokes equation (7.15). Since the magnetic and gravity forces are conservative (their curl is zero), this operation yields the following equation for *vorticity* ($\boldsymbol{\Omega} \equiv \nabla \times \boldsymbol{v}$):

$$\frac{\partial \boldsymbol{\Omega}}{\partial t} = \frac{\eta}{\rho} \nabla^2 \boldsymbol{\Omega} + \nabla \times (\boldsymbol{v} \times \boldsymbol{\Omega}) \qquad (7.21)$$

This equation is of the same form as (7.20) with η/ρ set equal to k_m and $\boldsymbol{\Omega}$ to $\boldsymbol{H}$. Indeed a theorem similar to the frozen-in-field theorem, sometimes referred to as the Helmholtz theorem for vorticity, can be derived from this equation. This theorem proves that the vorticity is carried with the fluid in a homogeneous inviscid fluid. Several important insights into the origin of the earth's magnetic field can be gained from considering fluid mechanical analogues. One example is provided by the analogy of the magnetic induction equation with the above equation. This analogy also shows why k_m in (7.20) is sometimes referred to as *magnetic viscosity*. However, the analogies can be carried only so far and even experts in fluid mechanics have been led astray by the similarities of *some* of the equations. An example of one important difference between fluid mechanics and magnetic theory is that the boundary conditions at the core–mantle interface are entirely different for $\boldsymbol{H}$ and $\boldsymbol{\Omega}$.

7.4.3 Free Decay of the Magnetic Field

In the absence of a velocity field, the magnetic induction equation (7.2) becomes:

$$\frac{\partial \boldsymbol{H}}{\partial t} = k_m \nabla^2 \boldsymbol{H} \tag{7.22}$$

As has been noted, this is a vector diffusion equation for the magnetic field. The field will decay away to $1/e$ of its initial intensity in some time τ. A dimensional analysis can be used to obtain a rough estimate of that time. Let $t \sim \tau$ and $\nabla^2 \sim 1/L^2$ where L can be taken to be roughly 3×10^6 m for the radius of the earth's core. Then, from (7.22):

$$\tau \simeq \frac{L^2}{k_m} \simeq \frac{(3 \times 10^6)^2}{2} \text{ seconds}$$

The free decay time estimated by this method is of the order of 100,000 years, a figure that will be shown to be too large by one order of magnitude.

It is useful to use this approximate dimensional analysis to see if laboratory dynamo models can be constructed to model the conditions of the earth's core adequately. The subscript l will be used to denote laboratory conditions and the subscript c to denote core conditions. The above dimensional analysis yields:

$$\frac{\tau_l}{\tau_c} = \frac{L_l^2 \sigma_l}{L_c^2 \sigma_c}$$

where the relationship $k_m = (\sigma\mu)^{-1}$ has been used.

Suppose the dimension of a laboratory model is 1 m, and it contains liquid mercury, then:

$$\tau_l = 10^{-13} \tau_c$$

For a reversal of the earth's magnetic field, $\tau_c = 3 \times 10^{10}$ to 3×10^{11} s, giving a value of $\tau_l = 3 \times 10^{-2}$ to 3×10^{-3} s. It is difficult to examine in detail the properties of a reversal that occurs so quickly. As mercury has one of the highest conductivities readily available for a liquid model, it is doubtful that one will be able to model in a laboratory some of the interesting phenomena occurring in the core. However, other phenomena might be modelled with more success. For example, Lowes and Wilkinson (1963, 1968) have used rotating solid magnetic cylinders to increase μ_0 and therefore τ_l. Their model achieves reasonable scaling for the time factor, but at the expense of eliminating the fluid aspects of the problem.

The methods for solving (7.22) to obtain better estimates of free decay depend on the initial and boundary conditions. P. Roberts (1967) suggested a problem that provides valuable insight into the free decay of the earth's magnetic field. Figure 7.4 illustrates the geometry of the problem.

At time $t = 0$, the field $\boldsymbol{H}$ can be represented by:

$$\boldsymbol{H}_0 = (H(z), 0, 0)$$

The boundary conditions are $H = 0$, at $z = 0$ and $z = d$ (for all time t). $H(z, t)$ can be represented by:

$$H(z,t) = \sum_{n=1}^{\infty} H_n(t) \sin \frac{n\pi z}{d}$$

$H(z, t)$ clearly satisfies the boundary conditions. Substitution of this last equation into (7.22) gives:

$$H_n(t) = H_n(0) \exp\left[-k_m t\left(\frac{n\pi}{d}\right)^2\right]$$

This shows that the smaller wavelength features (larger n) decay faster than the longer wavelength features. Moreover, the fundamental mode, $n = 1$, decays to $1/e$ of its initial value in a time:

$$\tau = \left(\frac{d}{\pi}\right)^2 k_m^{-1}$$

This time is π^2 smaller, or roughly one order of magnitude less, than obtained by the previous dimensional analysis. Calculations involving spherical geometry provide similar results and indicate that the free decay time of the dipole-part of the earth's magnetic field is of the order of 10,000 years (Elasser, 1946). Usually the free decay time of the non-dipole components are still smaller. With reasonable assumptions the higher the degree of the harmonic representing the non-dipole component, the more rapid its decay. The relaxation time is related to the square of the dimensions of the feature involved. Although other more sophisticated estimates for free decay times can be given, the greatest uncertainty in all estimates relates not to the

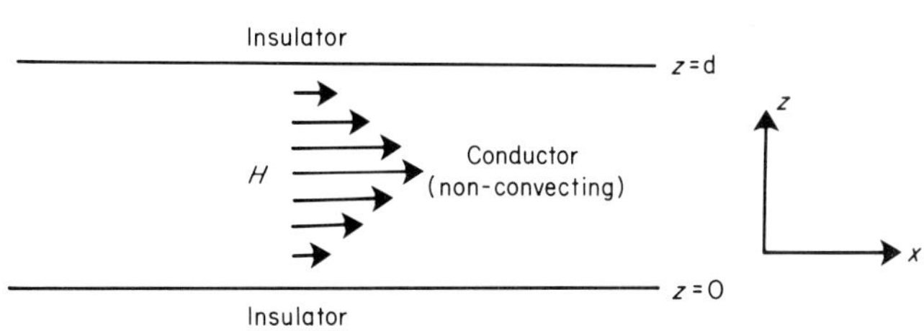

Fig. 7.4. Magnetic field decay in a conductor. A magnetic field $H(z)$ in the x direction is confined to a planar conducting region between two insulators. The boundaries of the insulators are at $x = 0$ and $x = d$. If there is no convection in the conductor, the magnetic field will decay away exponentially with time.

theory, but to the imprecise knowledge of the relevant core parameters (Tables 7.1 and 7.2).

7.5 General Concepts in Dynamo Theory

7.5.1 Dimensionless Numbers

The value in using dimensionless numbers in dynamo problems can be illustrated by considering the fluid mechanical Reynolds number. For simplicity, consider the case of an infintely long cylinder placed in a fluid initially flowing uniformly between two plates (see Fig. 7.5). The corresponding Navier-Stokes equation (7.15) is:

$$\rho\left(\frac{\partial v_i}{\partial t} + v_j\frac{\partial v_i}{\partial x_j}\right) = \frac{\partial P}{\partial x_i} + \eta\frac{\partial^2 v_j}{\partial x_j\,\partial x_i}$$

where i and j represent either of the two co-ordinates in the problem, x and y, with x in the direction of the initial flow. Summation is carried out over the repeated roman indicies. Let L be some representative length scale for the flow behind the obstacle and u some representative velocity. The following non-dimensional parameters are used: $v' \equiv v/u$, $t' \equiv tu/L$ and $x_i' \equiv x_i/L$. In addition, let $P' \equiv (P - P_0)/\rho u^2$, where P_0 is some representative value of the modified pressure. Under these conditions, the Navier-Stokes equation becomes:

$$\frac{\partial v_i'}{\partial t} + v_j\frac{\partial v_i}{\partial x_j} = \frac{\partial P'}{\partial x_i} + \frac{1}{R_e}\frac{\partial^2 v_j^2}{\partial x_i\,\partial x_j} \tag{7.23}$$

where $R_e \equiv \rho Lu/\eta$ is the *Reynolds number*.

The Reynolds number provides a great deal of information on the class of non-unique solutions of (7.23). Any three of the four parameters, velocity, density, length scale or viscosity, can be varied arbitrarily, while the fourth is varied in such a manner to keep R_e constant. The solutions to (7.23) are

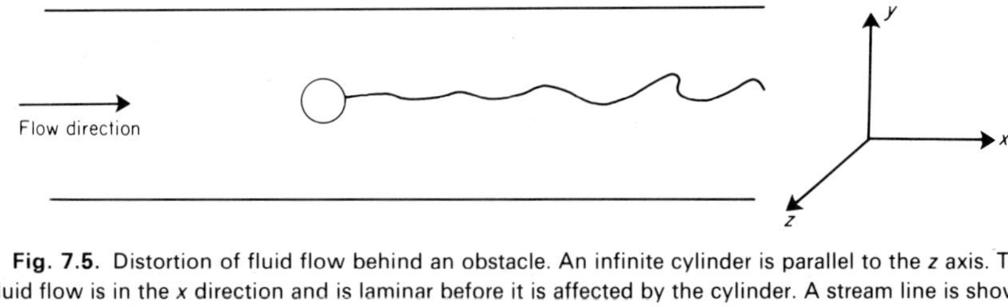

Fig. 7.5. Distortion of fluid flow behind an obstacle. An infinite cylinder is parallel to the z axis. The fluid flow is in the x direction and is laminar before it is affected by the cylinder. A stream line is shown behind the cylinder.

clearly independent of these variations, so long as R_e is constant. All fluid flows having the same value for R_e in (7.23) are said to be *dynamically similar*. Changes in R_e will be reflected in changes in the dynamical flow of the system. In the case of the infinite cylinder the flow is laminar for $R_e = 25$ and is turbulent for $R_e = 250$.

At first it might seem an easy task to determine whether the earth's outer core is turbulent or not by determining the appropriate Reynolds number. Note that even if the equations used in the earth's core are more complicated (7.23), the Reynolds number is effectively the ratio of the inertial to viscous forces and the larger R_e becomes the more complicated the flow structure will be in any given situation. Unfortunately, on closer examination, the problem becomes far more complicated. The magnitude of the *critical Reynolds number*, that value above which the flow becomes turbulent, depends both on (7.23) and on the *boundary and initial conditions*. All too often a particular "critical value" for some characteristic dimensionless number determined for a particular boundary value problem will be erroneously used to characterize a different boundary value problem. For example, a very strong magnetic field in the x direction could result in laminar flow in the above example, even for very high values of R_e. However, once one has a solution for a particular problem, then the use of an appropriate dimensionless number can be very helpful.

A great deal of effort is expended to determine which dimensionless number is the most useful to characterize a particular problem and hence the literature is filled with a variety of such numbers. A few of the common ones used in dynamo theory are given in Table 7.4, along with a brief description of their meaning. The list is far from complete: for example, the dynamo number has already been mentioned in §7.4.2.

Most of the parameters needed to estimate these numbers have been given in Table 7.4. One exception is the velocity, which is difficult to determine and is critical for most estimates. Some investigators assume the velocity to be equal to the westward drift of the non-dipole field $\approx 10^{-3}$ to 10^{-4} ms^{-1} but if hydromagnetic waves are responsible for the observed drift, then this value may be radically wrong (see Chapter 9). A value for thermal expansion (α) of the order of 10^{-5} K^{-1} seems reasonable. A final word of caution is needed in reference to the physical intuition provided in Table 7.4. This intuition for the first five dimensionless numbers is not necessarily appropriate if the magnetic forces in the core are significant, as may well be the case.

7.5.2 The α-effect

Both the observational and theoretical evidence suggest that the core motions are reasonably complex, so that correlations in the velocity field

TABLE 7.4

Some common dimensionless numbers used in dynamo theories

$R_e \equiv \dfrac{uL}{\nu}$	*Reynolds number.* This gives the ratio of inertial to viscous effects. Flow becomes turbulent for large R_e and is laminar for small R_e. See text of this section for more discussion
$R_a \equiv \dfrac{g\alpha\beta L^4}{k\nu}$	*Rayleigh number.* A fluid, heated from below will become unstable at the critical Rayleigh number, above which convection occurs. The size of this critical number depends on the specific circumstances, but is often of the order of 10^3
$R_b \equiv \dfrac{u}{\Omega L}$	*Rossby number.* The time it takes a fluid element moving with speed u to traverse a distance L is L/u. If that time is less than the period of rotation of the core then the fluid cannot sense the rotation. In short, if $R_b \gg 1$, rotation can be neglected, while rotation plays a central role if $R_b \ll 1$
$P_r \equiv \dfrac{\nu}{k}$	*Prandtl number.* This is a constant for a given material and does not depend on the properties of the flow. The viscous response will be instantaneous relative to the thermal response for a very large Prandtl number and vice versa for a very small one
$E \equiv \dfrac{\nu}{\Omega L^2}$	*Ekman number.* This describes the relative importance of frictional effects. If E is small, frictional effects will be small and the Ekman boundary layer thickness at solid–liquid boundaries will be small
$P_e \equiv P_r Re = \dfrac{uL}{K}$	*Peclet number.* $P_e \gg 1$ implies that convective heat transport is much more important than conductive heat transport
$R_m \equiv \dfrac{uL}{k_m}$	*Magnetic Reynolds number.* This gives the relative importance of the fluid modification of the magnetic field to the diffusion of that field. R_m must be large for the frozen-in-field concept to be a reasonable approximation. See §7.4 for further discussion
$A \equiv \dfrac{B}{u(\rho\mu_0)^{1/2}}$	*Alfven number.* The magnetic field has a major effect on the fluid flow for $A \ll 1$. The magnetic field has a major effect on the fluid flow for $A \gg 1$. In particular, turbulence will be inhibited if A is very large
$Q \equiv \dfrac{v_A K}{2}$	*Magnetic Rossby number.* This describes the relative importance of rotation on hydromagnetic waves; if Q is large then rotational effects are negligible. Probably $Q \ll 1$ in the earth's core and rotational effects are very important there

Notation:

u = estimate of the magnitude of some characteristic velocity
L = some representative length scale
ν = kinematic viscosity $= \eta/\rho$
ρ = density
g = gravitational field
α = thermal expansion
β = temperature gradient in *excess* of the adiabatic temperature gradient
k = thermal diffusivity
v_a = Alfven velocity
K = wave number ($2\pi/\lambda$ where λ = wavelength)
Ω = angular rotation of the core
k_m = magnetic diffusivity (see §7.6)
B = magnetic induction
μ_0 = magnetic permeability of free space

over moderate dimensions may be small or negligible. If the length scale over which these correlations occur is very small, then the fluid is said to be turbulent. However, the use of the word "turbulent" often causes unnecessary confusion. The important step in so-called "turbulent dynamos" is simply applying statistical approaches to obtain a mean-field approximation of the earth's magnetic field.

Turbulent dynamics have become very fashionable with important contributions from Krauss, Radler and Steenbeck (as translated by P. Roberts and Stix, 1972), P. Roberts (1971), Gubbins (1974), Krauss (1977) and Moffat (1961, 1970, 1978). All turbulent models depend to some degree on the so-called *α-effect* to magnify the magnetic field.

Ohm's law can be written in terms of some current density, $\boldsymbol{J}$:

$$\boldsymbol{J} = \sigma \boldsymbol{E} + \sigma \boldsymbol{v} \times \boldsymbol{B}$$

where $\boldsymbol{E}$ is an externally applied electric field. An effective "internal electric field", is defined by:

$$\boldsymbol{E}_i \equiv \boldsymbol{v} \times \boldsymbol{B}$$

Suppose a turbulent system is partially described by $\boldsymbol{v} = \boldsymbol{v}_0 + \boldsymbol{v}'$, and $\boldsymbol{B} = \boldsymbol{B}_0 + \boldsymbol{B}'$. The subscript, 0, denotes a steady part of the $\boldsymbol{v}$ or $\boldsymbol{B}$ field, while the prime is used to indicate the fluctuating part of that field about the steady value. The average of the internal field is then:

$$\langle \boldsymbol{E}_i \rangle = \boldsymbol{v}_0 \times \boldsymbol{B}_0 + \langle \boldsymbol{v}' \times \boldsymbol{B}' \rangle \tag{7.24}$$

since $\langle \boldsymbol{v}' \rangle = \langle \boldsymbol{B}' \rangle = \boldsymbol{0}$.

An additional e.m.f. associated with $\langle \boldsymbol{v}' \times \boldsymbol{B}' \rangle \equiv E'$ occurs when v' and B' are correlated. Krause (1977) shows one simple example in which E' (or effectively, an e.m.f.) can be written as

$$\boldsymbol{E}' = \alpha \boldsymbol{B}_0 \tag{7.25}$$

In this case the internal electric field produced from the fluctuating parts of the $\boldsymbol{v}$ and $\boldsymbol{B}$ fields is directly related through the constant α to the steady field $\boldsymbol{B}_0$. This is the so-called *α-effect*. α need not be a scalar, but can be a second order tensor.

Figure 7.6 helps illustrate one way in which the *α-effect* given by (7.25) can occur. Suppose $\boldsymbol{v}' = \boldsymbol{v}_1 + \boldsymbol{v}_2$, where $\boldsymbol{v}_2$ is along the z axis and $\boldsymbol{v}_1$ represents a right-hand rotation about z. In fluid mechanics, $\boldsymbol{v}'$ is referred to as a right-handed *helical velocity*. Take $\boldsymbol{B}_0$ to be uniform in the x direction. $\boldsymbol{v}_1 \times \boldsymbol{B}_0$ will produce a current density, $\boldsymbol{J}_1$, flowing in planes perpendicular to the y axis in Fig. 7.6. Associated with $\boldsymbol{J}_1$ will be a magnetic field $\boldsymbol{B}'$ directed in the y direction. The interaction of $\boldsymbol{B}'$ with $\boldsymbol{v}_2$, $(\boldsymbol{v}_2 \times \boldsymbol{B}')$, will produce an electric field $\boldsymbol{E}'$ parallel to $\boldsymbol{B}_0$, as given by (7.25).

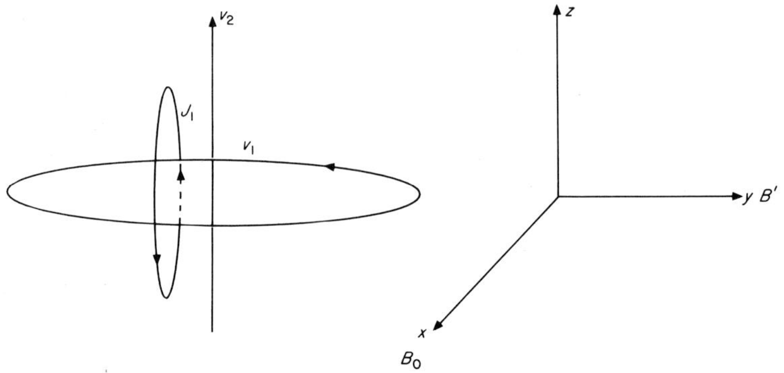

Fig. 7.6. To illustrate the α-effect. Consider a right-handed helical velocity field depicted by $\boldsymbol{v}_1$ and $\boldsymbol{v}_2$ in the presence of a field $\boldsymbol{B}_0$ aligned along the x axis. This will produce current loops such as J_1 lying in the xz plane. Associated with the current loop J_1, is a field B' aligned parallel to the y axis. This new field, $\boldsymbol{B}'$ interacts with $\boldsymbol{v}_2$ to produce an electric field parallel to the x axis.

Equation 7.25 represents the α-effect in the theories of turbulent dynamos. Because a mean electric field is produced in a turbulent system, analyses of this sort are often referred to as *mean field electrodynamics* (see §8.3.1). The existence of the α-effect was verified experimentally by Steenbeck *et al.* (1967) in a rectangular box of liquid sodium. This internal electric field (or equivalently, the e.m.f.) can drive a current, and if the current has the proper geometry it could reinforce an initial magnetic field. Figure 7.7 illustrates how this could occur.

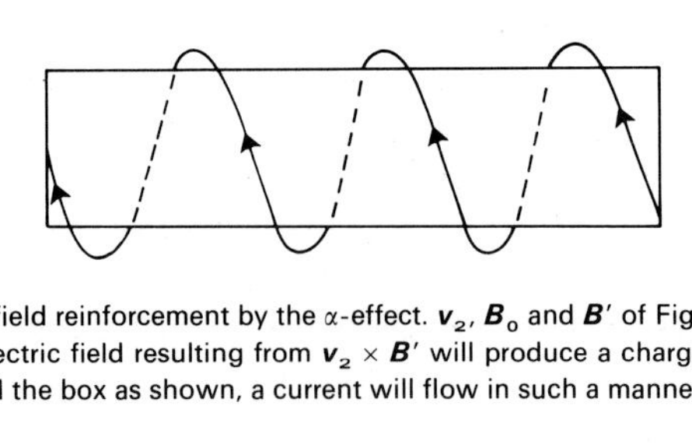

Fig. 7.7. Magnetic field reinforcement by the α-effect. $\boldsymbol{v}_2$, $\boldsymbol{B}_0$ and $\boldsymbol{B}'$ of Fig. 7.6 are shown in a large rectangular box. An electric field resulting from $\boldsymbol{v}_2 \times \boldsymbol{B}'$ will produce a charge separation as shown. If wires are added around the box as shown, a current will flow in such a manner to magnify the initial $\boldsymbol{B}_0$ field.

The experimentally measured quantity used to verify the α-effect was the e.m.f. (Steenbeck *et al.*, 1967; Krause, 1977). There is no evidence that reinforcement of the initial B field occurred. Even though wires, such as given in Fig. 7.7, could be used to obtain reinforcement of the initial B field, the analogue in the earth is not clear, since this "reinforcement current" would then have to be close to the core–mantle boundary. More sophisticated treatment than given above is thus required to examine the α-effect in the earth (§8.3). Note that a necessary condition for the presence of an α-effect is a correlation between vorticity, $\nabla \times \boldsymbol{v}$, and the velocity (i.e. there is some helicity, $\boldsymbol{v} \cdot (\nabla \times \boldsymbol{v}) \neq 0$). The α-effect is essentially a mechanism by which turbulent energy is converted to electrical energy.

The α-effect is fundamental to two of the most fashionable dynamo types, the $\alpha\omega$ and α^2 dynamos. Briefly, the *$\alpha\omega$ dynamos* involve a toroidal magnetic field (see §7.5.3) associated with a large variation in the fluid velocity in a plane perpendicular to the rotation axis. This contrasts with the *α^2 dynamos* (§8.3.2) in which no large velocity shear is required. In the latter, both a toroidal magnetic field and a poloidal magnetic field are amplified by an α process.

7.5.3 A Heuristic $\alpha\omega$ Dynamo

Parker (1955) has presented a dynamo model that does not make use of any mathematics, nor does it require much understanding of the previous sections. It helps to provide a visual picture of the dynamo process.

The basic idea of a dynamo is that some initial magnetic field is altered through interactions with an assumed (kinematic) or derived (hydromagnetic) velocity field in such a way that the magnetic field is reinforced. In the absence of a velocity field in the core, the magnetic field would decay with time (§7.4.3). The velocity field interacts with the magnetic field through Lenz's law. Any attempt to move an electrically conducting material into a magnetic field, will induce currents in the conductor to oppose its motion. These induced currents will alter the initial magnetic field. Figure 7.3 shows what happens in one simplified case. The magnetic field lines are compressed in front of a moving conductor and the field is thus intensified there. Figure 7.3 also illustrates the frozen-in-field concept that any magnetic field lines already in the conductor will be carried with it. If the conductor is not perfect, the field can diffuse into and out of the conductor. In fact, some of this must be occurring because it would be impossible to get the field lines into a perfect conductor in the first place. The process of moving a conductor to intensify the magnetic field illustrates how kinetic energy can be converted into magnetic field energy.

Figure 7.8 illustrates how some initial magnetic field, represented by a single field line can be magnified by moving (conducting) core fluid. The initial magnetic field has a radial component and is referred to as a S_1^0 poloidal magnetic field. The notation used here refers to the vector spherical harmonic notation developed in §8.1. Consider a simple toroidal velocity field, T_1^0, as shown in Fig. 7.8a (simple shear; the definition of a *toroidal field* implies that there is no radial component). Figure 7.8b shows that this toroidal velocity field, T_1^0, interacts with the magnetic field S_1^0, to generate a

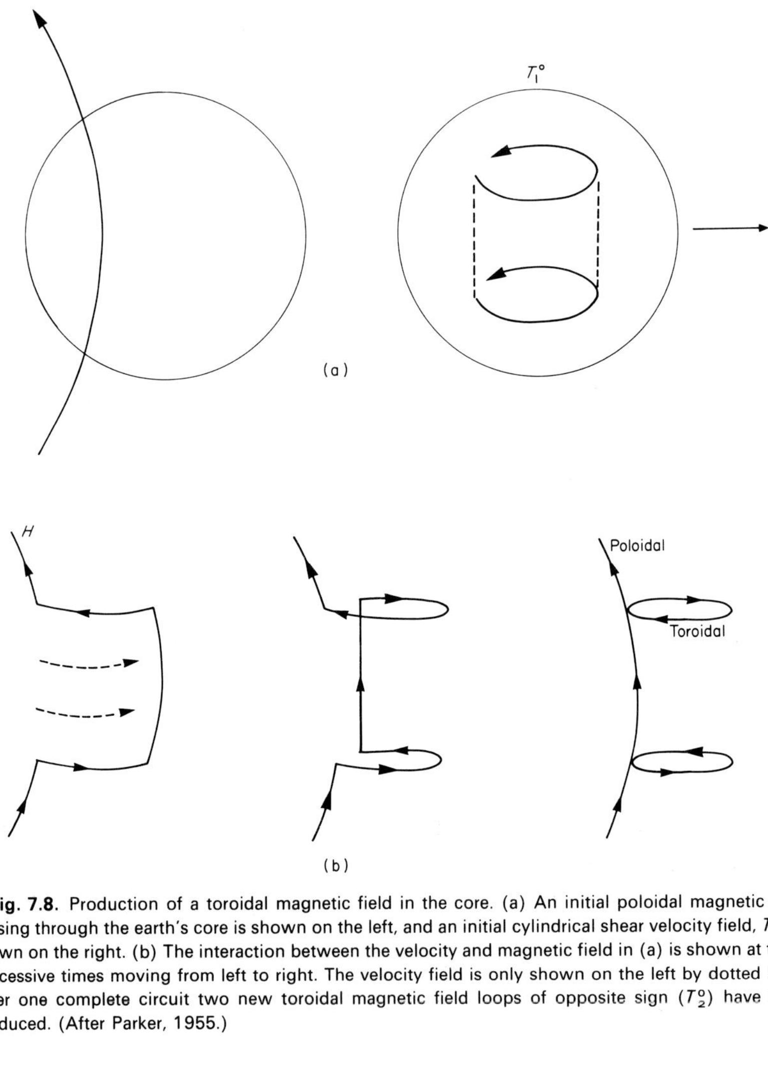

Fig. 7.8. Production of a toroidal magnetic field in the core. (a) An initial poloidal magnetic field passing through the earth's core is shown on the left, and an initial cylindrical shear velocity field, T_1^0, is shown on the right. (b) The interaction between the velocity and magnetic field in (a) is shown at three successive times moving from left to right. The velocity field is only shown on the left by dotted lines. After one complete circuit two new toroidal magnetic field loops of opposite sign (T_2^0) have been produced. (After Parker, 1955.)

more complex toroidal magnetic field, called a T_2^0 field. Note that this latter toroidal field has opposite signs in the two hemispheres. In simple terms:

$$S_1^0 \text{ magnetic field} + T_1^0 \text{ velocity field} \rightarrow T_2^0 \text{ magnetic field}$$

This process is referred to as an ω *effect*.

Another velocity field with a radial component (poloidal) must be added in order to reinforce the original field. This is done by assuming that convection exists in the core. Figure 7.9 illustrates how a toroidal field would be affected by an uprising of fluid in the northern hemisphere. A bulge occurs in the field line, taken to represent a segment of a toroidal field line as it moves with the fluid. The Coriolis force in the northern hemisphere will act to produce a counter-clockwise rotation in the region of fluid upwelling. The field lines will be twisted with this rotation and a poloidal magnetic loop (Fig. 7.9) will be produced after every 180° rotation. Figure 7.10 sketches how various poloidal loops of this sort could come together to produce a large poloidal field. This process involves the α-effect. Because downwelling regions produce poloidal loops of opposite sign to upwelling regions, it must be

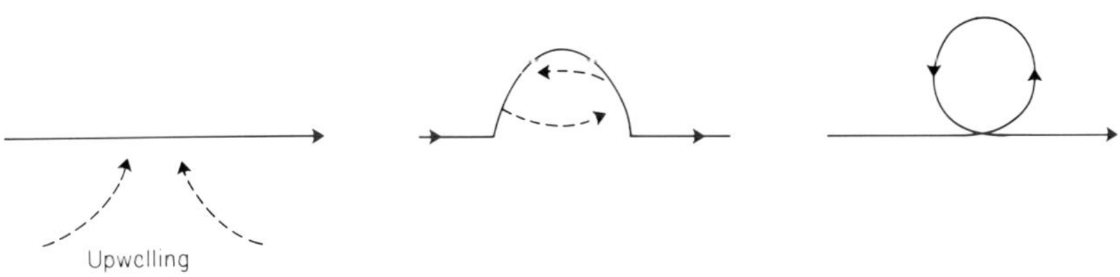

Fig. 7.9. Production of a poloidal magnetic field. A region of fluid upwelling, illustrated by dotted lines on the left interacts with a toroidal magnetic field (solid line). Because of the Coriolis effect (northern hemisphere) the fluid exhibits helicity, rotating as it moves upward (dotted lines centre). The magnetic field line is carried with the conducting liquid and is twisted to produce a poloidal loop as on the right. (After Parker, 1955.)

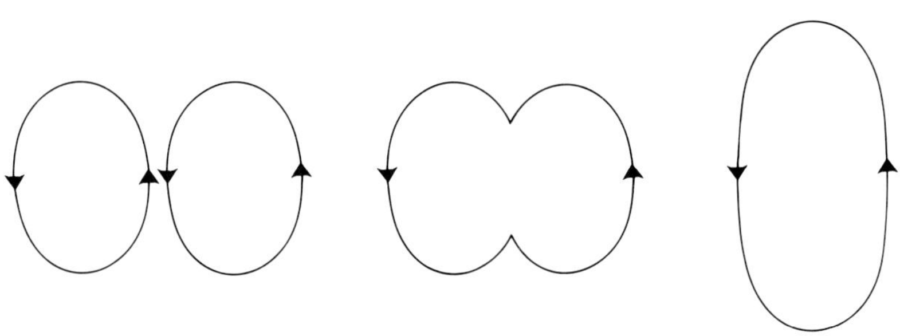

Fig. 7.10. Convergence of two poloidal loops as produced in Fig. 7.9 results in a larger poloidal loop. (After Parker, 1955.)

supposed that some inhomogeneity is present. The above heuristic dynamo combining the α and ω processes, is an example of an $\alpha\omega$ *dynamo*.

Little attention has been given to the origin of the initial weak poloidal field. Perhaps the initial field resulted from the trapping of a solar or galactic field that may have been present at the time of formation of the earth's core. In this model, the earth would have had a magnetic field since the time of core formation. Alternatively, even though it would be very small, the initial magnetic field could have arisen in the earth (or the sun) from separation of charge due to temperature or pressure gradients (§7.2.2).

Although the above heuristic model is intuitively appealing, a more mathematical examination of it indicates that it probably does not work. Lilley (1970a) suggests that it fails because more complicated initial fields are required for dynamo action. This appears to be correct (Bullard and Gubbins, 1977). Nevertheless the intuition provided by this model is still extremely valuable, even if the real situation involves more complex fields. However, this intuition is not necessarily applicable to alternative dynamo models (see Chapter 8).

7.5.4 Energy Sources

The major uncertainties in estimating the power required to sustain the geodynamo can be traced to the poor knowledge of the electrical conductivity of the core and the configuration and strength of the magnetic field in the core. Uncertainties in the magnitude of the toroidal field alone involve two to three orders of magnitude (see discussion in §8.3.4) and therefore the energy in the field at any given time is extremely poorly known (recall that the energy goes as the square of the magnetic field). Because of such uncertainties, the approach has been to seek a probable lower bound for the power and subsequently to make guesses about more reasonable values. This is presently a subject of intense research and more extensive discussion than is given here can be found in Gubbins (1977), Loper and P. Roberts (1979), Häge and Muller (1979), Verhoogen (1980) and Olson (1981a, b).

The velocity field can either magnify or destroy the magnetic field (§7.4.2). Statistically, the magnification process must dominate the destructive process for a dynamo to exist. If the velocity field acts to destroy the magnetic field in a given locality for a short period of time, then elsewhere the magnifying processes must be large enough to sustain a dynamo. Therefore, by setting the velocity equal to zero in the magnetic induction equation, one can estimate the *minimum* power to sustain the field (§7.4.3). This is minimum because it assumes that no dynamic destruction of the magnetic field ever occurs. If the magnetic field is mapped downwards into the earth, the higher degree spherical harmonics increase much faster than lower degree ones (Chapter 2). In addition, the higher the degree of the harmonic the more rapid its decay (§7.4.3), and thus substantial amounts of power may be required to sustain harmonics other than those associated with the dipole field. To obtain a minimum estimate of the power, consider only the dipole field, ignoring both the toroidal field and the higher degree terms in the poloidal field.

The magnetic field energy density is given by $B^2/2\mu_0$. Now continue the dipole field downwards to the core and a value for B near 4×10^{-3} T is obtained, or a magnetic field energy density roughly of 10 J m^{-3}. Using this value for the energy density throughout the core (volume equal to 10^{20} m^3), a probably lower bound for the magnetic field energy in the core of 10^{21} J can be obtained. Although the uncertainty in the value of the conductivity makes estimating the free-decay time of the magnetic field difficult, a value of 10^5 years (3×10^{12} s) is probably a reasonable upper limit for it. Therefore, to maintain the field against decay, power must be supplied at the minimum rate of 3×10^8 W. A similar estimate for power is obtained for the loss of energy in the dipole field that is external to the core (obtained from $(\int_{r_c}^{\infty}(B^2/2\mu_0)\,\mathrm{d}V)\tau^{-1}$ where τ is the decay time and r_c is the radius of the core). Therefore, 10^8–10^9 W seems to be a likely minimum estimate for the power required to maintain the geodynamo. This estimate is consistent with other minimum estimates made by different methods (e.g. Gubbins, 1977; Verhoogen, 1980).

To obtain a guess as to what is the actual power, assume (controversially—see Chapter 8) a toroidal magnetic field that is one order of magnitude larger than the dipole field and (less controversially) a free decay time near 10^4 years (§7.4.2). By similar methods to those employed above, this leads to estimates near 10^{11}–10^{12} W consistent with a value of 4×10^{11} W obtained from a more complicated calculation by Braginskiy (1976), who favours a strong toroidal field (§8.3.4).

Assuming that thermally driven convection powers the geodynamo, Stacey (1977b) used a different approach to estimate the power required. Under these conditions the temperature gradient in the outer core is closely approximated by the adiabatic temperature gradient, $(\mathrm{d}T/\mathrm{d}r)_a$. The minimum core heat flux $F_{\min}$, from the adiabatic temperature gradient can then be estimated (Stacey, 1977b):

$$F_{\min} = -k4\pi r_c\left(\frac{\mathrm{d}T}{\mathrm{d}r}\right)_a \approx 3 \times 10^{12}\ W$$

where r_c is the outer core radius and k is the thermal diffusivity.

The calculation above ignores the problem of thermodynamic efficiency. Stacey (1977b) estimated this to be 10% by equating the efficiency to $\Delta T/T$, where ΔT is the estimated drop in the temperature across the outer core and T is the estimated temperature in the core–mantle boundary. The question of efficiency, and in particular whether a thermal or gravitational driven dynamo is more efficient, is a complicated one that will not be extensively dealt with here. Backus (1975) has established upper and lower bounds on the efficiency, but these bounds are too large to be of much practical value. Considerable controversy exists on the efficiency of particular energy sources (e.g. Gubbins, 1977; Lopez and Roberts, 1978; Verhoogen, 1980).

Olson (1981b) has examined the energy source problem by making reasonably broad assumptions about the dynamo processes acting in the core. In particular he assumes the earth's magnetic field is generated by small-scale rotational turbulence consisting of propagating inertial waves (Chapters 8 and 9) and that this turbulence is supported by a flux of buoyancy originating at one surface of the core. He shows that these assumptions can be used to derive a relationship between the rate of buoyancy production γ and dipole moment D given by:

$$\gamma = cD^2$$

where c is a constant that involves several factors, including conductivity, rotation, the density contrast between the inner and outer core, the volume of the core, etc. This analysis suggests that there is a simple relationship between the strength of the magnetic field and buoyancy production. If so, long-term variations in the intensity of the field are due to long-term variations in buoyancy production. Olson's analysis of the dynamics of the problem yields estimates of efficiency and power similar to those given before.

The following energy sources have been suggested to drive the dynamo.

(i) *Radioactivity*. Most geochemical models for the earth's interior have suggested that heat production from heavy radioactive elements, such as uranium and thorium, can be neglected in the core.

The most commonly suggested radioactive element in the core has been K, although even its presence there is debated (Oversby and Ringwood, 1972a, b; Hall and Murthy, 1971; Stacey, 1977a; Ringwood, 1979). Nevertheless, only a small amount of radioactive material is needed to power the geodynamo. For example, Verhoogen (1973) has shown that only 0.1% K in the core gives rise to a thermal power of 4×10^{12} W, a value sufficient by itself to power the geodynamo.

(ii) *Latent heat of crystallization*. Verhoogen (1961) used the Claussius-Clapyron equation to estimate the latent heat associated with crystallization:

$$\frac{dP}{dT_m} = \frac{\Delta Q}{T_m \, \Delta V}$$

where P = pressure, T_m = melting temperature, ΔV = volume change between the liquid and solid, and ΔQ is the latent heat. The use of this equation assumes that the inner–outer core boundary represents a first-order phase change and that the inner (solid) core is slowly growing due to cooling of the core. Several additional assumptions are required to estimate the parameters appearing in the above equation. Although many of these parameters are uncertain, Verhoogen took a cooling rate of 10–50 K per Ga at the initial inner–outer core boundary and estimated a power around 10^{11}–10^{12} W.

(iii) *Gravitational energy.* Braginskiy (1963) suggested that core convection is powered gravitationally either by the upward migration of a light component or by growth of the inner core by the freezing out of a heavier component. Gubbins (1976, 1977), Loper (1978a, b) and Loper and P. Roberts (1978, 1979) have renewed interest in this gravitational mechanism by claiming it is far more efficient than thermally driven convection, but this is contested by Verhoogen (1980).

The power available from gravitational energy, P_g (force x momentum) is given by:

$$P_g = \int (\nabla \phi_g) \cdot \rho \mathbf{v} \, dV$$

where ϕ_g is the gravitational potential and ρ the local density of core material. The integration is carried out over the volume of the core. Integrating by parts and using Stokes' theorem yields:

$$P_g = \int \phi_g \rho \mathbf{v} \cdot \mathrm{d}\mathbf{s} + \int \phi_g \nabla \cdot (\rho \mathbf{v}) \, \mathrm{d}V \tag{7.26}$$

where s is the surface of the core. Gubbins (1977) assumes that there is no net volume change in the core and that the first term on the right can be neglected. Following Gubbins and using (7.26) and the continuity equation (7.16) then,

$$P_g = \int \phi_g \frac{\partial \rho}{\partial t} \, \mathrm{d}V$$

The power released depends critically on the rate of growth of the inner core. For comparison sake, Gubbins takes a value of 25 $m^3 \, s^{-1}$ of material frozen at the inner core boundary, a value used by Verhoogen (1961) in calculating latent heat energy (ii). Estimating P_g numerically from reasonable seismic models gives:

$$P_g \approx \Delta p \times 2 \times 10^{11} \text{ W}$$

Gubbins (1979) analyses eigenvibration data to conclude that $\Delta \rho$, the density contrast across the inner–outer core boundary, lies between 0.25 and 1.25.

Note that all values in this range are greater than those for the 1981 PREM Model (Table 7.1). The analyses of Loper (1978a, b) give slightly larger estimates for the gravitational energy available (4×10^{11} W) based on theoretical calculations that lead to estimates of $\partial\rho/\partial t$. The radius of the inner core would have had to increase (non-linearly) by roughly 5–10% in the last 4 Ga to give between 10^{11} and 10^{12} W for the power. This may be a very important source of energy for the geodynamo.

(iv) *Precessional energy.* Malkus (1963, 1968) suggested that differential precession might drive the dynamo. Briefly, the ellipticity of the entire earth differs slightly (by about 0.5%) from that of a rotating hydrostatic body with the same mass distribution. Assuming the dynamic ellipticity of the core is due to a rotating liquid, external torques on the core will produce a different precession rate than that of the solid earth. The earth's rotation axis precesses about the pole of the ecliptic with a period of about 25,700 years. However, it has been argued that the energy associated with the dissipative coupling resulting from this differential precession does not exceed 10^8 W (Rochester *et al.*, 1975; Stacey, 1977a, b). In addition, a serious objection to precession is that it is not clear how it can drive a dynamo.

(v) *Primordial heat.* Because the earth has a very large heat capacity, a substantial amount of heat must be lost for even a small decrease in the earth's mean temperature (Schubert *et al.*, 1980; Stacey, 1980). Starting with an initial hot earth due to conversion of gravitational energy during formation (including that associated with core formation), a range of values for the change in total heat flux with time can be obtained. Even when mantle-wide convection is assumed, there may be a significant amount of "primordial heat" remaining to help drive the dynamo.

The problem of the energy source for the earth's dynamo is not well constrained. It is not even certain whether the earth is presently heating up or cooling down (e.g. McKenzie and Richter, 1981). The magnitude of the field in the core today or in the past is also still debated. It is doubtful that the determination of an adequate energy source will ever be a serious obstacle to finding viable dynamo mechanisms for the earth. However, it is important to determine the nature of the energy source to distinguish between competing classes of dynamo models.

Chapter Eight

Origin of the Earth's Magnetic Field 2: Introduction to Advanced Dynamo Theory

8.1 Vector Spherical Harmonics

8.1.1 Helmholtz Scalar Equation

The Helmholtz vector equation,

$$\nabla^2 \boldsymbol{H} + k^2 \boldsymbol{H} = 0 \tag{8.1}$$

commonly occurs in many branches of physics and geophysics. For example, suppose decay of the magnetic field is assumed to be represented by $\boldsymbol{H}e^{-at}$, where a is a constant and t is time. Substitution of this into the magnetic diffusion equation (7.22) yields (8.1) with $k = (a/k_m)^{1/2}$.

Solutions of (8.1) are usually obtained in terms of vector spherical harmonics. These harmonics play a major role in many dynamo theories and also they can be directly linked with the terminology presently used to describe the earth's magnetic field. A more elegant and complete development of these harmonics can be found in Morse and Feshbach (1953). It is hoped that the treatment that follows will be helpful to those encountering them for the first time.

The development of vector spherical harmonics is most conveniently done by first analysing (8.1) in its scalar form:

$$\nabla^2\psi + k^2\psi = 0 \tag{8.2}$$

This last equation is referred to as the *Helmholtz scalar equation*. Solutions to Laplace's equation in spherical co-ordinates are given in terms of surface harmonics y_l^m (§2.2). Therefore, a solution of (8.2) is attempted by the substitution:

$$\psi = \sum_{l=0}^{\infty} \sum_{m=l}^{\infty} f_l(r) y_l^m(\theta, \phi) \tag{8.3}$$

where $f_l(r)$ is some undetermined function of r.

Substitution of (8.3) into (8.2) leads to an equation for $f_l(r)$:

$$\left[\frac{\mathrm{d}^2}{\mathrm{d}r^2} + \frac{2}{r}\frac{\mathrm{d}}{\mathrm{d}r} + k^2 - \frac{l(l+1)}{r^2}\right] f_l(r) = 0$$

Let

$$f_l(r) = \frac{1}{r^{1/2}} u_l(r) \tag{8.4}$$

and substitute this value of $f_l(r)$ into the preceding equation to obtain:

$$\left[\frac{\mathrm{d}^2}{\mathrm{d}r^2} + \frac{1}{r}\frac{\mathrm{d}}{\mathrm{d}r} + k^2 - \frac{(l+\frac{1}{2})^2}{r^2}\right] u_l(r) = 0 \tag{8.5}$$

Equation 8.5 is well known to mathematicians as one form of Bessel's equation. Solutions of it are:

$$u_l(r) = \xi_{l+1/2}(kr) \tag{8.6}$$

and

$$u_l(r) = N_{l+1/2}(kr)$$

where $\xi_l(R)$ is a Bessel function given by:

$$\xi_l(R) = \left(\frac{R}{2}\right)^{l+1/2} \sum_{n=0}^{\infty} \frac{(-1)^n}{n!\Gamma(n+l+\frac{3}{2})} \left(\frac{R}{2}\right)^{2n} \tag{8.7}$$

$\Gamma(z)$ is a gamma function, given by $\Gamma(z) \equiv \int_0^\infty e^{-t} t^{z-1}\,\mathrm{d}t$. $N_{l+1/2}$ is a Neumann function, given by

$$N_{l+1/2}(r) = \frac{\xi_{l+1/2}(r)\cos l\pi - \xi_{-(l+1/2)}(r)}{\sin l\pi} \tag{8.8}$$

The solution of the Helmholtz scalar equation is therefore given by (8.3) where $f_l(r)$ is obtained from (8.4) through (8.8).

8.1.2 Helmholtz Vector Equation

Take ψ to represent a solution of the Helmholtz scalar equation given in the last section. Then direct substitutions of $\nabla\psi$ into the Helmholtz vector equation (8.1) shows that $\nabla\psi$ is a solution. Note that $\nabla\psi$ is irrotational, or curl free ($\nabla \times \nabla\psi = 0$). To find a general solution, two other vector solutions must be found that are orthogonal to $\nabla\psi$ and to each other. With this objective in mind the cross product of the terms in the Helmholtz vector equation can be taken with r to obtain:

$$[\nabla^2 \nabla\psi] \times \boldsymbol{r} + k^2 \nabla\psi \times \boldsymbol{r} = 0$$

Since $\nabla^2 r = 0$, the above can be written:

$$\nabla^2(\nabla\psi \times \boldsymbol{r}) + k^2(\nabla\psi \times \boldsymbol{r}) = 0 \tag{8.9}$$

Therefore,

$$\boldsymbol{T} = \nabla\psi \times \boldsymbol{r} \tag{8.10}$$

is a second solution of the Helmholtz vector equation. For completion a third vector solution, $\boldsymbol{S}$, is needed, where $\boldsymbol{S}$ is orthogonal to $\boldsymbol{T}$ and $\nabla\psi$. $\boldsymbol{S}$ is given by:

$$k\boldsymbol{S} = \nabla \times \boldsymbol{T} \tag{8.11}$$

This can readily be proved by taking the curl of (8.9) with $\boldsymbol{T}$ substituted for $\nabla\psi \times \boldsymbol{r}$. It is also easy to show that,

$$k\boldsymbol{T} = \nabla \times \boldsymbol{S}$$

by multiplying both sides of (8.11) by k and using the fact that $\nabla \cdot \boldsymbol{S} = 0$. Therefore, $\boldsymbol{S}$ and $\boldsymbol{T}$ are related to each other by identical equations. Moreover, $\boldsymbol{S}$ and $\boldsymbol{T}$ are clearly orthogonal to each other and to $\nabla\psi$.

The three orthogonal vectors $\boldsymbol{U}(\boldsymbol{r}) \equiv \nabla\psi$, $\boldsymbol{T}$ and $\boldsymbol{S}$ are the solutions to the Helmholtz vector equation. The spherical harmonic representations of these vectors are:

$$\boldsymbol{U} = \frac{\partial\psi}{\partial r}, \quad \frac{1}{r}\frac{\partial\psi}{\partial\theta}, \quad \frac{1}{r \sin\theta}\frac{\partial\psi}{\partial\phi}$$

$$\boldsymbol{T} = 0, \quad \frac{1}{\sin\theta}\frac{\partial\psi}{\partial\phi}, \quad -\frac{\partial\psi}{\partial\theta}$$

$$\boldsymbol{S} = k_m r\psi + \frac{1}{k_m}\frac{\partial^2(\psi r)}{\partial r^2}, \quad \frac{1}{k_m r}\frac{\partial^2(r\psi)}{\partial r\,\partial\theta}, \quad \frac{1}{k_m r \sin\theta}\frac{\partial^2(r\psi)}{\partial r\,\partial\theta}$$

Therefore, any solenoidal, or divergence free vector, such as $\boldsymbol{B}$, can be written:

$$\boldsymbol{B} = \nabla \times \psi\boldsymbol{r} + \nabla \times (\nabla \times \Psi\boldsymbol{r}) = \boldsymbol{T} + \boldsymbol{S}$$

Ψ has been used instead of ψ in the second term to achieve more generality (e.g. Ψ could be a constant times ψ). The orthogonality conditions for $\boldsymbol{T}$ and $\boldsymbol{S}$ can be shown to be:

$$\begin{aligned}
\int \boldsymbol{T}_a \cdot \boldsymbol{T}_b \, \mathrm{d}\Omega &= \int T_a L^2 T_b \, \mathrm{d}\Omega \\
\int \boldsymbol{T}_a \cdot \boldsymbol{S}_b \, \mathrm{d}\Omega &= 0 \qquad (8.12) \\
\int \boldsymbol{S}_a \cdot \boldsymbol{S}_b \, \mathrm{d}\Omega &= \int \nabla(rS_a) \cdot \nabla(rL^2 S_b) \frac{\mathrm{d}\Omega}{r^2}
\end{aligned}$$

for all a and b. $\mathrm{d}\Omega$ is the increment of solid angle ($\mathrm{d}\cos\theta\,\mathrm{d}\phi$), and the operator L^2 is defined by:

$$L^2 = -\left[\frac{1}{\sin\theta}\frac{\partial}{\partial\theta}\left(\sin\frac{\partial}{\partial\theta}\right) + \frac{1}{\sin^2\theta}\frac{\partial^2}{\partial\phi^2}\right]$$

The vector $\boldsymbol{T}$ has no radial component and is called a *toroidal vector*. $\boldsymbol{S}$ is called a poloidal vector in geomagnetic theory and a spheroidal vector everywhere else (for example, spheroidal modes in eigenvibrations in seismology). Thus a magnetic field represented by $\boldsymbol{T}$ is referred to as a *toroidal magnetic field* and one represented by $\boldsymbol{S}$ is referred to as a *poloidal* (spheroidal) *field*. Since $\boldsymbol{B}$ is always divergence free ($\nabla \cdot \boldsymbol{B} = 0$), it can always be represented by a toroidal and poloidal magnetic field.

8.2 Kinematic Dynamos

8.2.1 Toroidal and Poloidal Fields

To gain more insight into the toroidal magnetic field the special case of a velocity field in cylindrical co-ordinates (R, θ, z) given by:

$$\boldsymbol{v} = (0, R\omega, 0)$$

will now be considered. For illustrative purposes it is assumed that the conductivity is sufficiently high that the decay term in the magnetic induction equation can be neglected. Then the rate of build-up of the toroidal magnetic field is given by the magnetic induction equation:

$$\frac{\partial H_\theta}{\partial t} = R\left[H_R\frac{\partial\omega}{\partial R} + H_z\frac{\partial\omega}{\partial z}\right] - \omega\frac{\partial H_\theta}{\partial\theta} \qquad (8.13)$$

The subscripts of H refer to the appropriate components of H in the R, θ and

z directions. Note that H_θ is a toroidal field, since it has no R component. Symmetry dictates that the above magnetic field can be represented in spherical co-ordinates (r, θ, ϕ), by the toroidal vector: $(0, 0, T_\phi)$. Equation 8.13 shows how a toroidal magnetic field can be created from shear of a poloidal magnetic field ($\partial\omega/\partial R$ and $\partial\omega/\partial z$ terms) or rotation associated with a magnetic field that varies as a function of θ ($\partial H_\theta/\partial\theta$ term). This equation also allows one to calculate the rate of build-up of the toroidal magnetic field, a rate that can be relatively high (depending on the assumed shear, rotation rate, magnetic field strength, and field inhomogeneity).

Additional insight into toroidal and poloidal magnetic fields can be gained from Fig. 8.1, which gives representations of a few simple toroidal and poloidal fields. The following "rules" can easily be shown to be valid.

(i) for S_l^0, the number of surface latitudinal nodal lines is given by l.
(ii) for S_l^m, the number of longitudinal nodal lines is $l - m$.
(iii) for T_l^0, the number of latitudinal nodal lines is given by $l - 1$.

Therefore, an axial geocentric dipole can be represented by S_1^0. S_2^0 represents an axial geocentric quadrupole, etc.

Because a toroidal magnetic field in the earth's core has no radial component, if any part of a field line is in the core, the entire field line will be in the spherical core. Figure 8.2 helps show why no toroidal field could exist in the earth's mantle, providing the mantle is assumed to be an insulator. This figure shows a toroidal field loop that is entirely in the mantle. Since $\nabla \times \boldsymbol{H} = \boldsymbol{J}$, a current must pass vertically through this loop, thus contradicting the assumption that the mantle is an insulator, so that no toroidal loop is possible there. A similar, but more complicated explanation, can be made for a mantle toroidal loop that lies in the plane of the equator. If the mantle is partially conducting, then currents can exist there as can a toroidal field. The presence of toroidal fields in the partially conducting mantle can be imagined as originating through outward diffusion of a toroidal field produced in the core. However, the conductivity and permeability of the mantle are such that virtually no toroidal field should be manifested at the earth's surface. Thus the core's toroidal field must be inferred from theoretical models and its magnitude is very controversial (§8.3.4).

8.2.2 Bullard–Gellman Models

Bullard and Gellman (1954) presented a brute-force kinematic dynamo model that had pronounced effects on the geomagnetic community, in part because it was the first model that produced actual numerical values. A valuable review and analysis of this model has been given by Gibson and P.

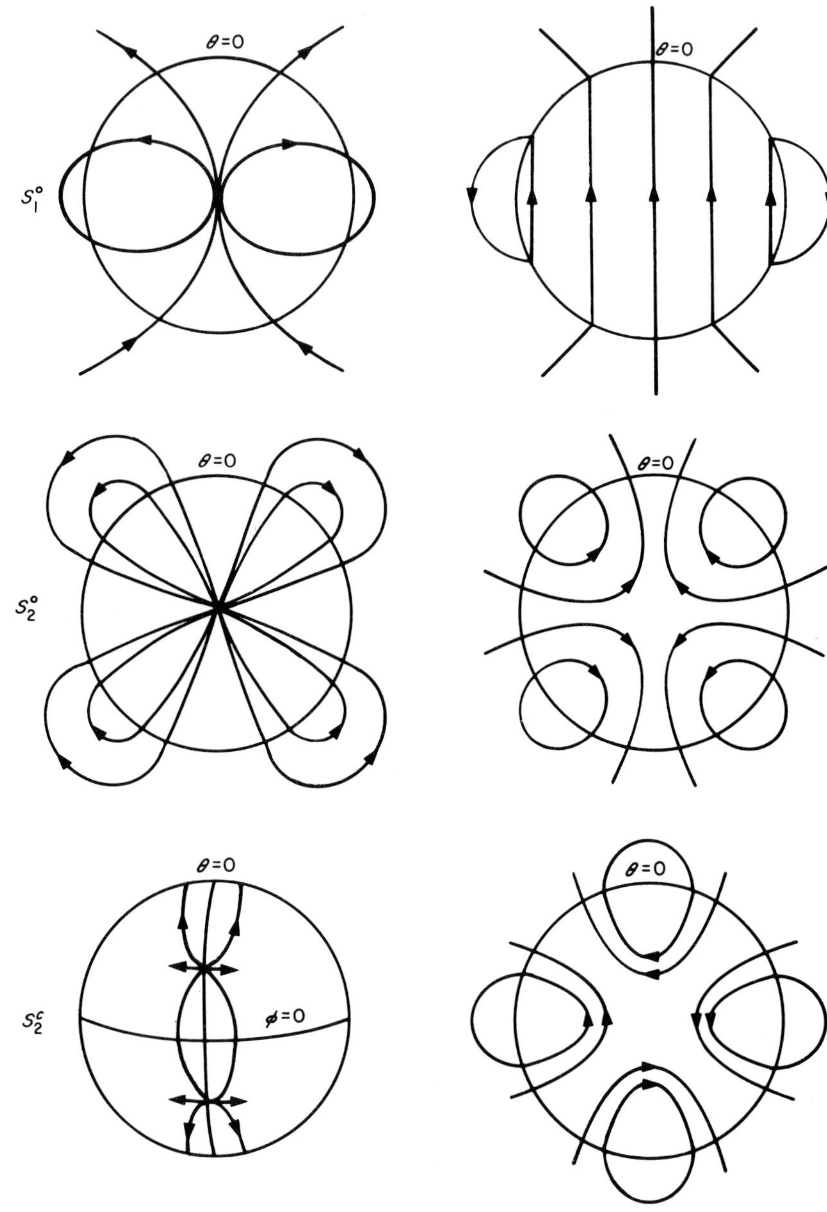

Fig. 8.1. Examples of (a) poloidal fields and (b) toroidal fields. Superscripts c and s respectively stand for cosine and sine and indicate that part of the ϕ component in the surface harmonic ψ_l^m that is retained. (After Bullard, 1949a.)

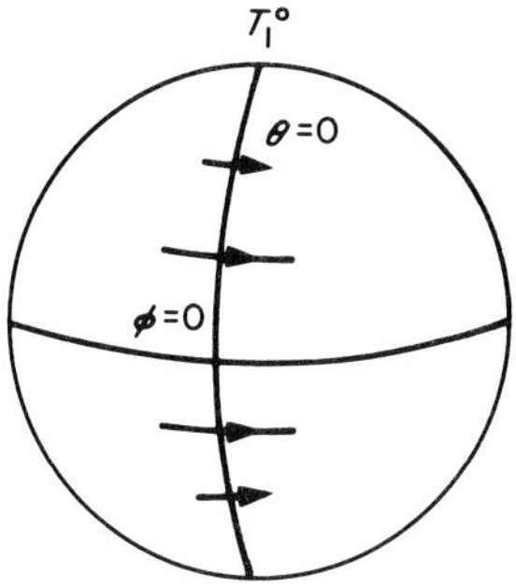

$\psi_1 = f(\gamma)\cos\theta$
$T_\gamma = 0$
$T_\theta = 0$
$T_\phi = R\sin\theta$

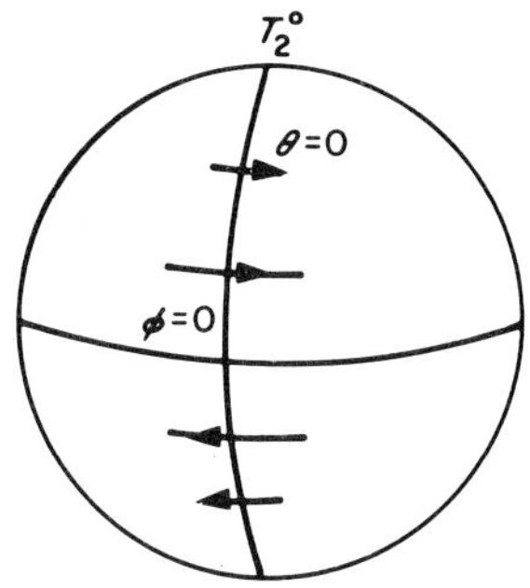

$\psi_2 = f(\gamma)(3\cos^2\theta - 1)$
$T_\gamma = 0$
$T_\theta = 0$
$T_\phi = \frac{3}{2}R\sin 2\theta$

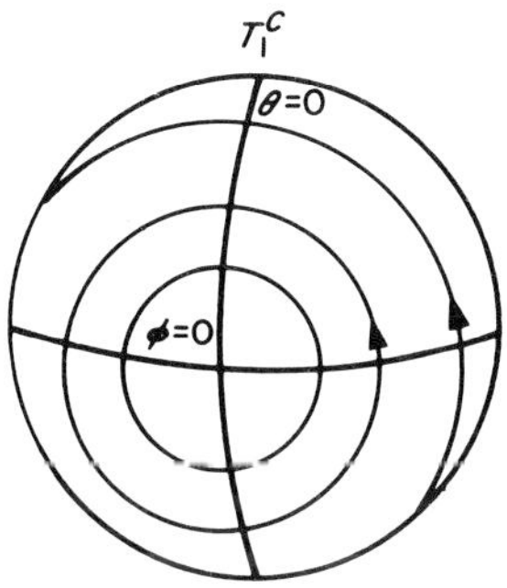

$\psi_1^c = f(\gamma)\sin\theta\cos\phi$
$T_\gamma = 0$
$T_\theta = R\sin\phi$
$T_\phi = R\cos\theta\cos\phi$

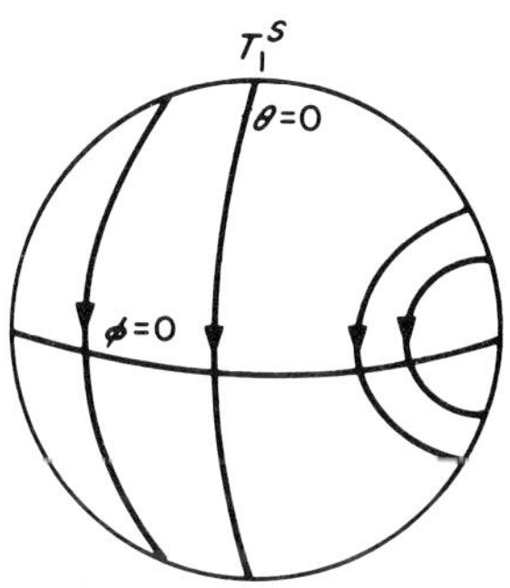

$\psi_1^s = f(\gamma)\sin\theta\sin\phi$
$T_\gamma = 0$
$T_\theta = R\cos\phi$
$T_\phi = -R\cos\theta\sin\phi$

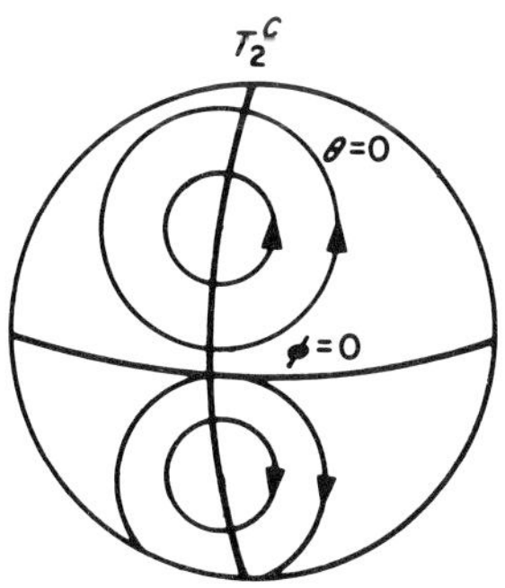

$\psi_2^c = \frac{3}{4}f(\gamma)\sin 2\theta\cos\phi$
$T_\gamma = 0$
$T_\theta = -\frac{3}{2}R\cos\theta\sin\phi$
$T_\phi = -\frac{3}{2}R\cos 2\theta\cos\phi$

(b)

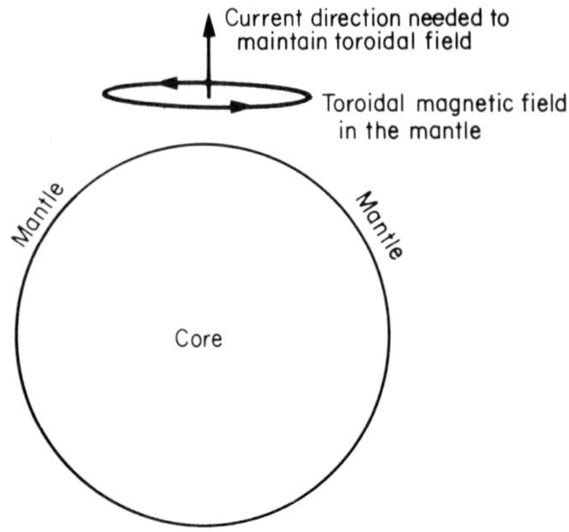

Fig. 8.2. A toroidal magnetic field is assumed to exist in the earth's electrically insulating mantle. The current needed to maintain this field must pass through the toroidal loop, thus contradicting the assumption of an insulating mantle.

Roberts (1969). The approach taken by Bullard and Gellman was to convert the magnetic induction equation to an eigenvalue equation by the following transformations:

initial length scale = new length scale × a

Therefore, $\nabla_{\text{initial}} = a^2 \nabla_{\text{new}}$

Similarly, $t_{\text{initial}} \rightarrow \sigma\mu_0 a^2 t_{\text{new}}$

Let $\boldsymbol{v}$ be represented by $U\boldsymbol{u}$. Then using the above, $\boldsymbol{v}_{\text{initial}} = U\boldsymbol{u}/\sigma\mu_0$. The magnetic induction equation can be rewritten:

$$\frac{\partial \boldsymbol{B}}{\partial t} = \nabla^2 \boldsymbol{B} + U\nabla \times (\boldsymbol{u} \times \boldsymbol{B}) \tag{8.14}$$

For a given set of boundary conditions, (8.14) will have solutions for particular values (eigenvalues) of U, the magnitude of the velocity field. Because the Laplacian has mixed derivatives in spherical co-ordinates, it is convenient to use the following form of the magnetic induction equation:

$$\frac{\partial \boldsymbol{B}}{\partial t} = U\nabla \times (\boldsymbol{u} \times \boldsymbol{B}) - \nabla \times (\nabla \times \boldsymbol{B}) \tag{8.15}$$

To solve this equation, Bullard and Gellman expanded both the velocity field and the magnetic field in poloidal and toroidal vectors:

$$\boldsymbol{u} = \sum_{\alpha} (\boldsymbol{s}_\alpha + \boldsymbol{t}_\alpha) \tag{8.16}$$

and

$$\boldsymbol{B} = \sum_{\beta} (\boldsymbol{S}_\beta + \boldsymbol{T}_\beta) \tag{8.17}$$

where the short-hand notation of Gibson and P. Roberts (1969) which has been used $\left(\sum_{\alpha}, \sum_{\beta}, \text{etc.}\right)$ is a notation for summation of the appropriate vector spherical harmonics, y_l^m, over $\sum_{l=-\infty}^{\infty} \sum_{m=-l}^{l}$. Substituting (8.16) and (8.17) into (8.15) yields:

$$\sum_{\gamma} \left[\frac{\partial \boldsymbol{S}_\gamma}{\partial t} + \frac{\partial \boldsymbol{T}_\gamma}{\partial t}\right] = U \sum_{\alpha, \beta} \nabla \times [(\boldsymbol{s}_\alpha + \boldsymbol{t}_\alpha) \times (\boldsymbol{S}_\beta + \boldsymbol{T}_\beta)] - \sum_{\gamma} \nabla \times [\nabla \times (\boldsymbol{S}_\gamma + \boldsymbol{T}_\gamma)] \tag{8.18}$$

where γ is yet another dummy index (that is, the summation is over "γ").

Now take the scalar product of $\boldsymbol{S}_\gamma$ with the terms in (8.18) and integrate over the solid angle to eliminate the θ and ϕ dependence and to make use of the orthogonality properties of the spheroidal and toroidal vectors (see 8.12). This yields:

$$\left[\nabla^2 - \frac{\partial}{\partial t}\right] \boldsymbol{S}_\gamma = U \sum_{\alpha, \beta} \{\boldsymbol{s}_\alpha \boldsymbol{S}_\beta \boldsymbol{S}_\gamma\} + \{\boldsymbol{t}_\alpha \boldsymbol{S}_\beta \boldsymbol{S}_\gamma\} + \{\boldsymbol{s}_\alpha \boldsymbol{T}_\beta \boldsymbol{T}_\gamma\} \tag{8.19}$$

where

$$\{\boldsymbol{s}_\alpha \boldsymbol{S}_\beta \boldsymbol{S}_\gamma\} \equiv \iint \frac{y_l^m \boldsymbol{r} \cdot [\nabla \cdot (\boldsymbol{s}_\alpha \times \boldsymbol{s}_\beta)] \, d\Omega}{l_\gamma (l_\gamma + 1)(y_l^m)^2 \, d\Omega} \tag{8.20}$$

and

$$\nabla^2 \boldsymbol{S}_\gamma = \left[\frac{\partial}{\partial r^2} + \frac{2}{r}\frac{\partial}{\partial r} - l_\gamma \frac{(l_\gamma + 1)}{r^2}\right] \boldsymbol{S}_\gamma$$

Equations similar to (8.20) exist for $\{\boldsymbol{t}_\alpha \boldsymbol{S}_\beta \boldsymbol{S}_\gamma\}$ and $\{\boldsymbol{s}_\alpha \boldsymbol{T}_\beta \boldsymbol{T}_\gamma\}$. In addition, similar equations can be generated for $\boldsymbol{t}_\alpha$, by taking the scalar product of $\boldsymbol{T}_\gamma$ with (8.18) and integrating.

The basic procedure used by Bullard and Gellman (1954), as has been followed by several others (see review by Gubbins, 1974), is to assume some initial magnetic and velocity field and to substitute these into the above equations. Bullard and Gellman assumed an initial dipole field, S_1^0, and initial toroidal velocity field, T_1^0 (uniform shear), and poloidal velocity field, S_2^{2C}. (C and S as superscripts refer to the cosine or sine part of the harmonic respectively—see Fig. 8.1.) They looked for steady-state solutions $\partial \boldsymbol{S}_\gamma/\partial t = \partial \boldsymbol{T}_\gamma/\partial t = 0$ in (8.19). However, because of terms like $\{\boldsymbol{s}_\alpha \boldsymbol{S}_\beta \boldsymbol{S}_\gamma\}$ in (8.19), this procedure leads to an infinite series of toroidal and poloidal vectors. That is, the interaction of even very simple velocity fields and magnetic fields leads to the generation of magnetic and velocity fields that are represented by an infinitely large number of toroidal and poloidal terms. In particular, the assumption made by Bullard and Gellman generates terms like T_2^0, T_2^{2C}, T_2^{2S}, S_3^0, S_3^{2C}, S_3^{2S}, etc. They truncated the series at $l = 12$ to obtain a finite system that could be solved numerically using a computer. For $U = 2 \times 10^{-4}$ m s^{-1} (a velocity at the outer core–mantle boundary that is comparable to the westward drift velocity of the non-dipole field, 0.2°/year), and for S_1^0 at the earth's surface around 10^{-4} T (comparable to the present field intensity) they found that the lowest order toroidal magnetic field, T_2^0, has an order of magnitude value of 10^{-2} T. It should be noted that the toroidal field in this model contains most of the magnetic energy. Because the toroidal field cannot be observed directly (§8.2.1), the Bullard-Gellman estimate for the toroidal field magnitude has been widely quoted.

The question of convergence of the infinite series used in the expansion by Bullard and Gellman was not examined adequately until Gibson and P. Roberts (1969) showed that it does not converge. Therefore, the numerical results of Bullard and Gellman may be radically wrong. More complicated initial flows have been assumed to circumvent this problem. An instructional dynamo model by Lilley (1970a) demonstrated that the initial velocity field $\boldsymbol{T}_1^0 + \boldsymbol{S}_2^{2C} \rightarrow \boldsymbol{S}_2^{2S}$ produces much more rapid convergence. However, even in this case the convergence still does not appear to be rapid enough (Gubbins, 1974). The most likely reasons for this lack of convergence are discussed in §8.2.3.

Backus (1958) suggested one way to modify the above approach to escape from the convergence problem by noticing that convergence would occur if the velocity field were time-dependent. In particular, allowing the velocity field to be zero for long time periods eliminated the effects of the higher-order harmonics. This follows from §7.4.2, where it was shown that higher-order harmonics, which are associated with shorter wavelengths, decay more rapidly than lower-order harmonics. One could simply "turn off" the velocity field for as long as needed to allow the higher-order harmonics to become negligible. In this way the troublesome convergence problem could be eliminated, but at the expense of retaining terms like $\partial S_\gamma/\partial t$ in (8.18).

8.2.3 Necessary Conditions

Cowling (1934, 1957) argued that a steady poloidal magnetic field with an axis of symmetry cannot be maintained by motion symmetric about that axis. Subsequently numerous efforts have been expended on generalizing and examining the rigour of *Cowling's Theorem* (e.g. Backus and Chandrasekhar, 1956; Backus, 1957; Braginskiy, 1964a, b; James *et al.*, 1980; Hide, 1981; Hide and Palmer, 1982). Backus (1957) was able to generalize the Cowling Theorem to show that non-steady axisymmetric magnetic fields cannot be maintained by symmetric fluid motions. A complete proof of Cowling's Theorem using a Bullard-Gellman type formulation was proposed by James *et al.* (1980), in which the fluid was assumed to be incompressible with no sources or sinks and its velocity to be bounded. Hide (1981) showed that the Theorem is still applicable when thermoelectric and thermomagnetic effects are present, casting doubt on Hibberd's (1979) use of the Nernst-Ettinghauser effect in generating the earth's magnetic field. Hide and Palmer (1982) have recently argued that Cowling's Theorem is applicable even when the fluid is compressible.

Only a heuristic argument will be given here to illustrate the validity of the Cowling Theorem (Cowling, 1957). Assume there are axisymmetric magnetic fields such as shown in Fig. 8.3. At the two limiting points, 0_1 and 0_2 (Fig. 8.3), the magnetic field is zero, but an electric current exists, since $\nabla \times \boldsymbol{H} = \boldsymbol{J} \neq \mathbf{0}$. The current at 0_1 and 0_2 flows around the symmetry axis and therefore cannot be maintained by electrostatic forces. Also, the magnetic field is zero at 0_1 and 0_2, so the currents cannot be maintained by the magnetic field. Therefore, the currents must decay with time by Ohmic dissipation, and likewise so must their associated magnetic fields.

In addition to Cowling's Theorem, a necessary condition for positive dynamo action is that the magnetic Reynolds number be greater than one

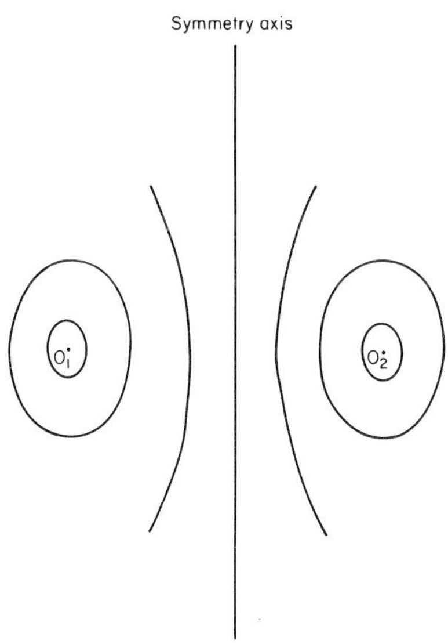

Fig. 8.3. An axisymmetric magnetic field is illustrated. The curl of this field is finite at the two limiting points 0_1 and 0_2 but the field itself is zero there. This leads to contradictions (§8.2.3) and was one of the first "proofs" of Cowling's theorem. (After Cowling, 1957.)

(§7.4.2). It can also be shown by a development similar to that given in §7.4.2 that for the build-up term to dominate the decay term in the magnetic induction equation $\lambda_M L^2/k_m$ must exceed unity, where L is the length scale involved and λ_M is the maximum eigenvalue for velocity shear (Childress, 1978). Still another necessary condition for dynamo action can be obtained by writing the field in terms of its toroidal and poloidal parts (Busse, 1975):

$$\max(\boldsymbol{v}\cdot\boldsymbol{r}) > \left[\frac{2E_s}{E_m}\right]^{1/2} k_m$$

where E_m is the total magnetic field energy, E_s is the magnetic energy in the poloidal magnetic field and k_m is the magnetic diffusivity. This last condition implies, among other things, the physically appealing result that, in steady-state conditions, a poloidal magnetic field in the core requires that there be some radial velocity motion.

Finally, there is the intriguing question of the role of helicity in dynamo theories. Although it can be shown that no dynamo action can occur unless the volume integral of helicity over the earth's core vanishes (i.e. mean helicity is zero), Ruzmaikin and Sokoloff (1980) have also shown that no dynamo action is possible if the helicity vanishes everywhere. So helicity and its distribution are important components of viable dynamos. Considerably more work on the necessary conditions associated with helicity will probably occur in the future, since helicity is a critical ingredient of mean field electrodynamic approaches to dynamo theory. Other necessary conditions are required for particular classes of dynamo models, such as illustrated in §7.4.2 in which the dynamo number N of Parker-Levy dynamos must exceed zero.

Following the development by Cowling, opponents of dynamo models expended a considerable effort to find a general anti-dynamo theorem. Proponents initially called on the Coriolis force to produce the needed asymmetry, but until the last decade they were unable to demonstrate the existence of a single viable kinematic dynamo. Some psychological support for kinematic dynamos came from analogue models, particularly from the work of Lowes and Wilkinson (1963) and Rikitake (1966). However, these models were not totally convincing, because of scaling and other problems (§7.4). Pioneering work by Childress (1969, 1970) and G. Roberts (1969, 1970, 1972) on the existence of kinematic dynamos gave a strong boost to dynamo theory. In particular, G. Roberts (1970) proved the important theorem that almost all spatially periodic motions give rise to kinematic dynamos in unbounded (infinite size) conductors. ("Almost all" means that the number of spatially periodic motions that do not give rise to dynamo action is countable.) This does not assure one that the earth has a dynamo, but it dispells doubt that any general mathematical anti-dynamo proof exists.

Subsequently Soward (1975) has given an excellent, but mathematically involved, discussion on how random waves can produce dynamo action. An important part of the proof by G. Roberts (1970) is to demonstrate that large-scale magnetic fields can be maintained in a system in which all fluid motions are periodic with small-scale wavelengths ("small" relative to the scale of the magnetic field). The proof of this is essentially the same as that for Bloch's Theorem dealing with electrons in a solid with a periodic potential (see Ashcroft and Merman, 1976). This result is important not only for the existence of kinematic dynamos but it also provides insight into turbulent dynamos (§8.3).

The one remaining problem treated here concerns the numerical modelling problem discussed in §8.2.2. The question can be asked that if almost all periodic motions give rise to kinematic dynamos, why is there a problem in finding one that works for the earth? This question has been investigated by Bullard and Gubbins (1977) and it appears that the answer may have to do with the geometry assumed. The existence theorems usually treat periodic motions in an *infinite medium*, while the generation of the earth's magnetic field is confined to a finite region in the earth's outer core. Bullard and Gubbins (1977) consider two cases, a conducting fluid core bounded by an insulating mantle and a conducting fluid core bounded by a perfectly conducting mantle. Here only the first case will be discussed, in which no current can flow out of the core. This causes a concentration of current near the boundary, a concentration which can have a catastrophic effect on the nature of the magnetic field in the core. Note that this concentration of current does not occur in the infinite dynamos, because nowhere is there an impediment to the current flowing between the various small-scale cells in which the velocity field is periodic. Bullard and Gubbins (1977) argue that Ohmic dissipation, which depends on the square of the current density, can even be sufficient to inhibit dynamo action in some cases. In particular, the existence of large T_2^0-type magnetic fields over very long times, such as utilized in §7.5.3 might be impossible.

The results of G. Roberts and Childress are still encouraging because when a large number of convection cells are present, the magnetic field in the interior will not be substantially altered by the concentration of current near the core's boundary. Of course, this will depend on the precise geometry of the boundary current. In any case, it appears that the core's magnetic field might be more complex than has been assumed in many numerical models. This suggestion provides a motivation for considering the possibility of turbulent dynamos.

8.3 Turbulent and Hydromagnetic Dynamos

8.3.1 Mean Field Electrodynamics

The evidence of Bullard and Gubbins (1977) is consistent with the earlier results of G. Roberts (1970, 1972) that several fluid eddies are probably required in the earth's core for a dynamo to be possible. If sufficiently large numbers of eddies are present, then correlations in the velocity field over moderate dimensions may be negligible and the statistical approaches used in mean field electrodynamics become desirable.

Turbulence is best treated by a statistical approach in which one considers some mean field value $\langle \boldsymbol{F} \rangle$ and some fluctuating value $\boldsymbol{F}' \equiv \boldsymbol{F} - \langle \boldsymbol{F} \rangle$, where $\boldsymbol{F}$ denotes whatever field is being considered (e.g. the velocity and magnetic fields.) Clearly:

$$\langle \boldsymbol{F}' \rangle = \mathbf{0}; \qquad \langle \boldsymbol{F} + \boldsymbol{G} \rangle = \langle \boldsymbol{F} \rangle + \langle \boldsymbol{G} \rangle; \quad \langle \langle \boldsymbol{F} \rangle \cdot \boldsymbol{G}' \rangle = 0.$$

However, $\langle \boldsymbol{F}' \cdot \boldsymbol{G}' \rangle$ is not necessarily zero, but will be zero if no correlation exists between $\boldsymbol{F}'$ and $\boldsymbol{G}'$, that is, if $\boldsymbol{F}'$ and $\boldsymbol{G}'$ are statistically independent variables. However, if $\boldsymbol{F}'$ and $\boldsymbol{G}'$ are correlated, then $\langle \boldsymbol{F}' \cdot \boldsymbol{G}' \rangle$ need not vanish.

The average can be taken over space or time or both. If the averages do not depend on time, the turbulence is said to be *steady*. When the turbulence does not depend on space, the turbulence is said to be *homogeneous*. The turbulence is *isotropic* if it is independent of the orientation of the co-ordinate system used, otherwise it is *anisotropic*.

The Einstein summation convention will be used in the following to reduce the equations and terms to a manageable level: summation occurs over repeated roman subscripts. In addition, δ_{ij} is the Kronecker tensor and ε_{ijk} is the permutation tensor in which $\varepsilon_{123} = 1$ for a right-handed co-ordinate system. Thus $\boldsymbol{v} \times \boldsymbol{B}$ in tensor notation is written $v_i B_j \varepsilon_{ijk}$.

$\langle \boldsymbol{v}' \times \boldsymbol{B}' \rangle$ defined in §7.5.2 will now be investigated under several simplifying assumptions to illustrate some of the concepts associated with mean field electrodynamics. First assume that $\boldsymbol{B}$ is independent of time and only weakly dependent on position, so that $\langle \boldsymbol{v}' \times \boldsymbol{B}' \rangle$ (in tensor notation) can be approximated by the Taylor expansion:

$$\langle v_j' B_k' \varepsilon_{ijk} \rangle \approx a_{ij} \langle B_j \rangle + b_{ijk} \frac{\partial \langle B_j \rangle}{\partial x_k} \tag{8.21}$$

where the higher-order terms are neglected. The second-order tensor a_{ij} and the third-order tensor b_{ijk} are functionals of $\langle \boldsymbol{v} \rangle$ and $\langle \boldsymbol{v}' \rangle$.

The properties of a_{ij} and b_{ijk} will be obtained only for some simple cases. Suppose $\langle \boldsymbol{v} \rangle = 0$, then a_{ij} and b_{ij} depend only on $\boldsymbol{v}'$. Further assume there is a *homogeneous* turbulent field, so that fluid averages are independent of position. It follows that a_{ij} and b_{ijk} are constant in space. If $\boldsymbol{v}'$ is isotropic, then a_{ij} and b_{ijk} must be invariant under any rotation about the axes (x_1, x_2, x_3) of a Cartesian co-ordinate system. With such assumptions, it is relatively easy to show that all second- and third-order isotropic tensors are linearly proportional to δ_{ij} and ε_{ijk}, respectively; i.e.

$$a_{ij} = \alpha\delta_{ij}$$

$$b_{ijk} = \beta\varepsilon_{ijk}$$

where α and β are scalar functionals of $\boldsymbol{v}'$ that are independent of position. The above two equations can be referred to respectively as the α- and β-effects. Both the α- and β-effects can be thought of as in terms of an effective electric field (or EMF), as seen from (8.21). Also note that if the conditions assumed above are relaxed (less symmetry), then the problem is far more complicated, since a_{ij} and b_{ijk} cannot generally be linearly related to δ_{ij} and ε_{ijk}. In this case several "induced EMFs occur" and the simple physical insight given in Chapter 7 on the α-effect requires substantial modification. More general relations have been discussed by P. Roberts and Stix (1972) and Moffat (1978). Note further that for each higher-order term retained in (8.21), there is, effectively, a "new effect". Generally, more rapid decay of higher-order features in the magnetic field (i.e. features in which $\boldsymbol{B}$ varies more rapidly in space) is expected (see §7.4) relative to lower-order features, and it is usually assumed that only the first few terms in (8.21) need be retained.

In the oversimplified case of isotropic turbulence considered above, (8.21) in vector notation can be reduced to:

$$\langle \boldsymbol{v}' \times \boldsymbol{B}' \rangle = \alpha\langle \boldsymbol{B} \rangle - \beta\nabla \times \langle \boldsymbol{B} \rangle \tag{8.22}$$

where mirror symmetry has also been assumed. This implies the flow possesses non-zero helicity, $\boldsymbol{v} \cdot [\nabla \times \boldsymbol{v}] \neq 0$. Equation 8.22 explicitly shows the α- and β-effects of mean field electrodynamic dynamos. Contrary to that often assumed, both the α- and β-effects can contribute to dynamo action. Hereafter only simple examples of the α-effect will be considered.

8.3.2 α^2 Dynamos

The heuristic dynamo described in §7.5.3 is an example of an $\alpha\omega$ dynamo. A large shear in the fluid is used to generate the toroidal magnetic field and the α-effect is called on to convert this toroidal field back to a poloidal field. In α^2 dynamo models, both the toroidal and poloidal magnetic fields can be obtained from α-effects. Two examples of α^2 dynamos are given here to illustrate the principles and potential of mean field electrodynamic approaches to dynamo theory. The notation of §7.5.2 is used throughout.

Example 1

This follows that given by Childress (1978) who begins by using a first-order smoothed version of the magnetic induction equation (see discussion in §7.5.2):

$$\frac{\partial \boldsymbol{B}'}{\partial t} - k_m \nabla^2 \boldsymbol{B}' = \nabla \times (\boldsymbol{v} \times \boldsymbol{B}_0) = (\boldsymbol{B}_0 \cdot \nabla)\boldsymbol{v}' - (\boldsymbol{v}' \cdot \nabla)\boldsymbol{B}_0 \tag{8.23}$$

Because v' can be assumed to vary spatially much more rapidly than B_0, the last term can be omitted from further discussion.

Now consider a periodic velocity field in Cartesian co-ordinates:

$$\boldsymbol{v}' \equiv u(0, \cos Kx, \sin Kx)$$

Hence from (8.23):

$$\frac{\partial \boldsymbol{B}'}{\partial t} - k_m \nabla^2 \boldsymbol{B}' = \boldsymbol{B}_{0x} Ku\ (0, -\sin Kx, \cos Kx)$$

where B_{0x} is the x component of the mean field $\boldsymbol{B}_0$. A steady-state solution of this last equation for the particular case that $\boldsymbol{B}_0$ is constant is:

$$\boldsymbol{B}' = B_{0x} \frac{u}{k_m K} (0, \sin Kx, -\cos Kx)$$

In this case the effective electric field derived from the fluctuating parts (eqn. 7.24) is:

$$\langle \boldsymbol{v}' \times \boldsymbol{B}' \rangle = \left(B_{0x} \frac{u^2}{k_m K}, 0, 0 \right) = \bar{\boldsymbol{\alpha}} \cdot \boldsymbol{B}_0$$

where

$$\bar{\boldsymbol{\alpha}} = \frac{u^2}{k_m K} \begin{pmatrix} 1 & 0 & 0 \\ 0 & 0 & 0 \\ 0 & 0 & 0 \end{pmatrix}$$

Note if $\boldsymbol{B}_0$ is in the x direction, then the electric current density, $\boldsymbol{J}_0$, produced would also be in the x direction. However, this current does not work to magnify B_{0x}. This simple α-effect does not produce dynamo action and thus the magnetic field would decay away! In §7.5.2 this problem of reinforcement was bypassed by putting boundaries into the problem. Here a more satisfactory solution is obtained by examining alternative velocity fields.

The fluid velocity used above is periodic, but does not possess helicity. A simple velocity field that does possess helicity is:

$$\boldsymbol{v}' = u(\sin Ky, \cos Kx, \sin Kx + \cos Ky)$$

Substitution into (8.23) with similar assumptions yields:

$$\langle \boldsymbol{v}' \times \boldsymbol{B}' \rangle \equiv \bar{\boldsymbol{\alpha}} \cdot \boldsymbol{B}_0$$

as before, but here,

$$\bar{\boldsymbol{\alpha}} = \frac{u^2}{k_m K}\begin{pmatrix} 1 & 0 & 0 \\ 0 & 1 & 0 \\ 0 & 0 & 0 \end{pmatrix}$$

α_{11} interacts with $\boldsymbol{B}_{0x}$ constant field to produce a current J_{0x}, while α_{22} interacts with a $\boldsymbol{B}_{0y}$ constant field to produce J_{0y} currents. Combining J_{0x} with $\boldsymbol{B}_{0y}$ and J_{0y} with $\boldsymbol{B}_{0x}$ makes dynamo action possible in this case. This kind of interaction is called the α^2-effect. Obviously, far more complex forms of α_{ij} are possible.

Example 2

An example given by Krause and Steenbeck (1967) and repeated by P. Roberts (1971), provides another illustration of an α^2 dynamo. Here α is assumed to be (unrealistically) constant. Imagine a conducting sphere of radius a, bounded by an insulator. The fluid inside the sphere is homogeneous, isotropic, and turbulent with constant helicity. Under these conditions, the mean field magnetic induction equation is:

$$\frac{\partial \boldsymbol{B}}{\partial t} = k_m \nabla^2 \boldsymbol{B} + \nabla \times (\alpha \boldsymbol{B}) \tag{8.24}$$

Note that although the instantaneous motions may be far more complex than in the Bullard-Gellman dynamo (§8.2.2), the magnetic induction equation is easier to solve, since the $\boldsymbol{v} \times \boldsymbol{B}$ term has been substituted by $\alpha\boldsymbol{B}$ (a *mean* field value). $\boldsymbol{B}$ is separated into its toroidal and poloidal magnetic fields by using the vector potentials $\boldsymbol{A}_T$ and $\boldsymbol{A}_S$ respectively for the toroidal and poloidal fields; that is:

$$\boldsymbol{B} = \nabla \times \boldsymbol{A}_T + \nabla \times (\nabla \times \boldsymbol{A}_S) \tag{8.25}$$

Substitution into (8.24) gives:

$$\frac{\partial \boldsymbol{A}_T}{\partial t} = k_m \nabla^2 \boldsymbol{A}_S + \alpha \boldsymbol{A}_T \tag{8.26}$$

The boundary conditions are that the magnetic fields must vanish when r equals a and when r approaches infinity. Solutions of this boundary value problem can be obtained by substituting the appropriate spherical harmonic expansions, such as

$$\boldsymbol{A}_T = \sum_{l=1}^{\infty} \boldsymbol{A}_T(r) Y_l(\theta, \phi)(\exp - \sigma_l t) \tag{8.27}$$

into the above equations.

As in the Bullard-Gellman dynamo (§8.2.2), solutions involve linear combinations of the Bessel function, $\xi_{l+1/2}$. Application of the boundary conditions necessitates that solutions exist only for certain eigenvalues of σ_l. From (8.27) a dynamo mechanism can occur only if there is a real part to σ_l and that part is less than zero. For the case at hand, σ_l turns out to be real (P. Roberts, 1971) and the marginal case ($\sigma_l = 0$) is obtained when:

$$\xi_{3/2}(c) = 0 \tag{8.28}$$

where c is the dimensionless number:

$$c = \frac{a\alpha}{k_m}$$

To satisfy (8.28), $c = 4.49$ and thus an α^2 dynamo occurs for all $c > 4.49$ (P. Roberts, 1971; P. Roberts and Stix, 1971; Krause, 1977).

8.3.3 Dipole and Quadrupole Families

Most work in mean field electrodynamic models in which α is not taken to be constant has been restricted to solving the mean field magnetic induction equation in a sphere:

$$\frac{\partial\langle \boldsymbol{B}\rangle}{\partial t} = k_m \nabla\langle \boldsymbol{B}\rangle + \nabla \times [(\langle \boldsymbol{v}\rangle \times \langle \boldsymbol{B}\rangle) + \alpha\langle \boldsymbol{B}\rangle]$$

Meridionally dependent velocities often appear needed for steady solutions and to achieve this it is commonly assumed that α is given by a function like:

$$\alpha = \alpha_0 \cos\theta$$

P. Roberts (1971) points out that under these conditions the solutions separate into two classes. The first class, called *the dipole family*, is described by the poloidal magnetic models with odd-order spherical harmonics and the toroidal modes with even-order spherical harmonics. The second class, *the quadrupole family*, is characterized by poloidal magnetic modes with even-order harmonics and toroidal modes with odd order ones.

Insight into the origin of the dipole and quadrupole families is provided by Fig. 8.4. It shows that the simple toroidal velocity shear T_1^0 produces an antisymmetric toroidal magnetic field with respect to the equator when it interacts with a S_1^0 poloidal field, while it produces a symmetric toroidal magnetic field when it interacts with an S_2^0 field. Only the first terms are shown in Fig. 8.4 for illustrative purposes. (Recall that antisymmetric toroidal magnetic fields are described by even-order toroidal harmonics and vice versa.) With only a little reflection, it can be seen that the toroidal magnetic fields will always be antisymmetric for the simple shear considered, providing the subscript of the initial poloidal field is odd, while it will be symmetric if the subscript is even. This can then easily be extended to the consideration of more complicated symmetric or antisymmetric initial toroidal velocity fields (see Moffat, 1978). Also, initial poloidal velocity fields can be introduced into the problem without disrupting the independence of the dipole family from the quadrupole family. Insight as to why this last statement is true can be obtained by recalling that the magnetic induction equation is similar to one for the vorticity (eqn. 7.21).

Although the above might suggest that the independence of the dipole family and quadrupole family is a general property of dynamos, there are ways in which this independence can be disrupted. The first way was suggested by Yukutake (1977), who argues that the conditions for independence are not satisfied if the Lorentz force is large (introducing non-linearity into the problem). A second way of disrupting the independence of the families is to assume sufficient asymmetries in the boundary conditions. In

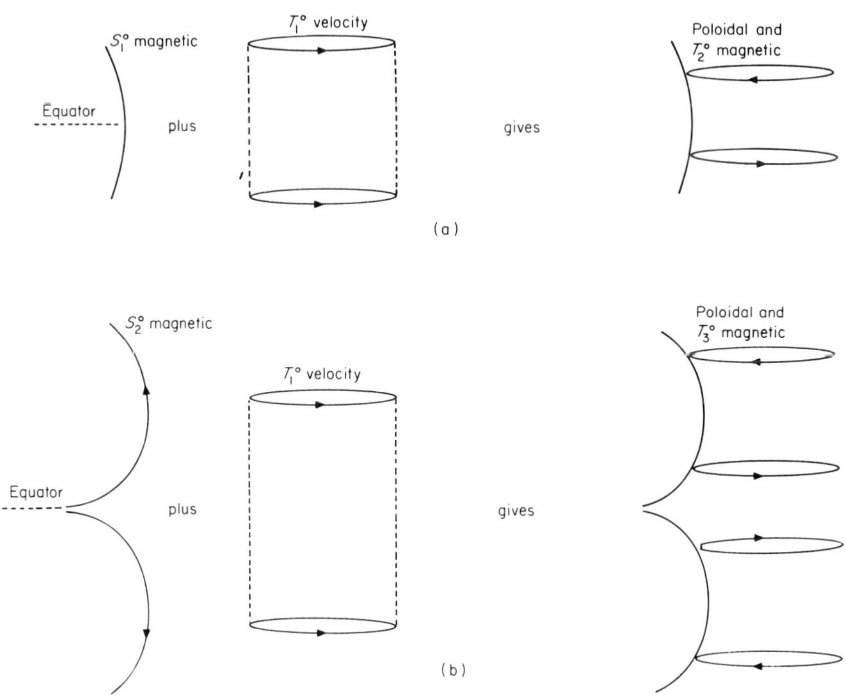

Fig. 8.4. Dipole and quadrupole families are generated by axially symmetric velocity fields. (a) An example of a dipole family, (b) an example of a quadrupole family. Notation for poloidal (S) and toroidal (T) fields is that given in §8.1.2.

any case, the palaeomagnetic evidence suggests that the strict independence between the two families does not occur. Most of the quadrupole poloidal field reverses sign when the dipole poloidal field reverses sign (§6.2).

8.3.4 Geostrophic Versus Magnetostrophic Balance

Busse (1970, 1975, 1977a, b, 1978) has constructed a hydromagnetic dynamo model for the earth and planets that has stimulated considerable discussion in geomagnetism. One of the important points to emerge from his work is the possibility that the toroidal magnetic field does not have to be larger than the poloidal magnetic field.

A major assumption in Busse's work is that the outer core is essentially in geostrophic balance. This means that the Coriolis force is balanced against the pressure force. All other forces, such as buoyancy, inertial, friction and Lorentz forces, are assumed only to perturb the basic geostrophic flow. This is a rather critical set of assumptions, any one of which could be wrong.

To achieve geostrophic balance, the Coriolis force is equated to the pressure gradient:

$$2\boldsymbol{\Omega} \times \boldsymbol{v} = -\frac{1}{\rho}\nabla P \tag{8.29}$$

where Ω is the angular velocity of rotation, P is pressure, and ρ is density. A right-handed co-ordinate system (x, y, z) is used, in which the z axis coincides with the rotation axis. In this case (8.29) can be rewritten in component form:

$$\frac{1}{\rho}\frac{\partial P}{\partial x} = 2\Omega v_y \tag{8.30}$$

$$\frac{1}{\rho}\frac{\partial P}{\partial y} = -2\Omega v_x \tag{8.31}$$

$$\frac{1}{\rho}\frac{\partial P}{\partial z} = 0 \tag{8.32}$$

Take $\partial/\partial y$ of (8.30), $\partial/\partial x$ of (8.31) and subtract the resulting (8.31) from the resulting (8.30) to obtain:

$$\frac{\partial v_y}{\partial y} + \frac{\partial v_y}{\partial x} = 0 \tag{8.33}$$

To conserve mass, $\partial v_z/\partial z = 0$. Equation 8.33 is a manifestation of the well-known Taylor-Proudman theorem of fluid mechanics, which states that the flow in the above situation is a steady one about the z-axis. No z dependence of the flow occurs, unless friction is inserted into the problem. The importance of friction can be described by an Ekman number, E, given

$$E = \frac{\nu}{\Omega L^2} \tag{8.34}$$

where ν is viscosity and L is a typical length scale in the system in the z direction. E is assumed to be small (remember that ν is one of the most poorly determined quantities in the core; §7.1), allowing Busse to model convection in the outer core by a cylindrical annulus. Convection in this case will thus be columnar with the column axes parallel to z (Fig. 8.5). Interestingly, these columns are not stationary, but drift about the rotation axis with time. This indicates how secular variation naturally falls out of hydromagnetic dynamo solutions (§9.1).

Using a no-slip velocity boundary condition ($v = 0$ at the boundary), an Ekman boundary layer will occur with thickness L given in (8.34), so that the thickness is proportional to $E^{1/2}$. In the boundary layer, the flow will be radially outward from the column's centre in the northern hemisphere for those columns having counterclockwise rotation (looking down, along the $-z$ direction) and radially inward for columns with clockwise rotation. This produces a secondary "Ekman" flow, the magnitude of which is crucial to the dynamo mechanism. Although the magnitude of the secondary flow can be adjusted through assumed changes in the value of E in (8.34), E cannot be taken too large, for then the cylindrical annulus approximation could not be used. This secondary Ekman flow produced both by the shape of the

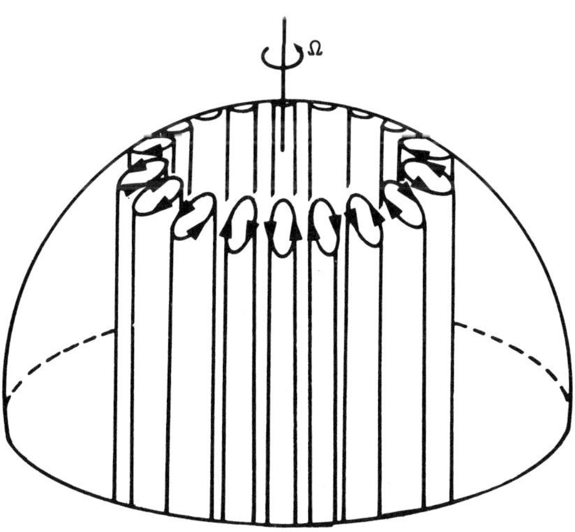

Fig. 8.5. Qualitative sketch of the onset of convection in a rotating fluid sphere following the experiments and results of Busse (1975). Convection occurs in columns parallel to the rotation axis.

boundary (spherical) and by friction, together with the primary geostrophic flow produces a net helical flow. Periodic helical velocity fields of sufficient magnitude and distribution lead to viable dynamo mechanisms (§7.5.2, §8.2.3, §8.3.2).

One of the major departures made by Busse in the above analysis is that the core is essentially in *geostrophic balance*, implying a balance between the pressure gradient and the Coriolis force. On the other hand, if the pressure gradient is balanced by both the Coriolis and magnetic force then the fluid is in *magnetostrophic balance*, mathematically given by:

$$2\rho\boldsymbol{\Omega} \times \boldsymbol{v} + \boldsymbol{J} \times \boldsymbol{B} = -\nabla P$$

The consequences arising from assuming geostrophic or magnetostrophic balance can be severe. This is seen in the following order of magnitude calculation in which the Coriolis force is balanced by the magnetic (Lorentz) force (*extreme magnetostrophic balance*):

$$|\boldsymbol{J} \times \boldsymbol{B}| \sim \frac{B^2}{\mu_0 l}$$

where l is taken as the dimension of the current loop that gives rise to the toroidal field. The Coriolis force gives us $\rho 2\Omega v$. Taking $l = 2\pi R_c$, where R_c is the core's radius and v as the westward drift velocity (4×10^{-4} m s^{-1}), a value for B (toroidal field) near 150×10^{-4} T is obtained. This value is close to that obtained in Bullard-Gellman type kinematic models (§8.2.2).

It is more difficult to obtain an estimate for the geostrophic balance case, but clearly the Lorentz force must be much smaller than that obtained above (by assumption!). For a very rough estimate, consider the Navier-Stokes equation (7.15) for an inviscid fluid. A geostrophic balance implies that the pressure term can be cancelled by the Coriolis term leaving:

$$\rho\left[\frac{\partial}{\partial t} + \boldsymbol{v}\cdot\nabla\right]\boldsymbol{v} = \boldsymbol{J} \times \boldsymbol{B} - \rho\nabla\phi_g$$

Ignoring the gravity term and assuming steady-state conditions, the approximate relationship obtained is:

$$|\boldsymbol{J} \times \boldsymbol{B}| \sim |\rho(\boldsymbol{v}\cdot\nabla)\boldsymbol{v}| \sim \frac{\rho v^2}{l} \tag{8.35}$$

Using $B^2/\mu_0 l$ as an estimate for the left-hand side of (8.35), a value for B (toroidal field) between 10^{-4} and 10^{-5} T is obtained. Thus the estimated toroidal magnetic field is 2 or 3 orders of magnitude smaller than analysed in the magnetostrophic balance case.

One of the consequences of Busse's work is that it appears the magnitude of the toroidal field is model dependent and not well known. Also, if Busse's model is right, then the magnitude of the toroidal field is about the same as the poloidal field. A second important finding is that no large net velocity shear in the outer core is apparent in his convection models. The consequence of this, if confirmed, is that the Bullard mechanism for westward drift is untenable (§9.1.2).

Olson (1977) has applied a "flux tube method" to carry out numerical estimates for the toroidal and poloidal fields. Olson finds that the secondary Ekman flow in Busse's model appears unable to produce a sufficient amount of helicity, a problem which might be solved by consideration of Lorentz force effects. Another objection to Busse's approach has been raised by Soward (1979), who argues that the Busse-type dynamos become unstable for reasonable size magnetic fields. The controversy over balance of terms (forces in the Navier-Stokes equation) is far from resolved. It is possible that this controversy will eventually be resolved by combining secular variation theory with observations (§9.1.6).

Chapter Nine

The Origin of Secular Variation and Field Reversals

9.1 Secular Variation

9.1.1 Introduction

A possible starting point for determining the changes in the core responsible for the observed secular variation is to downward continue the spherical harmonic representations of the earth's magnetic field to the core and to determine how the magnetic field in the core changes with time. Unfortunately, this approach has many pitfalls. Downward continuation is mathematically unstable and cannot be used to determine either the velocity or magnetic sources uniquely (§2.3). However, Backus (1968) was able to show that down continuation of the magnetic field could be used to determine null flux curves, which in turn could be used to determine some of the fluid motions in the core. *Null flux curves* are defined as locations (contours) on the earth's surface where the *radial* component of ***B*** is zero. Booker (1969) expanded on the non-uniqueness problem and defined much better which aspects of the core velocity field can be obtained from null flux curves. Even in an inviscid perfectly conducting core, only the component of the motion parallel to the core–mantle interface that is orthogonal to null flux curves will be reflected in changes of the geomagnetic field (Booker, 1969; Benton, 1979; Benton *et al.*, 1979). This is apparent physically, because any component of the velocity field parallel to a magnetic field line will not change the magnetic field in any manner. Thus, the constraints on theories of secular variation from direct and palaeomagnetic observations are few. Because of this, there are a wide variety of mechanisms proposed to explain secular variations, particularly the westward drift of the non-dipole field. The most complete of these are the explanations which emerge from the solutions of particular dynamo models (e.g. Busse, 1975; Krause, 1977). However, these usually depend on the details of the particular model solved and often do not provide much physical insight into the mechanism of drift. Probably more insight into westward drift can be obtained by the examination of two broad classes of westward drift models, referred to here as Bullard and hydromagnetic wave models of drift.

9.1.2 Bullard's Model

It is well known that an electromagnetic wave entering a conductor is damped to 1/e of its initial amplitude in a distance that is inversely proportional to the square root of the electrical conductivity. Because of this, Bullard *et al.* (1950) reasoned that the secular variation of the non-dipole field originates from magnetic eddies close to the core–mantle boundary. Magnetic eddies that lie deeper in the core could not produce the secular variation since they would be screened out by the outer part of the conducting core.

The westward drift of the non-dipole field was argued to originate because the outer part of the core rotated slower than the inner part. This shear in the rotation rate was believed to be the result of convection. Consider a core that is initially rotating uniformly, that is the angular velocity is constant at any R, where R is the distance from the rotation axis. A constant angular velocity, ω, throughout implies that the linear velocity, $v = \omega R$, is smaller for smaller values of R. Now consider any convection cell that transports material with a lower value of R to a higher value of R and vice versa (see Fig. 9.1). Material that was initially close to the rotation axis, and with small relative v is moved out into a region of high relative v, where statistically it will act to retard the rotation. Material moving inward will tend to increase the rotation rate and the total angular momentum of the system is conserved. The net effect is that the outer part of the core rotates more slowly than the inner part of the (outer) core. Because the varying part of magnetic fields originating in the inner part of the core are screened out, a westward drift of magnetic eddies in the outer part produces the westward drift of the non-dipole field. Although intuitively appealing, the relative slowing down of the outer part of the core due to convection has not been observed in the few experiments done in rotating fluid systems (e.g. Busse, 1975).

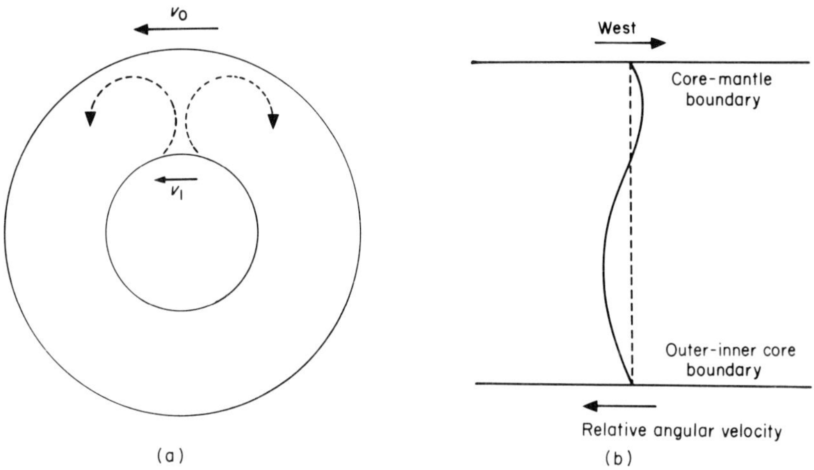

Fig. 9.1. Changes in angular velocity in the outer core produced by convection. (a) An equatorial section through the earth shows convection in the outer core (dashed lines). The linear velocity v_0 at the core–mantle interface is larger than that at the outer–inner core boundary, v_1, because the angular velocities there are equal. (b) The movement of material with lower values of v outward and material with higher values of v inward causes a slowing down of the rotation of the outer part of the core relative to the inner part. Dashed line shows line of constant angular velocity, with higher angular velocity to the left (not drawn to scale).

Interestingly, the slowing down of the outer core relative to the inner core would have the effect of taking a given initial uniform poloidal magnetic field and producing toroidal fields of opposite signs in the northern and southern hemispheres (§7.5.3). Boundary layers will be formed close to the inner and outer interfaces of the outer core, the size of which depends on various frictional effects (§10.3). The velocity across these interfaces will be constant under the expected "no-slip" boundary conditions, producing two additional velocity gradients to those considered by Bullard. Thus the applicability of a Bullard-type model depends on which forces balance each other in the Navier-Stokes equation (see §8.3.4) and on the nature of the boundary conditions, factors that are not yet well known.

9.1.3 MHD Waves

The screening discussion in the last section ignored the possibility that magnetohydrodynamic (MHD) waves occur in the core. If they do occur then it is possible for disturbances which originate at depth in a conducting liquid to give rise to waves that propagate outward to the core–mantle boundary. The possibility that waves could produce secular variation was first suggested by Hide (1966), while the first use of waves in dynamo models was by Braginskiy (1964a, b, 1967). Just like waves on a vibrating string, MHD waves can travel along field lines due to the presence of a magnetic stress. Consider an electric current density $\boldsymbol{J}$ in the presence of a field $\boldsymbol{B}$. The magnetic force on the current density is given by:

$$\begin{aligned}\boldsymbol{F} &= \boldsymbol{J} \times \boldsymbol{B} \\ &= (\nabla \times \boldsymbol{H}) \times \boldsymbol{B} \\ &= -\frac{\nabla(\boldsymbol{B} \cdot \boldsymbol{B})}{2\mu_0} + \nabla \cdot \left(\frac{\boldsymbol{BB}}{\mu_0}\right)\end{aligned}$$

$B^2/2\mu_0$ can be interpreted as a magnetic pressure, while $(\boldsymbol{BB}/\mu_0)$, a second-order tensor represented in dyadic form, can be interpreted as a stress (tension) along the field line. $(\boldsymbol{BB}/\mu_0)$ is a reduced form of the Maxwell stress tensor.

The velocity of waves in a vibrating string of density ρ and tension T_e is given by $(T_e/\rho)^{1/2}$. By analogy it might be expected (correctly) that a wave along a magnetic field line would have a velocity $\boldsymbol{B}(\mu_0\rho)^{-1/2}$. This velocity is called the *Alfven velocity* and the corresponding wave an *Alfven wave*.

This expectation is now justified for one simplified case. Assume there are waves in a fluid that is perfectly conducting, incompressible and inviscid. Suppose also that the initial magnetic field is uniform and denoted by B_0. Under these conditions the appropriate equations are:

$$\frac{\partial \boldsymbol{B}}{\partial t} = \nabla \times (\boldsymbol{v} \times \boldsymbol{B}) \tag{9.1}$$

(magnetic induction equation for a perfect conductor)

$$\begin{aligned}\rho\frac{\partial \boldsymbol{v}}{\partial t} &= -\nabla P + \boldsymbol{J} \times \boldsymbol{B} - \rho\nabla\phi_g \\ &= -\nabla\left(P + \frac{B^2}{2\mu_0} + \rho\phi_g\right) + \frac{1}{\mu_0}\nabla \cdot (\boldsymbol{BB})\end{aligned} \tag{9.2}$$

(Navier-Stokes equation; ϕ_g is the gravitational potential and P is pressure)

$$\nabla \cdot \boldsymbol{v} = 0 \tag{9.3}$$

(incompressible fluid)

$$\nabla \cdot \boldsymbol{B} = 0 \tag{9.4}$$

(one of Maxwell's equations)

A further simplification is made by assuming that the pressure field is always in equilibrium. This means that the first term on the right-hand side of (9.2) can be taken equal to zero and (9.2) becomes:

$$\rho \frac{\partial \boldsymbol{v}}{\partial t} = \frac{1}{\mu_0} \nabla \cdot (\boldsymbol{BB}) \tag{9.5}$$

This initial system, denoted by subscript 0, is now perturbed by a small amount and the perturbation values are denoted by the subscript 1:

$$\boldsymbol{B} \to \boldsymbol{B}_0 + \boldsymbol{B}_1$$

$$\rho \to \rho_0 + \rho_1$$

$$\boldsymbol{v} \to \boldsymbol{v}_1$$

The last equation assumes there is zero initial velocity. Substitution of these values into (9.1) and (9.5) gives for $\boldsymbol{B}_0$ taken in the z direction:

$$\frac{\partial \boldsymbol{B}_1}{\partial t} = \nabla \times (\boldsymbol{v} \times \boldsymbol{B}_0) = (\boldsymbol{B}_0 \cdot \nabla)\boldsymbol{v} \tag{9.6}$$

and

$$\frac{\partial \boldsymbol{v}}{\partial t} = (\boldsymbol{B}_0 \cdot \nabla) \frac{\boldsymbol{B}_1}{\mu_0 \rho} = \frac{B_0}{\mu_0 \rho} \frac{\partial \boldsymbol{B}_1}{\partial z} \tag{9.7}$$

Taking $\partial/\partial t$ of (9.6) and using (9.7) then:

$$\frac{\partial^2 B_1}{\partial t^2} = \frac{B_0^2}{\mu_0 \rho} \frac{\partial^2 B_1}{\partial z^2} \tag{9.8}$$

This is the wave equation and shows that there are (Alfven) waves propagated along field lines with a speed $B_0(\mu_0\rho)^{-1/2}$. For all laboratory speeds, the Alfven velocity is usually less than the sound velocity. However, great variability is found when solar and planetary bodies are considered. For example, the density of the sun's photosphere is very low (there are approximately 6×10^{10} hydrogen atoms $\mathrm{m^{-3}}$ there) and the Alfven velocity (v_A) is approximately $10^5\ B_0$. The general dipole field of the sun is about 10^{-4} T, while that in sunspots is roughly two orders of magnitude larger. v_A varies roughly from 10 m $\mathrm{s^{-1}}$ to 10^3 m $\mathrm{s^{-1}}$ as compared to 10^4 m $\mathrm{s^{-1}}$ for the estimated velocity of sound in the sun's photosphere.

A more general case for Alfven waves can be treated in which the field is not a perfect conductor, viscous effects are present, the second term of (9.2) is retained, and the field is compressible. For the same perturbations used in the last problem, it is relatively easy to show that (9.6) and (9.7) are replaced by (Jackson, 1962):

$$\frac{\partial \boldsymbol{B}_1}{\partial t} = \nabla \times (\boldsymbol{v}_1 \times \boldsymbol{B}_0) + \frac{\nabla^2 \boldsymbol{B}_1}{\sigma \mu_0} \tag{9.9}$$

and

$$\rho_0 \frac{\partial \boldsymbol{v}_1}{\partial t} = s^2 \nabla \rho_1 - \frac{\boldsymbol{B}_0}{\mu_0} \times (\nabla \times \boldsymbol{B}_1) + \eta \nabla^2 \boldsymbol{v}_1 \tag{9.10}$$

where s is the velocity of sound and η is the viscosity. Consider the possibility of plane wave solutions described by:

$$\boldsymbol{v}_1(r, t) = \boldsymbol{v}_1 e^{i(k \cdot r - \omega t)} \tag{9.11}$$

Using this with (9.9) and (9.10) the following dispersion relationship for Alfven waves that propagate parallel to the field can be derived:

$$k^2 v_A^2 = \omega^2 \left(1 + i \frac{k^2}{\sigma \mu_0 \omega}\right)\left(1 + i \frac{\eta k^2}{\rho_0 \omega}\right) \tag{9.12}$$

If the conductivity is large and the viscosity small, then the wave number is approximately

$$k \approx \frac{\omega}{v_A} + i \frac{\omega^2}{2 v_A^3} \left(\frac{1}{\mu_0 \sigma} + \frac{\eta}{\rho_0}\right) \tag{9.13}$$

The imaginary part of k is associated with attenuation. This can be seen directly for any plane wave, since

$$e^{i(\boldsymbol{k}_1 \cdot \boldsymbol{r} - i\boldsymbol{k}_2 \cdot \boldsymbol{r} - \omega t)} = e^{i(\boldsymbol{k}_1 \cdot \boldsymbol{r} - \omega t)} e^{-\boldsymbol{k}_2 \cdot \boldsymbol{r}}$$

where the wave vector $\boldsymbol{k}$ has been written in terms of a real part, $\boldsymbol{k}_1$, and imaginary part, $\boldsymbol{k}_2$. The factor $e^{-\boldsymbol{k}_2 \cdot \boldsymbol{r}}$ gives the decrease in amplitude of the wave with distance, r. Equation 9.13 shows that the attenuation increases rapidly with frequency and decreases with increase in v_A (or equivalently with B) and with increase in σ. This attenuation is clearly different from that obtained from the skin depth calculations for a solid conductor.

The real part of the hydromagnetic wave is non-dispersive. That is, $\omega/k = d\omega/dk = v_A$. The phase velocity and group velocity are both equal to the Alfven wave velocity. In the case under consideration, *longitudinal modes* (perpendicular to the field) are also allowed. The real parts of these longitudinal modes are also non-dispersive and the phase velocity is given by:

$$\frac{\omega}{K} = \sqrt{s^2 + v_A^2} \tag{9.14}$$

In the case when magnetic pressure goes to zero, $B \to 0$ and $v_A \to 0$. Thus, the velocity reduces to ordinary sound velocity. In the more general case, the longitudinal wave velocity is seen to depend on the sum of the hydrostatic and magnetic pressures.

Note that (9.14) can be used to estimate the effect strong magnetic fields have on the velocity of compressional (elastic) waves (P-waves) in the outer core. v_A is roughly 0.1 to 0.01 m s^{-1} for the core, a value roughly 10^{-5} to 10^{-6} of the velocity of P waves there. This shows that the velocity of compressional waves is virtually unaffected by the presence of the core's magnetic field. A more detailed analysis confirms this result (Knopoff and MacDonald, 1958).

9.1.4 Skiles' Model

Skiles (1972a, b) considered the effects that the core-mantle boundary (modelled as a plane) might have on the transmission and reflection of Alfven waves. Figure 9.2 provides valuable insight into what could occur at the core–mantle interface. Those same fields that provided so much insight into the heuristic dynamo model in §7.5.3 predict that the field lines will be pointing westward in both the northern and southern hemisphere. The angle the field lines make with the core depends on the relative magnitudes of the poloidal and toroidal field components. Traditional high values for the toroidal to poloidal magnetic field strengths arising from magnetostrophic models suggest that the field lines make very small angles with the boundary while geostrophic models suggest that the angles might be up to 45° or so (§8.3.4).

The component of the MHD waves considered by Skiles travelling across

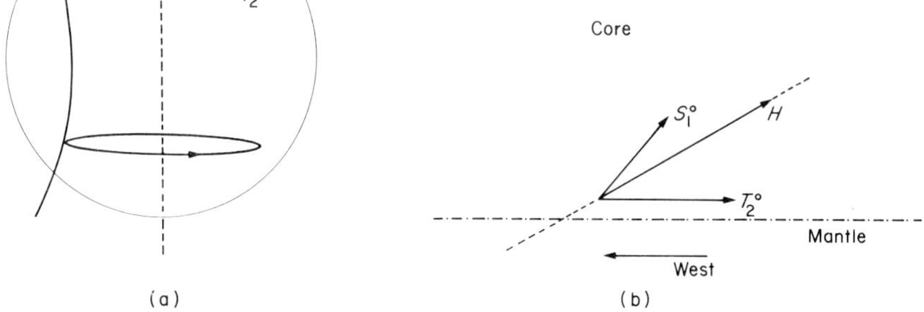

Fig. 9.2. Westward drift and MHD waves. (a) The S_1^o poloidal and a T_2^o toroidal magnetic field (as in §7.5.3). (b) H is the resultant of the S_1^o and T_2^o fields at the core–mantle boundary in the northern (upper) and southern (lower) hemispheres. The field line (dashed) through H coming out of the core is always westwards irrespective of the hemisphere. Alfven waves are preferentially transmitted along the field lines in the core and preferentially damped perpendicular to those lines.

magnetic field lines are damped more quickly than those that travel along field lines. This provides a selection mechanism that leads to westward travelling MHD waves being transmitted preferentially across the core–mantle interface. The boundary also will act as a filter in the sense that, depending on the angle of incidence, etc., waves of certain periods will be transmitted more effectively than those of other periods. These insights are verified by the formal solutions of Skiles. He calculated the reflection and transmission coefficients for Alfven waves using appropriate boundary conditions to obtain transmission and reflection coefficients as functions of angle of incidence and wave period. His results illustrate that the core–mantle boundary can act both as a filter on period and direction of MHD waves. Of course, MHD waves transmitted through the core to the mantle–core interface are converted to electromagnetic waves at the interface.

Unfortunately, Skiles' mechanism fails, simply because Alfven waves travel too quickly (roughly 0.1 to 0.01 m s^{-1}). The wavelength of the Alfven wave would have to be of the order of 10^8 km, a value clearly unacceptable, to explain a secular variation period of the order of 1000 years. Although this mechanism is not tenable, the idea that MHD waves might be transmitted more effectively westward than eastward across the core–mantle interface is well worth retaining. However, one must consider waves that travel at speeds far below those of pure Alfven waves.

9.1.5 Planetary Waves

Although pure Alfven waves should be considered poor candidates for secular variation (§9.1.3) there are other hydromagnetic waves that prove to be more promising. Planetary waves (Rossby waves) that exist in a rotating fluid in the absence of a magnetic field are considered first to gain insight into these other hydromagnetic waves.

Assume there is some initial flow of a fluid that is in geostrophic balance (pressure gradient balances the Coriolis force). This flow will then be perturbed by a small amount. To simplify the problem further it is assumed that all particle motion is horizontal and the viscosity is zero. Rectangular co-ordinates are used with the unit vectors $\hat{y}$ north, $\hat{x}$ east and $\hat{z}$ upward. Let v_1, v_2, v_3 represent velocities in the $\hat{x}, \hat{y}, \hat{z}$ directions, respectively.

For these conditions, the appropriate Navier-Stokes equations for v_1 and v_2 are:

$$\frac{Dv_1}{Dt} - fv_2 = -\frac{\partial P'}{\partial x} \tag{9.15}$$

$$\frac{Dv_2}{Dt} - fv_1 = -\frac{\partial P'}{\partial y} \tag{9.16}$$

where D/Dt is an operator given by $\partial/\partial t + \boldsymbol{v}\cdot\nabla$ and $P' = P/\rho$. f is the Coriolis force (equal to $2\Omega\cos\theta$), $\boldsymbol{\Omega}$ is the angular velocity vector, and θ is the colatitude. Assume $f(\theta)$ varies slowly as a function of θ and can be approximated by

$$f(\theta + \Delta\theta) = f(\theta) + \beta y$$

This approximation is called the β-plane approximation, where $\beta = 2\Omega\sin\theta/R$ and $R =$ the radius of the outer core (note that y increases northward). Technically, this approximation is only applicable to a thin plane. It is definitely not applicable to the entire outer core.

Differentiate (9.15) by y, (9.16) by x, and subtract using the β-plane approximation to get:

$$\frac{D}{Dt}\left[\frac{\partial v_1}{\partial y} - \frac{\partial v_2}{\partial x}\right] - \beta v_2 = 0$$

where the fact that $\partial v_1/\partial x = \partial v_2/\partial y$ has been used (the motions are purely horizontal). Now $\partial v_1/\partial y - \partial v_2/\partial x = \nabla^2\psi$, where ψ is the stream function and $v_2 = -\partial\psi/\partial x$. Physically, the lines of constant ψ are the paths the fluid particles follow. Substitution of these relationships into the last equation gives:

$$\frac{D}{Dt}(\nabla^2\psi) + \beta\frac{\partial\psi}{\partial x} = 0 \tag{9.17}$$

Let the initial velocity be given by $v_0\hat{x}$ and search for plane wave solutions of the form:

$$\psi = \psi_o e^{i(kx - \omega t)}$$

This implies that $\nabla^2\psi = k^2\psi$. In the perturbation approximation $D/Dt \approx \partial/\partial t + v_0(\partial/\partial x)$. By substituting these last two equations into (9.17), then solutions exist providing:

$$\omega = \frac{\beta}{k} + v_0 k \tag{9.18}$$

Thus there are dispersive plane waves called Rossby or Planetary waves, which satisfy (9.15) and (9.16). Since $\beta > 0$ the phase velocity in the reference frame of the fluid ($v_0 = 0$) is westward. However, the group velocity $d\omega/dk$ in this reference frame is to the east. If no rotation is present, $\beta = 0$ and the plane wave solutions exhibit no dispersion.

9.1.6 Hide's Model

Hide (1966) was the first to suggest that MHD waves similar to Rossby-type waves might explain the geomagnetic secular variation. Hide showed that valuable insight into the problem can be gained by considering the dispersion relation for a perfect conducting, incompressible, inviscid, homogeneous fluid of indefinite extent rotating uniformly in a magnetic field. The dispersion relationship for the case of plane waves in a *uniformly* magnetized field that is parallel to the rotation axis (Hide, 1966) is:

$$\omega = \pm[\Omega \pm (v_A^2 k^2 + \Omega^2)^{1/2}] \tag{9.19}$$

If rotation is negligible ($\Omega = 0$), (9.19) reduces to the case of simple non-dispersive Alfven waves, $\omega_+ = \omega_- = \boldsymbol{v}_A \cdot \boldsymbol{k}$. The plus root of (9.19) is denoted by ω_+ and the minus root by ω_-. A convenient way to determine the effects of rotation is to use the Rossby Number $Q \equiv v_A k/2\Omega$ (§7.5.1). When Q is small, rotational effects dominate and the four roots of (9.19) are approximately:

$$\left(\frac{\omega}{k}\right)_i \approx 2\Omega(1 + Q^2)/k \quad \text{(phase velocity)}$$

$$\left(\frac{\omega}{k}\right)_m \approx \pm v_A^2 k/2\Omega \quad \text{(phase velocity)}$$

$$\left(\frac{d\omega}{dk}\right)_i \approx \pm v_A^2 k/\Omega \quad \text{(group velocity)}$$

$$\left(\frac{d\omega}{dk}\right)_m \approx \pm v_A^2 k/\Omega \quad \text{(group velocity)}$$

The subscript i refers to the root corresponding to choosing the inner plus sign in (9.19), while m refers to the choice of the minus sign. The (i) subscript is associated with the normal inertial waves while the latter is referred to as a hydromagnetic inertial wave in which there is a magnetostrophic balance of forces (§8.3.4). In the absence of a magnetic field, $Q = 0$ and $\omega_i = \pm 2\Omega$. This result corresponds to a semi-diurnal oscillation involving no energy propagation. The presence of a weak field ($0 < Q \ll 1$) allows energy to propagate slowly along the magnetic field lines. Hide (1966) showed that very reasonable wave velocities can be expected for Magnetic Rossby waves, providing Q is not assumed to be too large, unlike the case for pure Alfven waves.

Hide (1966) constructed a model for hydromagnetic waves that used the β-plane approximation discussed in §9.1.5. His analysis leads to slow magnetohydromagnetic waves which propagate eastward (phase velocity), a result confirmed by the more rigorous mathematical treatment of Stewartson (1967). Because the β-plane approximation is only applicable to a thin shell, it was argued by Hide that westward travelling waves would exist in the outer core (thick liquid shell). The argument presented for this is intuitive rather than mathematical, and has been the subject of substantial debate. Even Hide once expressed doubts concerning its validity (Hide and Stewartson, 1972). Nevertheless, subsequent work by Acheson and Hide (1973) supported Hide's initial point of view. Acheson and Hide (1973) have also considered far more general conditions for Magnetic Rossby waves than assumed to obtain (9.19).

Subsequently a considerable amount of work has been carried out on instabilities giving rise to a wide variety of dispersion relations depending on the assumptions of the model. One of the most exciting results of these studies is derived from analyses that relate drift rates and directions to toroidal magnetic field strengths. Hide and P. Roberts (1979) argue that westward drift will occur for "strong" toroidal fields while eastward drift will occur for weak toroidal fields. In particular, they point out that the sense of propagation of hydromagnetic waves depends on the ratio of the toroidal to poloidal magnetic fields. Westward drift dominates when $B_T > B_P$ and eastward drift when $B_T < B_P$. Hide and Roberts (1979) argue that B_T is likely to be much larger than B_P, but is not so large that ohmic heating becomes excessive. Their range of tolerable values for B_T extends from 10^{-3} T to 5×10^{-2} T. These values appear incompatible with the Busse-type geostrophic models (§8.3.4).

9.2 Field Reversals

9.2.1 Physical Insight

Insight into magnetic field reversals can be obtained by considering the magnetic induction equation:

$$\frac{\partial \boldsymbol{H}}{\partial t} = k\nabla^2 \boldsymbol{H} + \boldsymbol{\nabla} \times (\boldsymbol{v} \times \boldsymbol{H}) \tag{9.20}$$

The velocity field $\boldsymbol{v}$ can be obtained by solving the Navier-Stokes equation, in which the magnetic field enters in only through the body force $\boldsymbol{J} \times \boldsymbol{B}$. In the MHD approximation (§7.3.2), this body force can be rewritten for core conditions as $(\nabla \times \boldsymbol{H}) \times \boldsymbol{H}/\mu_0$, where μ_0 is the permeability of free space (constant, independent of H). Therefore any solution $v(H)$ of the Navier-Stokes equation must be an even function of H, and thus the sign of the solution is independent of the sign of H. Substitution of this even function into the last term of (9.20) shows that this term must be an odd function of H. So is every other term in (9.20) and this implies for any solution H_0 of (9.20), there is an equal acceptable solution, $-H_0$. Thus reversed and normal states are expected from dynamo theory and the time-averaged properties are expected to be identical (Merrill *et al.*, 1979).

The remaining problem dealt with in the following sections is the mechanism of polarity transition. A simple kinematic dynamo explanation for transition is that convection ceases for a sufficient time that only weak (random) magnetic fields remain. During this time, the temperature gradient of the outer core would increase until convection resumes. The polarity of the new poloidal field would depend on the sign of the residual poloidal field at the time of the resumption of convection.

The above is an oversimplified example of one general class of reversal models, referred to here as the class of *free decay models*. Free decay models can either be kinematic or hydromagnetic, but they have the common property that the velocity field vanishes over sufficient time for the magnetic field to decay to a small value. This can be contrasted with *dynamic reversing models*, which also can be either kinematic or hydromagnetic, but which have the common property that the velocity is rarely, if ever, zero.

9.2.2 Levy–Parker Models

Parker (1969) and Levy (1972a, b, c) have produced models of reversals for kinematic dynamos in which the velocity field includes non-uniform rotation and cyclonic convection. Both Parker and Levy examine $\alpha\omega$-type dynamos in which large toroidal magnetic fields are generated in restricted regions. The velocity field associated with the production of the toroidal fields is assumed to have azimuthal symmetry, $\boldsymbol{v}_\phi$. The magnetic induction equation in spherical co-ordinates (r, θ, ϕ) for azimuthal velocity is

$$\frac{\partial B_\phi}{\partial t} = k_m\left(\nabla^2 - \frac{1}{r^2 \sin^2 \theta}\right)B_\phi = B_r\left(\frac{\partial v_\phi}{\partial r} - \frac{v_\phi}{r}\right) \tag{9.21}$$

The velocity shear is represented by a sphere of radius L rotating at a different angular velocity than that of an overlying spherical shell, $L < r \le R_c$, where R_c is the radius of the core (Levy, 1972a, b, c). The azimuthal velocity is given by:

$$v_\phi = C_1 r \sin\theta \qquad 0 \le r < L$$

$$v_\phi = C_2 r \sin\theta \qquad L < r \le R_c$$

where C_1 and C_2 are constants and $C_1 \ne C_2$. The velocity is yet undefined at $r = L$ and $(\partial v_\phi/\partial r - v_\phi/r)$ is zero everywhere but at $r = L$, where it is also undefined. This ambiguity is removed by taking,

$$\frac{\partial v_\phi}{\partial r} - \frac{v_\phi}{r} = (C_2 - C_1)L\,\delta(r - L)\sin\theta \tag{9.22}$$

where $\delta(r - L)$ is a Dirac delta function.

Both Parker and Levy consider turbulent dynamos so that they can treat convective cyclones (eddies which have counterclockwise rotation) in a statistical manner. It is the interaction of these cyclones with the toroidal magnetic field B_ϕ which generates a poloidal magnetic field. It is convenient to use the magnetic vector potential, $\boldsymbol{A}$, defined by $\boldsymbol{B} = \nabla \times \boldsymbol{A}$. The poloidal field B_r generated by interaction of cyclones with B_ϕ is then directly related to A_ϕ:

$$B_r = \frac{1}{r\sin\theta}\frac{\partial}{\partial\theta}(A_\phi \sin\theta) \tag{9.23}$$

The magnetic induction equation can be used to obtain the increase in A_ϕ with time due to the interaction of cyclones with B_ϕ:

$$\frac{\partial A_\phi}{\partial t} = k_m\left(\nabla^2 - \frac{1}{r^2\sin\theta}\right)A_\phi - \alpha(r, \theta, t)B_\phi \tag{9.24}$$

in which $\alpha(r, \theta, t)$ is a measure of the local strength of cyclonic motions.

The problem is reduced to solving simultaneously the above equations in which there is still the freedom of choosing $\alpha(r, \theta, t)$ (representing the cyclonic motions). The details of this solution will not be given here other than to point out that A_ϕ is expanded in an infinite series of spherical harmonic terms in a similar fashion to that discussed in §8.1 and §8.2. Statistical fluctuations in cyclonic bursts are simulated by Levy (1972b) by utilizing delta functions in space and time. That is, there are sharp peaks of cyclonic activity, whose locations can still be freely chosen, interspersed with periods of stasis in which no convection at all is assumed to occur. This guarantees both convergence of the infinite series and domination of long wave magnetic field features. Shorter wavelength features decay away exponentially at a rate proportional to the square of their wavelength (§7.4.2).

A heuristic explanation, differing slightly from the arguments of Parker and Levy, illustrates how reversals can occur by bursts of cyclones followed by periods of stasis. Suppose the *average* angular velocity of the earth's core is the same as that of the entire earth. Then any decrease in relative angular velocity in one part of the core must be balanced by an increase in angular velocity in another part of the core. Recalling how toroidal magnetic fields are generated (§7.5.3, §8.2.1, §9.1.2), the expectation is that there will be toroidal fields of opposite signs in the same hemisphere. These toroidal fields are argued to be located in different latitudinal zones. Parker and Levy assume that the main poloidal field is produced through cyclonic interactions with only one of these toroidal fields. Levy (1972b) presents two possible schemes in his reversal models. Both schemes (like those of Parker) call on fluctuations in the distribution of cyclonic cells in the core to reverse the otherwise stationary geomagnetic dipole. In this first scheme of Levy (1972b), the stationary field is maintained by cyclones at low latitude (Fig. 9.3). The

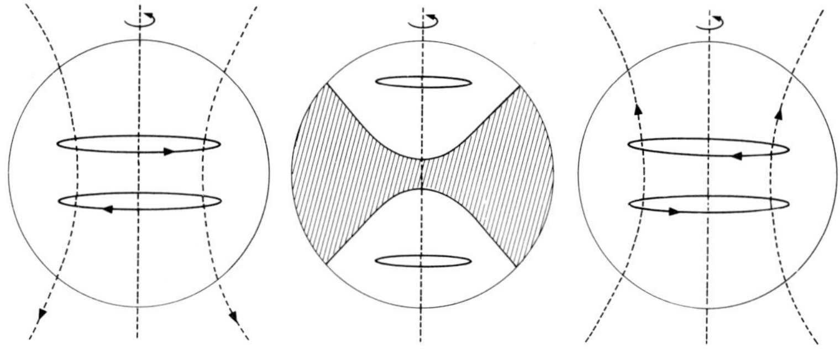

Fig. 9.3. A kinematic reversal model. (a) A poloidal (dipole) field (dashed lines) maintained by cyclones concentrated at low latitudes which interact with a "normal" toroidal field. (b) A burst of cyclones at high latitudes swamps the low latitude fields producing a region of "reverse" toroidal fields (shaded region). (c) Interaction between cyclones and this reverse toroidal field produces a reverse poloidal (dipole) field. (After Levy, 1972b.)

convection ceases and the field decays to a low value. A burst of cyclones then is hypothesized to occur in the high latitude zone and produces poloidal loops of the opposite sign. Diffusion of the resulting initial high latitude field to low latitudes will occur. The rotational shear then produces a main toroidal field of the opposite sign to that which was originally there (§7.5.3). Convective activity resumes in the lower latitude region, but now will interact with the new toroidal field (opposite sign) to produce poloidal loops of the opposite polarity. The second scheme is just the opposite. A burst of cyclones at low latitudes reverse the stationary field initially generated by cyclones at high latitudes.

Although these models have been criticized for being too contrived, they have provided valuable stimulation to palaeomagnetic research, largely because they indicate that a reversal may be initiated in one latitude zone (say low latitudes) and propagate to a different latitude zone. This possibility was initially exploited by Hoffman (1977) who showed that if the reversal is triggered at low latitudes and propagates to high latitudes, then different paths of virtual geomagnetic poles will occur, depending on whether the change in polarity goes from reverse to normal or vice versa (§5.4).

Although the models of Levy are described in terms of statistical fluctuations of cyclonic activity, the mathematical simulation of these statistical fluctuations involved critical periods of stasis. During these periods of stasis the process of free decay and diffusion of the magnetic field dominate and thus these models are examples of free-decay models of reversals.

9.2.3 Other Models

Besides the class of kinematic reversing models discussed in §9.2.2, there are hydromagnetic reversing models (e.g. Braginskiy, 1965; Gubbins, 1975a; Krause, 1977; Watanabe, 1981), statistical models (e.g. Cox, 1968; Kono, 1972; Phillips, 1977) and mechanical analogue models (e.g. Rikitake, 1958; Lowes and Wilkinson, 1967).

One way of examining dynamo models for reversals is to determine a solution to the dynamo problem at one time and then to generate successive solutions on a computer by advancing the time variable. The problem is complicated by the fact that steady or growing solutions are unacceptable. One way of doing this is exemplified by Watanabe (1981), who integrates forward in time the solutions of an $\alpha\omega$ dynamo obtained from the vector spherical harmonic expansion approach discussed in §8.1 and §8.2. Several assumptions are involved in this process that will not be discussed here. Watanabe obtains dynamic reversals with transition times comparable to those observed for the earth, but the intervals between reversals are far too short. This suggests that one or more of his assumptions need modification.

The most reasonable way to approach the mathematical formulation of such dynamic models is to extend the analysis of Chapter 8. As this is beyond the scope of this book and the purposes of this chapter, only a physical picture of dynamic reversing models will be given here. Although the picture to be given is based on insight gained from particular dynamo reversing models, it has not been completely justified mathematically and should be regarded as speculative.

The magnetic induction equation and the Navier-Stokes equation are coupled non-linearly to each other through the $\boldsymbol{v} \times \boldsymbol{B}$ term in the former equation and the $\boldsymbol{J} \times \boldsymbol{B}$ term in the latter. Because of this, small perturbations, for example, in the boundary conditions, can lead to a very different character of the magnetic field. It is such instabilities that are probably associated with reversals. This is illustrated in Fig. 9.4, following Gubbins (1975a, b) who presented numerical solutions showing that the core can switch from a steady state mode to an oscillatory one. Such switching could conceivably occur over and over again in the core with reversals occurring in the oscillatory mode. For another example, if the convection is thermally

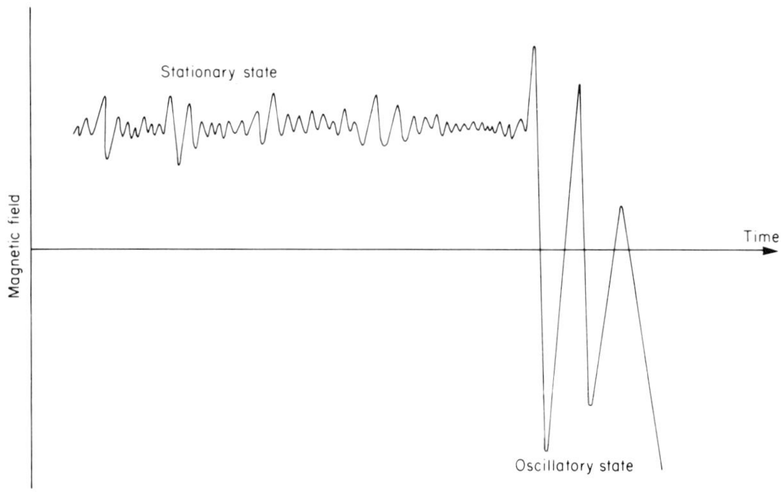

Fig. 9.4. A dynamic reversal process. One of several possible manifestations of the change in the magnetic field where there is a significant change in the velocity state. In this particular example the change in state is from a stationary state (left) to an oscillatory state. The oscillatory state would presumably be associated with the magnetic field during polarity transitions, while the stationary state would be associated with the field during non-transition times.

driven and if the temperature gradient became too steep during a steady-state mode the degree of turbulence could change markedly. Heat would be transferred more rapidly during the period of increased turbulence, lowering the temperature gradient towards the adiabatic one. Eventually the system is returned to the steady-state mode. The turbulent mode might be identified with times of polarity transition and the steady-state mode identified with normal and reverse polarity intervals. Reversals might, or might not occur at the time the state changes from a turbulent state to a steady state. Presumably whether reversals occurred or not (if not, one would have an aborted reversal or excursion) would depend on the particular configuration of the magnetic field at the time the state is returned to a steady-state mode. Although the particular examples given above may not stand the test of time, the palaeomagnetic evidence indicates that some sort of dynamic reversal is occurring in the core (see Chapter 10).

The above considerations and the apparent stochastic nature of reversals (§5.3) suggest that reversals might be modelled by invoking instabilities. For example, both Nagata (1969) and Lilley (1970b) suggested that such instabilities may occur when the magnetic field becomes too symmetric with respect to the rotation axis. Their suggestions are based on Cowling's Theorem (§8.2.3) which indicates that axially symmetric magnetic fields cannot be maintained by symmetric fluid motions. This idea is hard to test, because the fluid motions cannot be directly and uniquely obtained from even perfect magnetic field data. Nevertheless, the palaeomagnetic data (§10.2.3) suggest that this is not the instability leading to reversals.

The concept of instability is a useful concept for stochastic models of reversals. Such models are often "justified" by various physical arguments, but in fact are not complete dynamo models of reversals. Nevertheless, they are often useful statistically to characterize reversals and thereby provide constraints on more deterministic models. For example, Cox (1968, 1969b) suggested from the then existing palaeointensity data that the dipole field intensity varied in a simple sinusoidal fashion and that reversals occur when a rare statistical fluctuation in the non-dipole field of opposite sign to the dipole field happens to dominate the dipole field momentarily. Cox used a Poisson distribution to model this process and to characterize the field. It is important to realize that a Poisson distribution will occur simply if the probability of a reversal occurring in a given small interval of time is small. The success of using a particular distribution is not a strong argument supporting the physical insight used to justify the model. This is directly seen in the Cox model, since additional palaeomagnetic data (§4.1) illustrate that the sinusoidal varying dipole field assumption is invalid. Thus the physical arguments that led Cox to suggest a Poisson distribution are incorrect, but the use of a Poisson distribution or a Gamma distribution still appears to be a good way of statistically describing reversals in time (see §5.3 for further discussion of the statistical models).

Finally there are the mechanical analogue models for reversals (e.g. Rikitake, 1958; Lowes and Wilkinson, 1967), which have been briefly discussed in §7.3 and §7.4. Although very valuable at the time, because of scaling problems etc. (§7.4) and because both kinematic and hydromagnetic reversing dynamo models exist and are more satisfying theoretically, these models will not be discussed further here.

Chapter Ten

Palaeomagnetism and Dynamo Theory

10.1 Overview

Although there is little doubt that some dynamo process is responsible for the magnetic field of the earth, there is not good agreement on the details of the process. Part of the reason for this can be attributed to the mathematical complexity of the problem and part to the problem being poorly constrained. Actual constraints come from four general sources:

(i) constraints imposed by the theory itself;
(ii) constraints imposed by the physical properties of the earth, particularly the core;
(iii) constraints imposed by direct and indirect measurements of the earth's magnetic field through time;
(iv) constraints imposed by extra-terrestrial magnetic fields.

The first class of constraints (i) has already been discussed in §8.2.3 and will not be discussed extensively here. It is a very useful class of constraints that significantly reduces the number of reasonable models. For example, Hibberd's (1979) use of the Nernst thermomagnetic effect to generate the magnetic field encounters difficulties with Cowling's theorem (Hide, 1981; Hide and Palmer, 1982). Probably all thermoelectric or thermomagnetic effects are incapable of producing a magnetic field of suitable magnitude as is illustrated by the calculation in §7.3.3.

Constraints of the second class (ii) have also been discussed previously (§7.1), but have not served to reduce the numbers of dynamos significantly, partly because some critical parameters are not well known and partly because most dynamo models are not developed to the point of producing numbers (e.g. relating the mean intensity of the magnetic field to core size). Nevertheless, some physical properties are sufficiently well known to provide constraints, both in the earth and in other planets. The high temperatures in the cores of the earth and some planets probably rule out permanent magnetization there, even as a seed field which could be magnified by dynamo action. This and other constraints, such as subadiabatic temperature gradients, have already been discussed in §7.1. Similarly, constraints of the last class (iv) are discussed in §11.5 and §12.5 and also will not be extensively discussed in this chapter.

Constraints of the third class (iii) are usually stated in terms of the well-established aspects of the earth's magnetic field. These are that the field is largely dipolar, it is long-lived ($\simeq 3 \times 10^9$ years or longer), and it exhibits reversals and westward drift. However, this are many more constraints than these, although they are not generally appreciated by dynamo theoreticians or by palaeomagnetists. Table 10.1 summarizes some of these constraints, and they will be elaborated on in the remainder of this chapter.

10.2 Standing Field

10.2.1 Asymmetries in Polarity States

In §9.2 it was argued that the time-averaged properties of the reverse and normal polarity fields should be identical. The discussion there is incomplete because it neglected discussion of the boundary conditions, which have often been erroneously used to explain polarity asymmetries.

The only known way the magnetic field can affect the velocity field is through the body force ($\boldsymbol{J} \times \boldsymbol{B}$). This applies not only to the Navier-Stokes equation, but to any boundary conditions affecting the velocity field. The first set of equations in §9.2 (also see eqn. 10.1) illustrates that the magnetic field can be imagined as affecting the velocity field by changing the effective hydrostatic pressure and by adding a resistance to flow perpendicular to the magnetic field. However, these effects are clearly independent of the sign of the magnetic field, providing there is no permanent magnetization in the core (i.e. μ is not a function of B). Because it is highly unlikely that there is any permanent magnetization in the core, no polarity biases can arise from this effect. The main point here is that the velocity field cannot sense the sign of

TABLE 10.1

Summary of observed and inferred properties of the earth's magnetic field through time, with possible implications for dynamo theories

	Observation	Quality[a]	Comments	Possible implications (Chapters where discussed)
1.	Polarity reversal existence	CC		Not difficult to incorporate into dynamo theory, although precise origin remains unknown (5, 9, 10)
2.	Magnetic field is predominantly dipolar at earth's surface	CC	CC for the past 5×10^8 years, but probably only C for Precambrian times. Appears to be true for the other planets	Upward continuation and nature of dynamo processes. Not difficult to understand in terms of dynamo processes (2, 6, 7, 8, 10, 12)
3.	No simple periodicity of reversals	CC	Well established only for the past 2×10^8 years, and unlike the solar magnetic field	Solar and terrestrial fields may require different dynamo explanations. Has possible implications for the size of the Lorentz force (5, 6, 10, 12)
4.	Polarity transition time-scale is 10^3–10^4 years	CC	Well established for last 10^8 years. Times for transitions are probably variable with standard deviations not known	Slightly, but not significantly, smaller than the free-decay time estimates for reversals (5, 7, 10)
5.	Longevity of the magnetic field $\approx 3 \times 10^9$ years	CC		Not difficult to understand. Does suggest energy sources must be long-lived with a reasonable size core early in the earth's history (6, 7)
6.	Time-averaged palaeomagnetic field coincides with axis of rotation	CC to C	CC for the last 5×10^6 years; C for the last 5×10^8 years	Shows rotation plays an important role in dynamos (6, 7, 8)
7.	Westward drift of the non-dipole field from direct measurements	CC to C	Historical data CC to C. Archaeomagnetic data C	Several alternative explanations. Unlikely to be a powerful constraint on dynamo theories (2, 4, 6, 9, 10)
8.	Scatter of VGP increases with latitude	CC to C	For the last few million years only. Scatter unlike the solar field	Highly non-unique models in terms of dipole and non-dipole fields. Maxwell stress explanation (6, 10)

Table 10.1—*Cont.*

	Observation	Quality[a]	Comments	Possible implications (Chapters where discussed)
9.	Long-term variations in the reversal frequency	C		Change in core boundary conditions over long periods of time (6, 10)
10.	Intensity of the dipole field varied in time by roughly $\pm 50\%$ during the last 10^4 years	C		Dynamic processes in the core (4, 7, 8, 10)
11.	Polarity transitions are non-dipolar	C	Restricted to past 5×10^7 years only	Dynamic reversal transition (5, 10)
12.	Mean lengths of polarity intervals are the same for both polarity states over times $\sim 10^7$ years	C to R	Restricted to past 5×10^7 years only	Expected from dynamo theory (6, 10)
13.	Decrease in intensity during polarity transitions	C to R	Good relative and absolute palaeointensity data across transitions are few. Data largely restricted to the last 5×10^7 years	Dynamic reversal transitions (5, 10)
14.	Asymmetry of time-averaged field between normal and reverse polarity states	C to R	Established for past 5×10^6 years	Slowly varying standing field (6, 10)
15.	Westward drift of the non-dipole field from palaeomagnetic measurements	R	Changes in the non-dipole field are clearly present in the palaeomagnetic record, but it is difficult to demonstrate they are due to westward drift	Several alternative explanations. Unlikely to be a powerful constraint on dynamo theories (2, 4, 6, 9, 10)

16.	Existence of excursions	R	Almost all palaeomagnetists expect they are present, but their presumed properties make them difficult to prove	Important in distinguishing between different dynamic models for reversals (4, 5, 10)
17.	Variable rates of westward drift	R	Probably occur, but there are still inadequate measurements to demonstrate this	Hydromagnetic wave phase velocity dependence on magnetic field intensity (2, 4, 9, 10)
18.	Systematic direction changes during polarity transitions	R	Considerable scatter	Boundary conditions, standing fields, manifestation of change in core convection, etc. (5, 10)
19.	Correlation between the intensities of the non-dipole and the dipole fields	R	Only suggested for the past few million years	Dynamic interaction between decay and source terms in the magnetic induction equation (4, 6, 10)
20.	Existence of hemisphere asymmetries for more than 10^6 years	R	Best case for north-south hemisphere asymmetry	Difference in boundary conditions (6, 10)
21.	An upper limit on palaeointensities that is often approached but rarely exceeded	R	Very incomplete record	If true then a pure geostrophic balance of forces in the earth's core seems unlikely (4, 6, 8, 10)
22.	Long-term variations with no simple periodicities in either the dipole or non-dipole field	R	Considering the other variations in the magnetic field record, this seems highly likely, but the record is very incomplete	Non-linear processes (4, 5, 6, 7, 8, 10)

[a] *Explanation of "Quality" Factor in Table 10.1.* CC, completely convincing; C, convincing; R, reasonable. These definitions are in the language of the palaeomagnetist used at scientific meetings. Most often the palaeomagnetist accompanies his arguments with statistics. In these terms CC corresponds to better than 99% confidence. C to better than 95%, and R to less than 95% but greater than 67%. Unfortunately, palaeomagnetists often make implicit, as well as explicit, assumptions in calculating their statistics. For the benefit of those who are used to well-controlled experiments, the following definitions are probably more appropriate. CC stands for probably correct and it would cause a major revolution in geomagnetism if proved otherwise. C stands for this is also probably right, but even if it is wrong, it will be hard to prove. R stands for best guess given very imprecise and little data, but the speculation is consistent with what is known (at least as far as the authors can tell at the time of writing). Other palaeomagnetists would probably order the quality of the data in this table in about the same way as the authors have. The authors' confidence in this last statement is indicated by the symbol R!

the magnetic field, regardless of whether one is considering the magnetic induction equation or the boundary of initial conditions. Note this causes serious problems with some suggestions, such as that the reverse polarity fields are associated with convection in a particular region of the core and normal polarity fields with convection in a different region. If one simply reverses the sign of the "initial" weak field that is amplified in a given region, then the opposite polarity field should be sustained in that region. Thus whatever the eigenstate solutions of the dynamo problem are, reverse and normal states appear to have equal probability of occurring.

This problem cannot easily be circumvented by consideration of the magnetic boundary conditions either since, using the MHD approximation (§8.4.2), those boundary conditions stem from the Maxwell equations:

$$\nabla \cdot \boldsymbol{B} = 0$$

$$\nabla \times \boldsymbol{H} = \boldsymbol{J}$$

The first is clearly independent of the sign of $\boldsymbol{B}$. It is possible to have a $\boldsymbol{J}$ at a boundary in the second of these (e.g. an infinite conductor-insulator boundary), but *if* $\boldsymbol{J}$ is associated solely with an internal magnetic field, its sign will change every time that of $\boldsymbol{B}$ changes and no polarity asymmetry occurs. However, magnetic fields of external origin can produce asymmetry in polarity states and in boundary conditions.

In spite of these predictions, the dipole to quadrupole field ratios for the time-averaged field over the past five million years are significantly different for the reverse and normal polarity fields (§6.2). There are three possible explanations for these differences referred to respectively as the *data artifact hypothesis*, the *standing field hypothesis*, and the *transition probability hypothesis* (Merrill *et al.*, 1979).

The analysis of Phillips (1977) initially suggested asymmetries in the length of normal and reverse polarity intervals, but with additional data this turned out to be an artifact (§5.3.3). As the dipole to quadrupole field ratios also vary in time, the question arises as to whether the time average over five million years (§6.2) is adequate? The appropriate time constant to investigate this question is the free decay time, around 10,000 years (§7.4). The mean time for normal and reverse polarity intervals during the past five million years is of the order of 2×10^5 years (§5.3). Because the observed systematic differences persist for time intervals yet another order of magnitude greater than this, it seems unlikely that the reported asymmetries are a data artifact. It has already been shown that they are unlikely to be produced from rock-magnetic sources (§6.2) so that the data-artifact hypothesis appears improbable.

The second hypothesis imposes asymmetry from magnetic field sources external to the outer core, a possibility not excluded from the boundary condition discussion. A standing magnetic field with origin partially or totally external to the core could produce the observed asymmetries. The magnitude and sign of this standing field could change with time, and indeed probably does (§6.2.3). The period of time between changes in the sign of the standing field must be long with respect to the mean time of polarity intervals to be consistent with the data. Thus the standing field hypothesis does not refer to a time invariant standing field, but to a *slowly-varying "standing" field*. This type of standing field should not be confused with the standing and drifting parts of the non-dipole field in the analysis of Yukutake in §2.4.

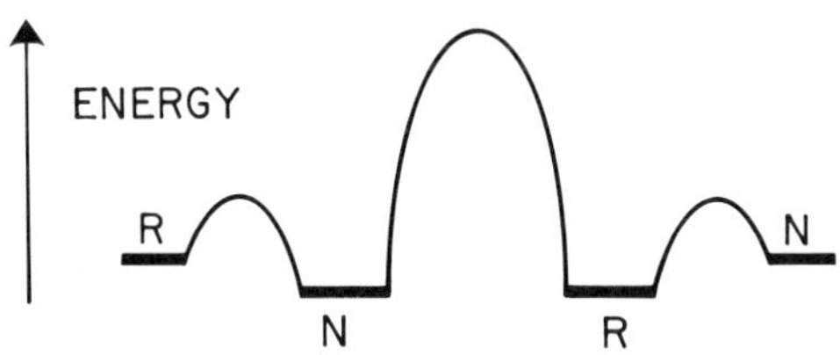

Fig. 10.1. The transition probability hypothesis. Suppose there are two normal polarity (N) eigenstate solutions of the magnetic induction equation that are characterized by different values of some parameter E (after energy in quantum mechanics). Then there must be two reverse polarity states (R) with E values equal to those of the N states as shown. However the probability of transition between eigenstates may be unequal as shown by the maxima. If the initial state happens to be the left N, then over a long period of time oscillations between N and R states may be confined to the left side of the central large maximum in the figure. The character (E) of the R and N states will be different. If the central maximum is ever traversed, then there is a high probability that there will be oscillation between the R and N states on the right side. Thus over long times, relative to a mean polarity time, there could be a change in the relative character of R and N states (e.g. in the dipole to quadrupole field ratios).

The third hypothesis is that although certain eigenstates have equal energies and equal probability of occurring in principle, not all have equal probability of occurring in practice, because transition probabilities between certain eigenstates are low or non-existent. This possibility is illustrated in Fig. 10.1 in which four eigenstates are assumed to exist for simplicity. The mechanism could (among other possibilities) allow reverse polarity and normal polarity fields to be associated with different parts of the core. The polarity of one state would have to be directly related to the seed field of the next polarity state for this latter speculation to be viable. The mechanism involving different probabilities of transition between allowed eigenstates is incompatible with free-decay models for reversal of the earth's magnetic field (§9.2), and is somewhat contrived.

10.2.2 Polarity Transitions

The possibility has been suggested that polarity transition data can be analysed to determine if a standing field exists (Hillhouse and Cox, 1976; also see Cox, 1975 and Fuller *et al.*, 1979). The basic idea is that during a transition, the core generated field is substantially reduced while the standing field remains unaffected, thus producing a bias in polarity transition paths. For example, suppose the standing field is reasonably large and free decay of the core field occurred to the extent that the standing field to core field ratio became significant. Then the standing field should certainly produce a strong bias in the palaeomagnetic directions during the transition. A closer look at these ideas suggests they may well be wrong.

The characteristic free decay time is of the order of 10,000 years, while the entire time of reversal transition (including build-up) is around 4000 years (§5.4 and §7.4). However, calculation of the free decay time depends on the (uncertain) values of some of the critical parameters in the core (Table 7.2), and the time for a reversal transition is reckoned for the directional changes only. There is some suggestion that intensity changes occur before directional changes are observed (Fuller *et al.*, 1979). Thus uncertainties in the estimates are such that it cannot be concluded from them that free decay is not occurring at reversal transitions. However, arguments against free decay during transitions are best provided by the nature of the field during the transition. The most significant observation in the palaeomagnetic transition data is that the field is highly non-dipolar during transitions (§5.4). This is precisely the opposite of that expected for free decay models in which higher order harmonics (shorter wavelength features) decay much more rapidly than lower order ones (§7.4). This implies that the dipole to non-dipole field ratio should *increase* during free decay, precisely the opposite to that observed. It can be concluded from the above that free decay models for reversals are not applicable to the earth (or sun; Chapter 12).

The use of universality theory (see Appendix B) shows that there are certain predictions that can be made about the behaviour of non-linear systems even if the detailed solutions are not available. This theory suggests a very different process may be occurring at transitions that precludes the possibility of seeing the standing field. In particular, an apparently trivial change in some critical parameter λ (Appendix B) can lead to increased bifurcations and an entirely different convection mode. Assuming that during this dynamic process λ increases, then the power in the non-dipole components should greatly increase (Appendix B). A change in heat transfer (and other properties) will most likely also occur with this change of convection mode and conceivably after a short time return the system to the original, and more typical convection mode. The particular polarity of the field would depend on the details of the magnetic field configuration at the instant the parameter characterizing the system passes back through its critical value. Excursions in this model would be nothing more than aborted reversals (§5.4.3). Such a model can be made consistent with the possibility of systematic directional changes during a reversal but if so, they will be difficult to prove. For example, if the field is quadrupolar during a reversal, then the ratio of quadrupole to higher order non-dipole field terms will be much smaller than the ratio of the dipole to non-dipole field at more usual (non-transition) times. This implies that the scatter associated with any VGP systematics will probably be large (see Appendix B). Interestingly, this class of reversal models includes the subclass in which convection is more dynamic during the polarity transition than during normal times, precisely the opposite of that predicted for free-decay models. Some sort of dynamic oscillatory process, such as occurs in the sun, is also compatible with the data. The consequences of such a process are the same as for the dynamic models and so will not be discussed further.

The consequences of such a dynamic model can be appreciated by recalling that a viable earth dynamo involves the magnification of an initial weak field through interaction with an electrically conducting and, most likely, convecting liquid. Any standing field that is manifested in the lowest order harmonics will be of global extent (§6.2). Thus such a standing field will be present in the core even if its origin is external to the core, and it will undergo magnification through positive dynamo action. It follows that if the asymmetries in the mean field data reflect a true standing field, then this occurs only after that standing field has undergone some, and very likely considerable, magnification. A change in convection mode that actively works to decrease the usual field, as is suggested above to occur at polarity transitions, will likely also have similar effect on the hypothetical weak standing field. Thus, it would be very difficult, if not impossible, to determine a standing field from the typically noisy palaeomagnetic data. Also, if in addition a reversal is preferentially initiated in some location in the core (§5.4 and §9.2.2), this may be the significant asymmetry in the problem, swamping any weak asymmetries from a hypothetical standing field. Therefore, it seems highly unlikely that the presence of a standing field could be detected in polarity transition data, as has been generally believed (Merrill *et al.*, 1984). Ironically, the transition probability hypothesis (§10.2.1) might be detected, because it is likely that if certain transitions are preferable to others that they will be associated with particular magnetic field configurations.

The palaeomagnetic data for rocks older than five million years are not of the quantity and quality to give completely reliable results for the lowest order zonal harmonics. However, they are adequate enough to indicate that there probably are major changes with time of the dipole to quadrupole field ratios for both N and R polarity states (§6.2). These changes include changes in sign and therefore rule out such possible candidates as thermoelectric fields for producing the standing field (Merrill *et al.*, 1984). Crustal magnetic anomalies have also been ruled out in the past due to the lack of large non-zonal terms in the mean field results (§6.2). This argument is in error because it neglects modification of the standing field by the dynamo processes. Figure 10.2 illustrates how axisymmetric toroidal magnetic fields are generated by symmetric velocity shear from either initial poloidal magnetic fields that are symmetric or non-symmetric with respect to the rotation axis. Of course Cowling's theorem must still be satisfied (§8.2.3), but this can be done in a variety of ways (e.g. §10.4.3). Figure 10.2 helps to show why the magnetic field of all the large objects in the solar system with active

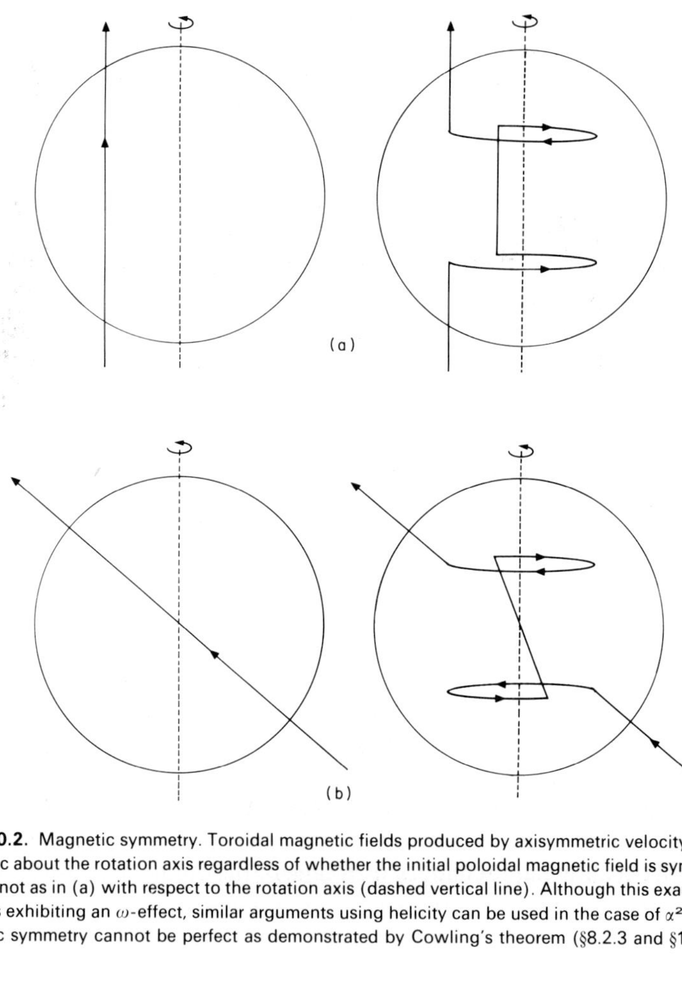

Fig. 10.2. Magnetic symmetry. Toroidal magnetic fields produced by axisymmetric velocity shear are symmetric about the rotation axis regardless of whether the initial poloidal magnetic field is symmetric as in (b) or not as in (a) with respect to the rotation axis (dashed vertical line). Although this example is for dynamos exhibiting an ω-effect, similar arguments using helicity can be used in the case of α^2 dynamos. Magnetic symmetry cannot be perfect as demonstrated by Cowling's theorem (§8.2.3 and §10.4.3).

dynamos have nearly axisymmetric fields (Table 12.1). Also, it shows why the observed lack of large non-zonal harmonics in the time-averaged palaeomagnetic field should not be used as an argument against magnetic anomalies of crustal origin as the ultimate source of the standing field. Non-axial symmetry is expected from such anomalies, but their palaeomagnetic manifestation may be highly symmetric.

For crustal magnetic sources to be responsible for the time-averaged polarity asymmetry they must have very long wavelength and large intensities to be significant in the core. One possible candidate involves the rocks producing marine magnetic anomalies associated with sea-floor spreading centres. These anomalies have their maximum near the axis of spreading and become smaller in intensity away from the axis (e.g. Johnson, 1979). Their wavelengths are short (few tens to a few hundreds of kilometres) in directions perpendicular to the spreading axis but are very long (up to thousands of kilometres) parallel to the spreading axis. There would have to be a net statistical bias for one of the polarities over a few millions of years for such anomalies to be a source of the field. Even though reversals occur much more frequently than this, this possibility cannot be ruled out because of random walk effects. It is also difficult to tell if there have been any relatively long-lived *continental* crustal magnetic anomalies which might contribute to the standing field. Cain (1975) concludes that all available evidence indicates that terms with $l = 8$ or smaller in a spherical harmonic analysis are of core origin, but the case is not closed because only relatively small crustal fields are needed. There may be more power at least in intermediate wavelength near surface fields than has been presumed although the origin of this power is controversial (Harrison and Carle, 1981; Schure and Parker, 1981). Analysis of MAGSAT data and comparisons with ground-based magnetometer studies may resolve these uncertainties. (The MAGSAT Project involves the measurement of the near-earth magnetic field utilizing a satellite with a vector magnetometer system estimated to have a roughly $\frac{1}{2}$ nT precision.) Any net long wavelength crustal anomalies would certainly reach the core, where they would be smoothed and amplified by dynamo processes. If they exist then the time-averaged results (§6.2) would give some indications of their magnification by the dynamo process and possibly even the size of the initial seed field for the earth dynamo (Merrill *et al.*, 1984). Significant changes in the locations of lithospheric plates occur on time scales of millions of years and thus the "standing field" would vary on similar time scales as seems to be observed (§6.2). Nevertheless preliminary analyses of MAGSAT data suggest that long wavelength crustal magnetic fields are very weak (e.g. Mayhew and Galliher, 1982).

10.3 Variations in the Geomagnetic Field

10.3.1 Magnetic Field Intensities

The amount and quality of the palaeomagnetic data decrease substantially with increase in age of the sample. Nevertheless there are several results that seem apparent from this sparse record. First there is no simple periodicity in palaeointensity as a function of age (§4.1), a result which in retrospect might have been anticipated from the lack of simple periodicities throughout the geomagnetic spectrum of internal origin, as illustrated most clearly in the stochastic reversal patterns (§5.3). There does appear to be a positive correlation between the magnitude of variations in intensity of the non-dipole field and the magnitude of the dipole field during most times (§6.3). However, the fluctuations on direction and intensity of the non-dipole field are very large during reversal transitions, a time when the dipole field is very low (§5.4). Thus the correlations between transition and non-transition times are opposite, supporting the suggestion of §10.2 that very different convective regimes are present during these times.

Although the data are sparse, the hypothesis cannot be rejected that there are no long-term (order of 10 Ma) trends in the intensity of the magnetic field during the past 3 Ga or so (§5.4). The majority of the palaeointensity data (excluding reversals) group around the present dipole moment of the earth's magnetic field. For example, the archaeomagnetic data for the past 10,000 years indicate moments with values that are essentially within ±50% of the present dipole moment (§4.1). Although the field might have been lower than this at times (§4.1.5), it seldom was much higher. In short, there appears to be an upper value for the field strength which is often approached but is seldom exceeded.

For a given velocity field in a kinematic dynamo (§7.3.3), an allowable solution is that the magnetic field approaches infinity in infinite time. Probably the reason this does not happen in the earth is that when the magnetic field becomes sufficiently large it significantly affects the velocity field through the body force ($\boldsymbol{J} \times \boldsymbol{B}$) in the Navier-Stokes equation (Table 7.3). This factor is not considered in kinematic models. The palaeomagnetic data suggest that this occurs for magnetic fields comparable in size to or slightly larger (e.g. 50% or so) than the present field. If so, this implies that geostrophic models (§8.3.4) are not good first-order approximations for determining the convective regime in the earth's core. It seems likely that when the magnetic field becomes too large, it impedes convection and decay of the magnetic field becomes dominant. The higher-order terms decay most rapidly and this provides an explanation for the correlation between variations in the intensity of the non-dipole field and the intensity of the dipole field during non-transition times (§6.4). As mentioned above this is the opposite to what happens at polarity transition times, indicating that polarity transitions probably do not involve significant periods of convection stasis.

Reversals of polarity in the sun are also associated with dynamic processes but not necessarily changes in convection mode (§12.2). In this case the Lorentz force is probably not as large a body force in the Navier-Stokes equation as it is in the terrestrial case. However, even in the sun there is observational evidence that granular convection is retarded in the presence of surface magnetic fields (Livingston, 1982). In the earth the relative size of the body force ($\boldsymbol{J} \times \boldsymbol{B}$) is probably even larger, accounting for the stochastic nature of the terrestrial field, such as found in the reversal patterns. Certainly the Navier-Stokes and magnetic induction equations are coupled *non-linear* equations, even in the first-order approximation in the earth. Although this greatly complicates the mathematics (Appendix B), it does provide a simple explanation for the palaeointensity observations.

10.3.2 Magnetic Field Directions

It is traditional in the analysis of secular variation to separate the dipole from the non-dipole field. However, there is no fundamental reason as to why this should be done. For example, any "real" current source in the core would give rise to dipole and non-dipole terms (§2.3). Suppose a current source were drifting westward, then so would the dipole and non-dipole terms and at the same rate. Evidently, the presently observed different rates of drift must be associated with changes in intensity of actual sources and/or actual differences in drift rates of the sources. One of the main reasons for dividing the field into the dipole and non-dipole fields is that it provides a convenient way of statistically separating large-scale features (dipole) from small-scale features (non-dipole). Nevertheless it has to be realized that this is somewhat artificial and it should not be taken to imply that the dipole and non-dipole fields have completely different current sources (§2.3).

There are several aspects of the palaeomagnetic record showing variations in magnetic directions which require explanation. In particular only recently have the following observations been explained (Merrill *et al.*, 1984):

(i) the scatter of virtual geomagnetic poles (directions) increases (decreases) with increase in latitude (§6.4);
(ii) westward drift rates of isoporic maxima and minima of inclination and declination indicate a drift rate that increases with latitude (§4.1.1);

(iii) there are different estimates of the drift rate arising from analyses of direct measurements of the earth's magnetic field (§2.4.3), from archaeomagnetic measurements (§4.1) and from analyses of sediments (§4.2 and §4.4);

(iv) fluctuations in the non-dipole field intensity are positively correlated with the magnitude of the dipole moment (§6.3).

The methods of analysis used to obtain (iv) involved a shorter window to obtain the mean scatter relative to that used to obtain the time-averaged dipole moment in a given interval (§6.3). The time-averaged magnetic field becomes more symmetric with respect to the rotation axis the longer the period of time (up to a few million years) used in the analysis. Therefore the methods used to determine (iv) emphasize the correlation in the fluctuations in intensity of the non-dipole zonal harmonics with the dipole moment. The explanation for the observed positive correlation was mentioned in the previous section. When the magnetic field intensity becomes too large it impedes convection and decay becomes more important. This decay may be free decay or dynamic (forced) decay. Apparently when the intensity becomes low enough the Lorentz force affects convection much less. It is the non-linearity that produces the stochastic behaviour observed at all levels in the earth's magnetic field (see Appendix B). The decay of zonal non-dipole components and their subsequent build-up is one possible explanation for the "standing non-dipole field" of Yukutake (§2.4.2). Note the Yukutake "standing non-dipole field" determined is not the standing field referred to in §10.2.

The explanation of (i), (ii) and (iii) requires the use of theoretical concepts developed in §9.1. Just as any continuous function of x can be represented by a Fourier series expansion, any continuous function of x and t (time) can be expanded in an infinite series of travelling waves with individual terms $e^{i(K_n x \pm \omega_n t)}$ in which n is the order of the term, K the wave number and ω the angular frequency. Thus secular variation can be represented in terms of waves regardless of its origin. This by itself does *not* imply the validity of the hydromagnetic wave explanations of westward drift (§9.1.6). Any disturbance in the magnetic field of the core is opposed by forces derivable from the Maxwell stress tensor (§9.1.3). All other factors equal, the magnitude of a disturbance will be related inversely to the square of the magnetic field. The propagation rate and direction of this disturbance does depend somewhat on the details of the theory invoked. However, both group and phase velocities in the hydromagnetic wave explanations are non-linear, but positively, related to the magnetic field strength. Possible explanations of (i), (ii) and (iii) follow from applications of the above (Merrill *et al.*, 1984).

Although the scatter in palaeomagnetic directions as a function of latitude as in (i) has been traditionally interpreted as due to the westward drift of the non-dipole field, these interpretations neglect any contributions from zonal harmonics, which by definition are longitudinally invariant. The free decay and build-up of the zonal terms would imply larger change in directions at low latitudes, or corresponding larger changes in VGP at high latitudes. Perturbations of the dipole field (i.e. *local* distortions of dipole field lines) will also be greater at low latitudes where the Maxwell stress is lowest. This follows because the Maxwell stress varies as B^2 and an axial geocentric dipole field has only half the intensity at the equator as it has at the poles. Note that these local disturbances of dipole field lines are very different from dipole wobble, in which the entire field changes simultaneously and the change in VGP is latitudinally invariant. Thus both the decay and build-up of zonal harmonics and the dipole perturbations give rise to an increase in scatter of VGP with latitude, providing a natural explanation for (i).

The methods used for obtaining (ii) indicate that westward drift of non-zonal harmonics are involved (§4.2.2). It is not the amplitude of the changes that is critical here but the drift rate. This again has a natural explanation in that hydromagnetic wave velocities depend non-linearly on B. As B generally increases poleward so should the drift velocity as is observed.

The non-linear dependence of drift rates on B also provides an explanation for (iii), because the intensity of the magnetic field varies with time. The direct type of analyses of westward drift (§2.4.3) of Bullard *et al.* (1950) uses data in a time interval of the order of 100 years (period of drift $\approx$ 2000 years). The Yukutake type analyses involve data in an interval of the order of 1000 years (period $\approx$ 1000 years; §4.1), while the palaeomagnetic data from sediments involve an interval greater than 10,000 years (broad peak in power; period consistent with values near 3000 years; §4.2 and §4.4). The average intensity of the field has changed significantly in these time intervals and is consistent with the notion that there is a positive correlation between mean westward drift rate and mean field intensity. With more precise data a better definition of the functional relationship between drift rate and field intensity may be possible, thus reducing the number of competitive models for westward drift (§4.1).

10.4 Core–Mantle Boundary

10.4.1 Evidence for Core–Mantle Coupling

The *length of day* (l.o.d.) varies roughly 3×10^{-12} rad s^{-1} over periods ranging from a few years to several tens of years (Lambeck, 1980). These are collectively referred to as *decade fluctuations* and their origins are not well known. However, it is generally conceded that the earth's core is the only sufficiently mobile part of the earth with sufficient mass to modify the rotation by the observed amounts on these time scales (Lambeck, 1980). There are four general ways of coupling the core to the mantle:

(i) *inertial coupling*, a function of the ellipticity of the core–mantle boundary;
(ii) *viscous coupling*, a function of the "friction" associated with a smooth boundary;
(iii) *topographical coupling*, a function of local topographical irregularities referred to as *bumps* on the core–mantle interface;
(iv) *electromagnetic coupling*, a function of magnetic field strengths and electrical conductivity in the core and mantle.

All of these are of interest to geomagnetists. Inertial coupling is related to precessional energy (§7.5.4), viscous coupling to the Ekman layer size (§8.3.4), topographical coupling to hydromagnetic waves (§9.1.6), and electromagnetic coupling to variations in the earth's magnetic field (§2.4 and §9.1).

Observational evidence for core–mantle coupling is very indirect and varies in quality from strongly suggestive to highly dubious. Lambeck (1980) has reviewed the problem in general and only a brief discussion will be given here. Some of the best evidence for core–mantle coupling can be found in Morrison (1979) who finds a correlation with lag, between magnetic declination and l.o.d. using data from the nineteenth century to the present. Perhaps the most debated observational data are those relating to topographical coupling. This is of particular interest to geomagnetists because Hide (1966, 1967, 1970) suggests that core–mantle bumps can serve as points of origin of hydromagnetic waves. In addition, Moffat (1978) has pointed out that helicity will be generated when fluid flows over a bumpy surface, suggesting that even large-scale dynamo processes might be significantly altered by bumps.

The radial dimensions of bumps cannot be much more than a few kilometres or they would be directly observable seismically. Indirect seismic evidence of bumps with "heights" of the order of 1 km is argued to be present in the observed precursors of the phase PKIKP, a phase corresponding to compressional elastic waves which travel through the earth's core. Scattering by these bumps has been argued to be the source of the precursors (Haddon and Cleary, 1974). Even more difficult to prove is the suggestion that bumps have gravitational manifestations (Hide and Malin, 1970, 1971a, b; Dziewonski *et al.*, 1977). Although lowest-order spherical harmonics of the *gravitational field* may reflect sources of density anomalies which statistically are of deeper origin than higher-order terms, there are some obvious exceptions. For example, a spherical shell with uniform density contributes only to the monopole term, regardless of the distance of this shell from the earth's centre. Hide and Malin (1970, 1971a, b) sought to circumvent this problem by considering correlations at the earth's surface between the magnetic potential associated with the non-dipole field ψ_m and the gravitational potential ψ_g. Such a correlation with lag was hypothesized because hydromagnetic disturbances would be manifested downstream from a bump on the core–mantle interface. The correlation coefficient used by Hide and Malin is given by,

$$R(\phi_0) = \frac{\langle \psi_m(\phi - \phi_0)\psi_g(\phi)\rangle}{\langle \psi_m^2 \rangle^{1/2} \langle \psi_g^2 \rangle^{1/2}}$$

where ϕ is longitude, and ϕ_0 is a phase difference between the bump as seen in the gravitational potential (geoid) and its magnetic manifestation. Using surface harmonics out to fourth order for ψ_m and ψ_g at the earth's surface at given values of latitude, Hide and Malin showed that $R(\phi_0)$ was maximum (0.84) for $\phi_0 = 160°$. That is, a correlation between the geoid and magnetic potential exists when the magnetic potential is rotated eastward by 160° relative to the fixed geoid.

Because it would be difficult to predict convincingly the theoretical lag between ψ_m and ψ_g, the empirical approach of Hide and Malin has merit. Nevertheless, the statistical significance of the "correlation" between ψ_g and ψ_m is certainly not clear (Khan, 1971; Lowes, 1971; Lambeck, 1980). Probably gravitational anomalies associated with sources in the mantle are such as to make any "correlations" non-unique. Moreover, mantle convection can also affect the geoid and alter the temperature of the core–mantle interface, thereby possibly affecting the geodynamo. Therefore even if a correlation is eventually demonstrated, its cause will not be uniquely determined.

10.4.2 Electromagnetic Coupling

Electromagnetic coupling appears to be the most promising form of core–mantle coupling to explain decade fluctuations because it is the most likely of the mechanisms given in the last section to vary significantly on such short time-scales. Perhaps the easiest way to envisage such coupling is to consider a magnetic field line crossing the core–mantle boundary and to assume the core and mantle are rotating with slightly different angular velocities. This field line, assumed not to be parallel to the rotation axis, is stretched by the shear at the core–mantle interface in a manner similar to that shown in Fig. 7.3.

There will be a tension associated with this stretching that can be relieved only by bringing the rotational velocities of the core and mantle to the same value. Mathematically the tension can be represented by the magnetic part of the Maxwell stress tensor, which in dyadic notation is (§9.1.3),

$$\bar{\boldsymbol{T}}_s = \frac{\boldsymbol{BB}}{\mu_0} - \frac{1}{2\mu_0}\bar{\boldsymbol{I}}B^2 \tag{10.1}$$

where $\bar{\boldsymbol{I}}$ is the unit tensor with components δ_{ij}. In the simplified example that the outer core rotates rigidly with respect to the mantle and all the torque is restricted to act at the core–mantle interface, the net electromagnetic torque $\bar{\boldsymbol{\Gamma}}_m$ is given by,

$$\bar{\boldsymbol{\Gamma}}_m = \int_S \boldsymbol{R}_c \times (\bar{\boldsymbol{T}}_s \cdot \boldsymbol{n})\,\mathrm{d}S \tag{10.2}$$

where R_c is the radius of the core, $\boldsymbol{n}$ is the outward directed normal at the core's surface, S. Substitution of (10.1) into (10.2) and using $\nabla \cdot \boldsymbol{B} = 0$ yields,

$$\bar{\boldsymbol{\Gamma}}_m = \int_S (\boldsymbol{R}_c \times \boldsymbol{B})\frac{B_R}{\mu_0}\,\mathrm{d}S \tag{10.3}$$

in which B_R is the radial component of $\boldsymbol{B}$. Bullard *et al.* (1950) first considered electromagnetic core–mantle coupling by assuming the core could be divided into two regions $R < R_1$ and $R_1 \le R \le R_c$, rotating with angular velocities ω_1 and ω_2 respectively. The mantle is rotating with angular velocity ω_3. The origin of these relative rotations was thought to arise from the convection mechanism described in §9.1.2. To conserve angular momentum

$$C_1\frac{\mathrm{d}\omega_1}{\mathrm{d}t} + C_2\frac{\mathrm{d}\omega_2}{\mathrm{d}t} = -C_M\frac{\mathrm{d}\omega_3}{\mathrm{d}t} \tag{10.4}$$

in which C_1, C_2 and C_M are the respective moments of inertia for the two parts of the core and the mantle. In the simplified version given here the core is assumed to rotate as a rigid body so that

$$C_c\frac{\mathrm{d}\omega_c}{\mathrm{d}t} = -C_M\frac{\mathrm{d}\omega_M}{\mathrm{d}t} \tag{10.5}$$

where C_c is the moment of inertia of the core, ω_c its angular rotation and ω_M that of the mantle. The torque can be calculated from (10.3) provided it can be decided what to use for $\boldsymbol{B}$. The change in angular velocity with time can then be calculated from (10.5).

Bullard *et al.* (1950) examined the three-layer case (eqn. 10.4), a case recently generalized further by Le Moüel and Courtillot (1982). The characteristic time constant derived by Le Moüel and Courtillot is given by a cumbersome algebraic expression involving the radii, the moments of inertia and the electrical conductivities of the rotating bodies. A critical assumption by these authors is that the characteristic time governing the process is derivable from the mechanical response of the system. That is, these authors examine the rate at which the electromagnetic torques bring the various rotating bodies (eqn. 10.4) into equilibrium.

In contrast Rochester (1968), Roden (1963) and P. Roberts (1972) assume that the characteristic time is the diffusion of a change (step function) in the magnetic field through the mantle. Following §7.4.3 an estimate for this diffusion time might be,

$$\tau_1 = \mu_0 \langle \sigma \rangle L^2 \tag{10.6}$$

where

$$\langle \sigma \rangle = \frac{1}{L} \int_{R_c}^{R_c + L} \sigma \, \mathrm{d}R \tag{10.7}$$

where L is the thickness of the conducting mantle shell and $\langle \sigma \rangle$ the average conductivity of this shell. Roden (1963) showed that (10.6) and (10.7) lead to an over-estimate of the characteristic time, as might be anticipated from the discussion and calculations of the free decay in §7.4. P. Roberts (1972) shows that a "compromise" diffusion time τ_3 is given by,

$$\tau_3 = \tau_1^{2/3} \tau_2^{1/3} \tag{10.8}$$

in which

$$\tau_2 = \frac{C_M C_c}{(C_M + C_c) 4\pi \langle \sigma \rangle R_c^4 L B_R^2} \tag{10.9}$$

Equation 10.9 is obtained by assuming that changes in the mantle rotation are so slow that the core always rotates as a solid body, that is (10.5) is applicable. Roberts shows that neither τ_1 nor τ_2 are appropriate, but that the compromise value τ_3 given by (10.8) is a reasonable first-order approximation. Le Moüel and Courtillot (1982) point out that both the mechanical and diffusion time constants must be combined in a complete description of the process.

There are several unknown factors that make any of the theories of doubtful application. Only the lowest-order poloidal magnetic field terms are available at the earth's surface and the downward continuation of these terms to the mantle–core boundary greatly magnifies errors (Chapter 2). The toroidal magnetic field strength and harmonic distribution are only very crudely estimated (§8.3.4) and so the magnetic field input into equations such as (10.3) can only be a rough guess. In addition, hydromagnetic waves with wavelengths and periods so small that they are not observable at the earth's surface might significantly affect the electromagnetic coupling (Hide, 1966). Finally, the mechanism producing differential rotation, and the magnitude and distribution of the velocity shear are all poorly known (§8.3.4). In spite of those difficulties it will be shown in the next section that electromagnetic coupling probably varies significantly over millions of years.

10.4.3 Variation in Time and Space

Gilman (1977) and Busse and Cuong (1977) have shown that convection in rotating spherical shells produces latitudinal variation in heat flux at the outer surface. In particular Gilman's study, carried out with the sun in mind, indicated a major maximum of heat flux in a narrow belt about the equator, a minimum at mid-latitudes and a minor maximum in the polar region. Although in both cases the authors have assumed that magnetic fields do not significantly alter the convection, long wavelength variations of heat flux at the core–mantle boundary are to be expected even if such effects were included. Also a characteristic of rotating shells is that the flow is not axisymmetric, possibly explaining why Cowling's Theorem is not violated in the earth's core (§8.2.3). The above indicates that spatial variation in the boundary conditions at the core–mantle boundary probably exist at any one given time.

Dzienwonski *et al.* (1977) have argued that body wave velocity data indicate heterogeneity of the lowermost mantle. In view of the above, some heterogeneity should probably be expected, although its magnitude and thus its detectability is conjectural. If spatial variations in core–mantle boundary conditions are large enough and long-lived, they may be manifested in the time-averaged magnetic field data. The first claim to evidence for long-lived spatial asymmetry involved the so-called dipole window in the Pacific Ocean beneath Hawaii. This turned out to be a data artifact, as has most evidence supporting long-lived spatial asymmetries (§6.3). Probably the best evidence for spatial asymmetry in the time-averaged magnetic field is the data showing asymmetry between the northern and southern hemispheres over the past five million years (Merrill and McElhinny, 1977; Lee and McElhinny, 1984). The difference is small but significant at the 95% confidence level (§6.2.2). Indeed there is evidence that this asymmetry was larger about 35 Ma ago (Coupland and Van der Voo, 1980; Lee and McElhinny, 1984) (see §6.2.3).

A far more convincing case can be made that the palaeomagnetic data necessitate variations over long times in the core–mantle boundary conditions. McElhinny (1971) first pointed out that time constants of the order of 10^8 years in the reversal frequency are present in the palaeomagnetic record. Phillips (1977) carried out detailed statistical analyses which supported McElhinny's view and also suggested that there were changes in the rate reversals occurring around 45 and 76 m.y. B.P. Phillips also suggested different stabilities for the normal and reverse magnetic field, a puzzling result in the light of his conclusions that the mean polarity times for reverse and normal fields are equal throughout the Cenozoic. This puzzle was removed when McFadden (1984) and McFadden and Merrill (1984) pointed out that Phillips' statistical analysis was very unstable to small changes in the reversal chronology (§5.3.3). Apparent differences in stability effectively vanish with the insertion of only two reversals in the entire Cenozoic era. However, long-term changes of the order of 10^8 years in the reversal frequency are clearly seen in the data and are not very sensitive to minor changes in the reversal chronology (§5.3).

The earth's magnetic field has a free decay time of the order of 10^4 years, four orders of magnitude smaller than the 10^8 years characteristic time for changes in reversal frequency. It seems likely that this longer characteristic time reflects changes in boundary conditions at the core–mantle interface as suggested by Jones (1977). Jones used an intermittent convection theory developed by Foster (1971) to argue that the roughly 150 km thick seismic D″ layer (Table 7.1) at the base of the mantle is a thermal boundary layer which intermittently breaks down by the formation of blobs or plumes. Figure 10.3 illustrates the situation as given by Foster (1971) applicable to a high Rayleigh number fluid ($\sim 10^7$) heated from below. There is a relatively rapid period of time in which a thermal boundary layer breaks down followed by a much longer period of time in which the temperature gradient increases by thermal conduction. Foster (1971) suggests that this asymmetry in build-up time of the thermal boundary relative to its breakdown time is probably a necessary condition for fully intermittent convection. McFadden and Merrill (1984) point out that the reversal data, which show a continuous increase in

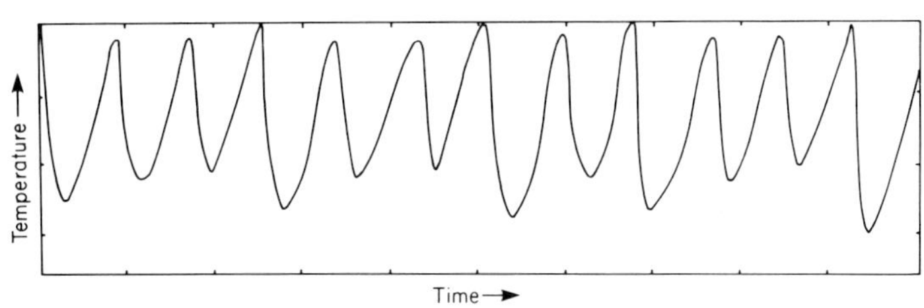

Fig. 10.3. Intermittent convection in a thermal boundary layer. Time variation of the temperature at the bottom of an intermittently unstable boundary layer for Rayleigh number 10^7 at a horizontal wave number L for optimum growth. Note the faster decrease in temperature when the thermal boundary layer breaks down relative to the increase in temperature during the thermal conduction phase. (After Foster, 1971.)

frequency since the long Cretaceous quiet interval, are incompatible with the changes expected if D″ is a simple thermal boundary layer as envisaged by Jones (1977). However, they may be compatible with more complex layer models for D″ (Ruff and Helmberger, 1982). Mantle convection is still very poorly understood. Suggested models include convection confined to the upper 700 km, convection in large mantle sized cells, and convection in various layers throughout the mantle separated by thin layers in which only thermal conduction occurs (Hager and O'Connell, 1978; Stevenson and Turner, 1979; Turcotte, 1980; Olson, 1981a; McKenzie and Richter, 1981). McFadden and Merrill (1984) point out that the palaeomagnetic data cannot distinguish between most of these, although the data do seem to require some convection, at least intermittently, deep in the mantle. In addition, some mantle layering appears to be necessary, although this conceivably could be restricted to D″. McFadden and Merrill (1984) also presented reasons why the temperature is probably maximum at the base of the mantle during very long polarity intervals, such as the nearly 30 Ma long normal polarity interval in the Cretaceous.

Electromagnetic coupling between the core and mantle depends on several interrelated factors, one of which is lower mantle electrical conductivity (§10.4.2). Because the electrical conductivity of semiconductors increases exponentially with temperature and because at least some of the properties of the magnetic field vary on the same time-scales as temperature, McFadden and Merrill (1984) have also speculated that electromagnetic core coupling may vary significantly with time-scales of the order 10^8 years. The important point however is the recognition that long period variations in the earth's magnetic field provide one of the few available tools using observational data to obtain information on heat transfer deep in the earth's interior in the past.

Chapter Eleven

Lunar Magnetism

11.1 Lunar Structure and Interior

Conversion of gravitational energy associated with accretion of the moon 4.6 Ga ago produced a magma ocean over the surface of the moon for which depth estimates range from one hundred to a few hundred kilometres. Subsequent cooling and differentiation of this magma ocean produced gabbros and anorthosites, rocks that are the major constituents of the lunar highlands. In Pre-Imbrian times (the period in which the highlands were formed), the moon was under heavy bombardment by planetesimals and large meteorites. This bombardment decreased considerably following a period of great impacts around 3.9–4.0 Ga ago. Subsequently the lunar maria formed when basalt melt (which earlier formed from the partial melting of rocks beneath the pre-Imbrian formed crust) flowed into large impact basins. The lunar samples returned from the maria have radiometric ages ranging from 3.1 to 3.9 Ga. Relative age relationships suggest that some maria, from which no samples were returned, may well have ages as young as 2.5 Ga (Head, 1976). It is doubtful that much volcanism, if any at all, has occurred during the last 2.5 Ga.

Although the major features of the moon have remained essentially unchanged during the past 2.5–3.0 Ga, the lunar surface has been subsequently broken up into layers of debris by the constant bombardment of meteorites, micrometeorites, cosmic rays, and solar wind particles. There are some exceptions—for example, Tycho crater was probably formed by a meteorite impact only 300 Ma ago. The surface layer of debris, which exists almost everywhere and is typically a few metres thick, is referred to as the lunar regolith. Essentially all returned samples come from this regolith and can be assumed to have been subjected to shock and radiation processes. In addition to forming the regolith, the constant rain of particles slowly mixes the upper parts of the regolith, a process known as "gardening".

The moon has a mean density of 3.34×10^3 kg m^{-3} and a moment of inertia of $0.391 \pm 0.002\ M_aR^2$ (M_a = total mass, R = radius (Blackshear and Gapcynski, 1977)). Together with the geological evidence, these values are consistent with the belief that some differentiation has occurred. However, both the extent of this differentiation and the present lunar structure are only crudely known.

As for the earth, seismic evidence provides the major information regarding the lunar interior. Four seismometers, and a gravity meter sometimes used as a seismometer, sent data back to the earth for several years until the instruments were turned off in the latter half of 1977. The seismic sources are moonquakes, meteorite impacts and controlled impacts from various Apollo spacecraft stages that were crashed onto the lunar surface. There are several criteria for distinguishing meteorite impacts from moonquakes, including the observation that the signal from meteorite impacts has a much smaller shear wave component. Most moonquakes have Richter magnitudes less than 3 and are found at a depth between 700–1000 km. A few near-surface moonquakes also occur and the highest recorded lunar magnitudes (between 3 and 4) have been associated with these surface events. Because of the locations of the instruments and the low magnitudes of the moonquakes, the seismic structure is essentially determined for the near side. All Apollo sites were on the near side, so that radio contact could be maintained with the astronauts.

The seismic data suggest that there is a crust that is roughly 70 km thick, an upper mantle that extends to a depth of 300 km (Nakamura *et al.*, 1976) or 500 km (Dainty *et al.*, 1976), below which there is a lower mantle with a decrease in the velocity of shear waves and probably compressional waves, as well as an increase in their attenuation. The different estimates of the extent of the upper mantle provide an indication of the problems of data resolution. Very little seismic evidence exists concerning the region extending from 1000 km to the lunar centre at 1738 km depth ($\sim$47 kb pressure). Evidence for a lunar core from seismic and other information is discussed in §11.3 and §11.5. Because of the non-uniqueness of the problem, there is a wide variation in possible lunar structures, not all of which have been given here (e.g. see alternative models summarized by Taylor, 1982).

There are some important asymmetries regarding the moon. These include a predominance of maria on the near side of the moon and their associated mascon gravitational anomalies, a 2 km offset of the centre of mass from the lunar centre towards the earth, and possibly a thicker crust on the far side of the moon (Kaula *et al.*, 1974). For a more detailed account of the lunar structure and interior, readers are referred to Taylor (1975, 1982).

11.2 The Present Magnetic Field

11.2.1 The Electron Reflection Method

Information concerning the lunar magnetic field has been obtained from direct measurements using surface magnetometers, satellite and subsatellite magnetometers and from indirect measurements such as the electron reflection method. Measurements made on samples returned to the earth attempt to provide information regarding the past magnetic field. The electron reflection method, first used by Howe *et al.* (1974), is relatively new and is not well known to most geomagnetists.

The gyromagnetic frequency, ω_0, is that frequency for which an electron, with charge e and mass m, rotates about a field $\boldsymbol{B}$. In equilibrium, the centrifugal force and Lorentz force are in balance:

$$e\boldsymbol{v} \times \boldsymbol{B} = \frac{mv^2\boldsymbol{r}}{r^2}$$

where $\boldsymbol{v}$ is the velocity of the particle and r is the radius of the orbit. Because $v = \omega_0 r$,

$$\omega_0 = \frac{qB}{m} \tag{11.1}$$

The moon has an abundant supply of electrons associated with the solar wind and with the geomagnetic tail of the earth's magnetosphere. Also it essentially has no atmosphere so that those electrons do not interact with atmospheric gases. In the absence of a surface magnetic field, electrons would be absorbed on collision with the moon. Indeed less than 5% of the electrons are subjected to coulomb back-scattering (Howe *et al.*, 1974). In the presence of a magnetic field the electrons are reflected back adiabatically if the spatial variations in the magnetic field are large compared with the gyro-radii of the electrons. The physics of this process of reflection, sometimes also referred to as mirroring, is described below. The non-adiabatic case, in which the spatial variations in the magnetic field are smaller than the gyro-radii of the electrons, requires numerical calculations on a computer and will not be treated here (see Howe *et al.*, 1974; Anderson *et al.*, 1976).

Figure 11.1 illustrates the adiabatic case and assumes cylindrical symmetry for simplicity. Electrons that spiral along the field lines will convert more and more of their translational energy into rotational energy until reflection occurs. As an example assume that an electron has a radial velocity component, v_r, and a z velocity component, v_z, where $v_r \gg v_z$. The total angular momentum of the particle can then be well approximated by $mr^2\omega_0$,

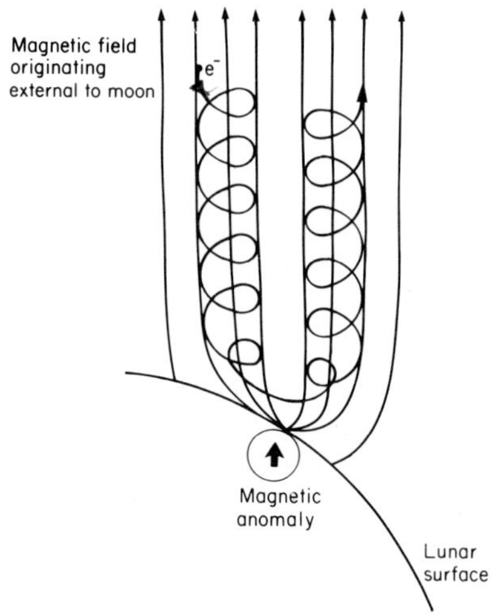

Fig. 11.1. The electron reflection method. An electron spirals in along magnetic field lines and is reflected adiabatically from a magnetic crustal source lying beneath the lunar surface. (After Anderson *et al.*, 1977.)

where ω_0 is the gyro-frequency. Equation 11.1 shows therefore that Br^2 must be constant. Br^2 is sometimes referred to as an adiabatic invariant and is equivalent to saying that the magnetic flux ($B\pi r^2$) contained within an electron orbit must be constant. The magnitude of the total velocity, v, is given by:

$$v^2 = v_r^2 + v_z^2 \tag{11.2}$$

Because $v_r = \omega_0 r = qrB/m$, the adiabatic invariant conditions can be expressed by saying that v_r^2/B must be constant. Let $B = B_0$ at some reference point $z = z_0$, then from (11.2)

$$v_z^2 = v_0^2 - v_{r0}^2 \frac{B}{B_0} \tag{11.3}$$

This last equation illustrates mathematically how electron reflection (mirroring) occurs. As an electron moves downward ($-z$) in a spiralling motion about the field lines, it encounters stronger B fields. Therefore, B/B_0 increases until v_z goes to zero in (11.3). The point is referred to as the mirroring point of the electron.

In more general cases, the incident electron makes some angle, α, with the direction of the magnetic field. In this case the theory can be extended to show that mirroring occurs when:

$$B = \frac{B_0}{\sin^2 \alpha}$$

Of course, any electron with a velocity that is parallel to the local magnetic field will not experience mirroring. Thus, there is a critical angle, α_c, below which mirroring ceases for a given velocity. The surface field, B_s, must then be given by

$$B_s = \frac{B_0}{\sin^2 \alpha_c} \quad (11.4)$$

The application of this technique to the moon involves counting the number of electrons that pass a satellite in a given direction with a given energy. For example, 0.5 keV and 14 keV electrons were used in the Rima Sirsalis study discussed below (Anderson *et al.*, 1977). The number of electrons that are mirrored back are counted, from which α_c can be determined since a sharp decrease in the electrons undergoing mirroring occurs at α_c. Measurement of B_0 with knowledge of α_c then determines the field at the moon's surface from (11.4). The best results occur for cases in which the field direction does not vary rapidly in space.

11.2.2 Magnetic Anomalies

The earlier spacecraft magnetometer measurements indicated that, at the present time, the moon only possesses a very small magnetic field (Dolginov *et al.*, 1961, 1966; Sonett *et al.*, 1967). Thus it came as somewhat of a surprise to find high values of remanent magnetizations from samples returned from the moon (see §11.4). The surface, satellite and subsatellite magnetometer data clearly indicate that surface magnetic fields range up to values that occasionally exceed 300 nT (Dyal *et al.*, 1970; Barnes *et al.*, 1971; Coleman *et al.*, 1972; Sharp *et al.*, 1973; Hood, 1981). At present, magnetic measurements are available for roughly 15% of the lunar surface. The observed fields are known to originate primarily from magnetization in the lunar crust. The observed magnetic field usually varies considerably, both in intensity and direction. As a result only an upper limit for a lunar magnetic dipole moment of the order of 1.3×10^{15} Am^2 can be obtained, compared with

8×10^{22} Am^2 for the present dipole moment of the earth (Russell *et al.*, 1974; Russell, 1978).

Although highly variable, there appears to be some general correlation between these magnetic measurements and surface features and chemistry (Lin *et al.*, 1977; Russell *et al.*, 1977; Soderblom *et al.*, 1977; Hood *et al.*, 1979; Anderson and Wilhelms, 1979; Hood, 1981). The highlands generally exhibit stronger magnetic anomalies than the maria. Because the highlands are relatively less abundant in the radioactive elements Th, U and K, there appears to be an inverse correlation between magnetic anomalies and rocks containing U, Th and K. Since the highlands are older than the maria, there is also a general correlation between age and magnetic anomalies. The rock magnetic properties also vary between maria and highlands (§11.4), so it is interesting to see if any age–magnetic relationships exist within the maria themselves. Preliminary studies suggest that there is such a relationship, statistically speaking, the strongest magnetic anomalies are exhibited by the oldest maria (3.5–4.0 Ga old) and seem to decrease with age with the smallest anomalies being associated with the youngest maria ($\simeq$2.5 Ga old).

Whether the above correlations imply some cause and effect relationship is controversial. The problem is complex in view of the number of factors that can alter magnetic anomaly strength, for example:

(i) the variability of magnetic properties within rocks;
(ii) the spatial and temporal variability of the inducing magnetic field;
(iii) the inhomogeneity in the magnetization brought about after formation (for example, the breaking up of rocks by impacting bodies);
(iv) spatial variability in the temperature distribution with depth (faster increase in temperature implies the Curie temperature will be reached at shallower depths).

Information on the spatial extent of magnetic anomalies is important because from this the minimum wavelength of the inducing field can be roughly estimated. Many of the results (Lin *et al.*, 1977; Metzger *et al.*, 1977; Russell *et al.*, 1977; Soderblom *et al.*, 1977) suggest that although the wavelengths of the magnetic anomalies are one to a few hundreds of kilometres in extent, small-scale fluctuations, typically of 10 kilometres (e.g. Lin, 1978), are usually imposed on the broader anomalies. Thus, it appears likely that there is a fairly rapid spatial variation in the remanent magnetization associated with the anomalies. These results have also been interpreted as implying that no large-scale inducing field, more than a few kilometres in extent, was present when the rocks formed. However, there are several factors, such as variations in the intensity of the remanence, that make unique interpretations impossible.

Strangway *et al.* (1973) and Anderson and Wilhelms (1979) have suggested

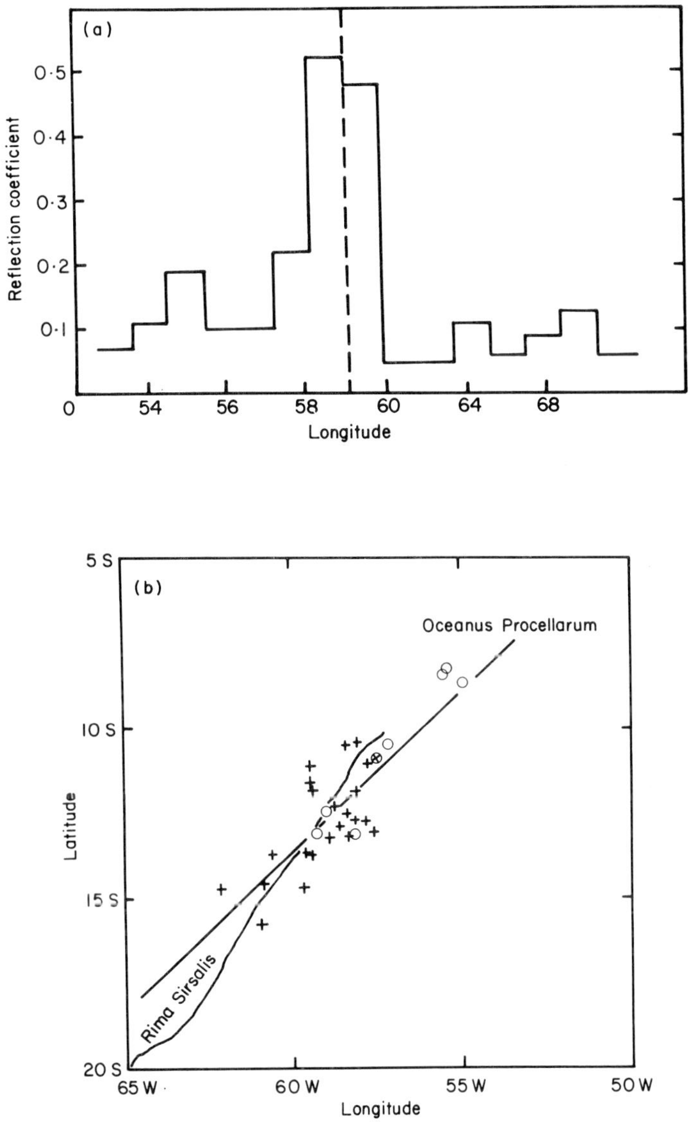

Fig. 11.2. The Rima Sirsalis magnetic anomaly. (a) 14 keV electron reflection data plotted versus longitude for a crossing of the Rima Sirsalis rille at 12.5°S (located by the dashed line). (b) 0.5 keV (+) and 14 kEV (○) electron maximum reflection peaks. A direct magnetometer measurement (⊗) is consistent with the electron reflection data. The least squares linear fit to the data is shown by the straight line. The location of the Rima Sirsalis rille is shown by the wavy line. Data from an Apollo 16 subsatellite flown in 1972. (After Anderson *et al.*, 1977.)

a correlation between magnetic anomalies and basin ejecta. Hood *et al.* (1979) found that significant anomalies were absent over several impact craters (e.g. Copernicus) but were present near craters with ejecta (e.g. Cayley Formation).

The most significant anomaly data are those for Rima Sirsalis, because the anomaly associated with Rima Sirsalis is probably the strongest indication that the moon once had a magnetic field with moderately large dimensions. Rima Sirsalis is a structural rille, a graben-like feature that is a few kilometres in width and depth and extends a few hundred kilometres in length (Fig. 11.2b). The magnetic anomaly is barely resolvable using satellite magnetometer data, but is reasonably well determined by the electron reflection method. Figure 11.2a gives some of the data used by Anderson *et al.* (1977) showing the large increase in electron reflections associated with Rima Sirsalis. Figure 11.2b shows the best least squares fit to the peak in electron reflection data along with the geographical location of Rima Sirsalis; the fit is clearly very good. The surface magnetic field strength is greater than 100 nT over a region of the order of 10 km in width and about 300 km in length (Anderson *et al.*, 1977). Anderson *et al.* (1977) show that these results can be modelled to a first approximation by assuming a uniformly magnetized lunar surface from which magnetized rock coinciding with a rille with a width of 10 km and depth of a few kilometres has been removed. These authors recognize the non-uniqueness of their modelling pointing out that an equally acceptable model at the other extreme is one in which a magnetic dyke coincides with the trace of the rille. This model is actually preferred by the authors since the magnetic anomaly continues on after the structural manifestation of the rille disappears under Oceanus Procellarum.

It is important to point out that although the magnetic models which fit the Rima Sirsalis data are very non-unique, all the models seem to necessitate some coherence in magnetization over a few hundred kilometres. Any hypothesis of lunar magnetism that exclusively invokes random magnetization processes is therefore difficult to sustain.

11.3 A Lunar Core?

11.3.1 Magnetic Sounding

Because the moon moves in and out of the geomagnetic tail (Fig. 11.3), the induced lunar magnetic field can be estimated. Assume that the geomagnetic tail is approximated as a uniform field, B_u, far removed from the lunar surface. In the magnetostatic case the moon has a uniform magnetic

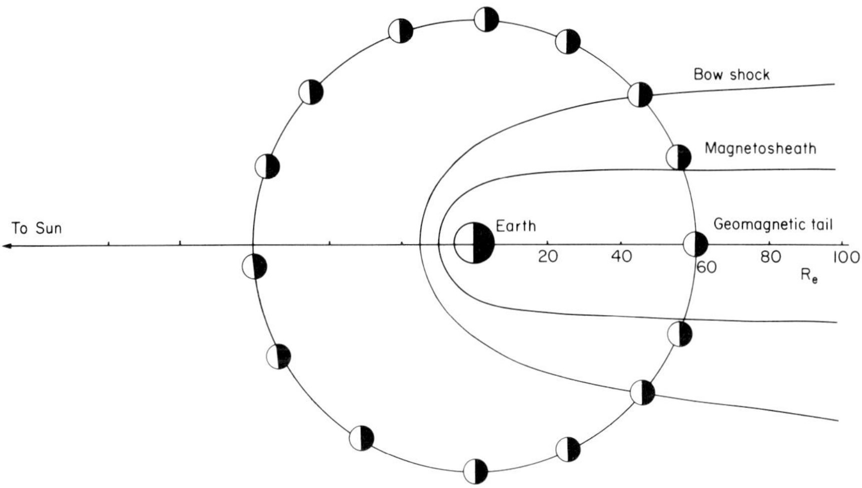

Fig. 11.3. The moon moves in and out of the earth's magnetosphere. The varying magnetic field present at the lunar surface induces a magnetization in the moon. Analyses of the induction effect provides an estimate of the lunar permeability.

permeability, μ_1. Then $\nabla \cdot \boldsymbol{B} = 0$ and $\nabla \times \boldsymbol{B} = \boldsymbol{0}$, implying that there is a potential, ψ, such that:

$$\boldsymbol{B} = -\nabla\psi \tag{11.5}$$

and

$$\nabla^2\psi = 0$$

For spherical co-ordinates (r, θ, ϕ) and assuming spherical symmetry, so that the ϕ part drops out, then

$$\psi = \sum_{n=1} \frac{a_n P_n}{r^{n+1}} + \sum_{n=0} b_n P_n r^n \tag{11.6}$$

where P_n is the *nth* order legendre polynomial. A boundary condition for the uniform field assumption is:

$$\psi(r, \theta)_{r\to\infty} = -B_u r \cos\theta + c \tag{11.7}$$

where c is constant.

An additional boundary condition, stemming from $\nabla \cdot \boldsymbol{B} = 0$, is:

$$\frac{1}{\mu_0}\left[\frac{\partial\psi}{\partial r}\right]_{\substack{r\to a \\ \text{outside}}} = \frac{1}{\mu_1}\left[\frac{\partial\psi}{\partial r}\right]_{\substack{r\to a \\ \text{inside}}} \tag{11.8}$$

where $a =$ the lunar radius.

Equations 11.6, 11.7 and 11.8 are easily solved for ψ and using (11.5) one obtains B everywhere:

$$B_r = B_u\left[1 + 2\frac{\mu_0}{\mu_1}\frac{a^3}{r^3}\right]\cos\theta$$

$$B = B_u\left[1 - \frac{\mu_0}{\mu_1}\frac{a^3}{r^3}\right]\sin\theta$$

Therefore, by measuring B_u (away from the moon) and B at the lunar surface, the mean magnetic permeability for the moon, μ_1, can be obtained. In practice a changing field for B_u is used and the change in B at the lunar surface is investigated since the unchanging part of B is essentially due to the remanent magnetization. This method was first applied by Dyal and Parkin (1971), who found a value of μ_1 near 1.013 μ_0.

Various shell models, which take into account the variability of μ (particularly above the Curie temperature of magnetic materials where μ drops drastically), have been used to estimate both a mean lunar permeability and the total free iron, Fe^0. By assuming that Fe^0 is primarily responsible for the observed induction effects, a value of 9% has been calculated (see Dyal *et al.*, 1974). This value has been widely used as a rigid constraint in many geochemical models for lunar origin, but it is not yet as well determined as commonly believed.

Russell *et al.* (1974) and Goldstein *et al.* (1976) have argued that other effects, such as a lunar core made of conducting material, could alter the above analyses. Because of the frozen-in-field theorem, a conducting core will produce a field opposite to the external variable field (see §7.4; moving a field into a conductor is equivalent to moving a conductor into a field). However, this effect is transient; the mean lunar permeability can reasonably be estimated a few hours after most large changes have occurred in the external magnetic field (Goldstein *et al.*, 1976).

The controversy over the existence of a lunar core is very complicated and not easily resolvable. Goldstein *et al.* (1976) concluded that a small lunar core is likely, disagreeing with Dyal *et al.* (1975) who claim that a core cannot be resolved within the uncertainty of the magnetic data. A different method of analysis, involving electromagnetic sounding, by Wiskerchen and Sonett (1977) and Sonett and Wiskerchen (1977) appears more consistent with the conclusions of Dyal *et al.* Some of these differences might also arise from:

(i) possible adverse effects from the presence of plasma in the tail of the earth's magnetosphere (producing diamagnetic-like effects);
(ii) possible calibration errors associated with data from Explorer 35 (King and Ness, 1977);
(iii) possible resolution and aliasing problems associated with magnetic

and electric properties (particularly the electrical conductivity) of the outer 1000 km of the moon.

It appears that the electromagnetic measurements cannot yet prove the existence or non-existence of a core, but they can provide a limit to the size of any conducting core. The maximum limit is around 400–500 km radius (maximum near 2%, by volume, of the moon). In addition, the mean permeability for the moon is not yet well known, but probably falls in the range of 1.006–1.018 μ_0. The uncertainties in the Fe^0 estimate are greater than the uncertainties in determining the mean lunar permeability, since further assumptions are required. The mean density of the moon (3.34×10^3 kg m^{-3}) is probably as good a constraint on the total amount of Fe^0 present, as is the constraint provided by the mean permeability estimate.

11.3.2 Other Evidence

The existence of a lunar core is very important in deciding whether the moon possessed a dynamo in the past. If proof for an ancient lunar dynamo were forthcoming, this would have several important consequences, including:

(i) that hydromagnetic dynamos could exist in very small bodies, and
(ii) the requirement of an unusual energy source, such as radioactive elements which have such short half-lives that they are no longer present (Runcorn, 1977; see §11.5).

Geochemical evidence for and against the existence of a lunar core is controversial. Brett (1973) suggested the presence of a small Fe-Ni-S core to explain the depletion, by factors ranging from 10^2 to 10^4, of siderophile elements (Au, Ag, Cr, Ni, etc.) relative to cosmic or terrestrial abundances (Anders, 1971). Alternate explanations for the depletion of the siderophile elements includes formation of a metallic core with a precursor planet or some sort of selective condensation-fractionation mechanism (Ringwood, 1978; Taylor, 1975). On the other hand, at least two geochemical arguments have been advanced which argue against a lunar core. The amount of ^{40}Ar at the lunar surface is apparently well above that which can reasonably be said to be coming from the sun (Hodges and Hoffman, 1975). This appears to be a reasonably strong argument against complete lunar differentiation, since ^{40}Ar in the upper regions of the moon is held tightly inside the crystal lattices and therefore cannot produce much of the observed ^{40}Ar at the lunar surface. Obvious evidence for this ^{40}Ar retention is the very old age of crustal rocks obtained in some cases by K-Ar dating. Presumably the observed ^{40}Ar originates in a primitive (not chemically altered) mantle at a few hundred kilometres depth. However, alternative core formation arguments that call on local differentiation and settling of Fe-FeS blebs for core formation (Goldstein *et al.*, 1976) cannot be ruled out by the ^{40}Ar observations. Ringwood (1978) believes that a lunar core is unlikely as a result of examining incipient melting in the laboratory for selected highland and mare samples. Ringwood thus argues that the mantle could never have been in equilibrium with an Fe-S melt.

A hot interior-core formation theory has been suggested by Runcorn (1976). However, this interpretation is in conflict with both the geochemical data given above and more careful thermal considerations which indicate that only the outer parts of the moon were very hot during the earliest times (Solomon and Head, 1979; Kaula, 1979). In particular Runcorn's model encounters difficulties with the lack of evidence for large-scale thermal contraction that would be expected if the entire moon were very hot for even a short period of time (Solomon and Chaiken, 1976; Solomon and Head, 1979). The present evidence, such as the restriction of grabens to rocks 3.4 Ga or older, indicates that only slight changes in the inner radius have occurred throughout lunar history (McElhinny *et al.*, 1978; Solomon and Head, 1979).

In principle, seismic data should be able to resolve the question of a lunar core. Indeed a large meteorite impact on the back side, was of sufficient magnitude to provide data on the lunar central region. The data from this single event have been interpreted as indicating rapid attenuation of seismic shear waves near the lunar centre and the possible existence of a molten region with radius of 170–350 km (Nakamura *et al.*, 1974). However, the resolution obtained from these data is clearly not good (Nakamura *et al.*, 1976; Dainty *et al.*, 1976). They are insufficient either to exclude the possibility of a larger core, or even to prove the existence of a core at all (Latham *et al.*, 1978; Goins *et al.*, 1977).

The moment of inertia of the moon ($0.391 \pm 0.002\, M_a R^2$) has been interpreted as evidence for a small Fe-S core with a few hundred kilometres radius (Solomon, 1974; Solomon and Head, 1979). Although the moment of inertia is less than that for a homogeneous sphere ($0.4\, M_a R^2$) and necessarily implies a density increase with depth, it still does not necessitate an inner core. Because the moment of inertia varies as the square of the distance from the axis about which it is calculated, small increases in density in the upper mantle have the same effect on the moment of inertia as do large density changes in the core. This ambiguity makes it impossible to conclude that a core exists from the moment of inertia alone.

In summary, three alternatives exist concerning a possible lunar core and central dynamo:

(i) a core exists and so did a central dynamo;

(ii) a core exists but not a central dynamo;
(iii) a core does not exist.

At present it is impossible on the basis of the evidence to rule out any of these alternatives. However, if a lunar core does exist, it must be very small—at least less than 500 km and probably less than 400 km in radius.

11.4 Magnetic Properties of Lunar Rocks

11.4.1 The Lunar Environment

Figure 11.4 shows the locations of the landing sites, including the American sites for Apollo 11 through 17, but excluding 13, from which samples were

Fig. 11.4. The location of the Apollo (A) and Luna (L) landing sites on the near side of the moon are shown relative to some of the mare (I = Mare Imbrium, C = Crisium, T = Mare Tranquilitatis, S = Mare Serenitatis and P = Oceanus Procellarum).

not returned. The first number used to describe a given lunar sample gives the mission number and thus the general location from which the sample was obtained. For example, rock 68415 comes from the location of the Apollo 16 mission. Very crudely, the rocks consist of soils, breccias, and igneous rocks. The chief magnetic carrier in these rocks is free iron (Fe^0), although nickel and cobalt are sometimes alloyed with the iron.

Ever since the first lunar samples returned to Earth were found to carry a remanent magnetization (Doell *et al.*, 1970; Nagata *et al.*, 1970; Runcorn *et al.*, 1970; Strangway *et al.*, 1970), a considerable effort has been made to determine the origin and properties of that remanence (see review by Fuller, 1974). Early concern that the remanence might be secondary in origin, particularly the possibility that the NRM was acquired after the collection of the sample on the lunar surface, was removed after the Apollo 12 mission carried an Apollo 11 basalt on a round trip to the moon and back through the entire receiving and sampling process. No significant secondary component was found for this basalt.

Although extensive and often successful efforts to determine the magnetic properties of lunar samples and to utilize this information to gain insight into the environments from whence they came has been undertaken, a detailed discussion of that work is inappropriate here (*The Proceedings of the Lunar Science Conferences* should be consulted for detailed work). Instead, some of the results will be discussed briefly to illustrate the lunar rock magnetic problems that must be considered for reliable palaeointensity determinations.

Some appreciation for the processes acting on the moon can be gained by considering the remanence in soils. The lunar regolith is formed from pre-existing hard rock at a rate of roughly 1 or 2 mm per million years. The gardening of the regolith (see §11.1 and Taylor, 1975) probably occurs at somewhat higher rates (statistically speaking, a few mm per million years). Because little chemical alteration of magnetic minerals might be expected in the near-perfect vacuum-like conditions at the lunar surface, it might be supposed that the magnetic carriers would be the same as those of the parent rocks. This is often not the case. Agglutinates, an important component of lunar soils, consisting of rock material welded together by glass, form during meteorite impacts into the soil (Kaula, 1971; Taylor, 1975). The heating associated with this welding occurs in a reducing environment due to gases such as hydrogen, which have been implanted on the surfaces of grains by the solar wind. Small superparamagnetic-single domain size Fe^0 grains probably form during the heating associated with the agglutinate-forming process, referred to as the Housley mechanism (Housley *et al.*, 1973). Because small uniformly magnetized Fe^0 grains can easily be detected and distinguished from larger multi-domain grains, ferromagnetic resonance techniques

(Morris, 1976; Housley *et al.*, 1976; Gose and Morris, 1977) have been used to identify agglutinate content in soil samples and to obtain some indication of soil maturity (more mature soils contain more small Fe^0 grains).

The above example shows how magnetic properties can be used to study the origin and evolution of lunar samples. It also serves to summarize the complex processes that can affect the magnetic minerals. In the case above, the initial magnetic material was fractured, mostly by meteorite and micrometeorite impacts, to form a primitive regolith. Subsequent impacts produced shock and heating effects in the initial magnetic material as well as new magnetic material by the Housely mechanism. Also additional Fe-Ni material is often introduced by the micrometeorites themselves. Afterwards, the samples are subjected to the lunar diurnal temperature cycle and to viscous magnetization effects.

Many other rocks also indicate complex magnetic histories. For example, breccias often show several generations of brecciation. Some breccias have recrystallized, probably at temperatures above 770°C, the Curie temperature of Fe^0. Other breccias probably have not been uniformly reheated to high temperatures. The presence of substantially different directions of remanent magnetization between different clasts from the same breccia sample has been demonstrated by Gose *et al.* (1978).

The magnetic classification of lunar rocks can be made in at least three different ways:

(i) the amount and composition of the magnetic minerals;
(ii) the grain size of those minerals—superparamagnetic, single domain, pseudosingle domain, or multidomain—and
(iii) the nature of the remanent magnetization (presence of secondary components, etc.).

Using a two parameter hysteresis classification (method ii), Wasilewski (1972) pioneered a method which could be used for quick identification of rock types. These procedures and some extensions have been reviewed by Fuller (1974) and discussed in Chapter 3. Broadly speaking igneous rocks and recrystallized breccias appear to consist primarily of multidomain grains, while almost all soil and soil breccia samples have magnetic grains that are in the superparamagnetic to single domain size range (except for the orange soils of Apollo 17).

The few examples given above illustrate how the study of magnetic minerals in lunar samples provides important data on the classification, origin and history of the moon. In particular, they indicate that the magnetic properties vary from one type of lunar terrain to another. The important question as to whether these rocks are suitable for palaeointensity estimates is discussed in the next section.

11.4.2 Rock Magnetic Problems

The basis for the most reliable palaeointensity techniques are outlined in Chapter 3. The nature of the NRM and the problems encountered in the use of these techniques to determine lunar palaeointensity estimates is outlined below. A hypothesized TRM component must first be isolated from a sample and then a reliable palaeointensity estimate must be determined from that component. Usually both are attempted at once by applying one of the consistency checks discussed in Chapter 3. The argument used is that only a TRM component that reliably records the field intensity would pass the consistency checks.

Most workers are quick to emphasize that they are only trying to obtain estimates of the field that are reliable within a factor of 2 or 3, or even within an order of magnitude. This argument has often been misused. For example, a variation by only 30% in the field estimated for a given sample should cause serious doubt regarding the consistency check employed. The NRM in such a case may not be a TRM at all and the resulting palaeointensity may be completely wrong, even though *only* 30% variation occurred in the estimates (see §3.4). Thus lunar palaeointensity estimates have been extremely controversial. Interpretations range from the extreme in which the NRM is believed to be essentially unrelated to TRM (Brecher, 1976) and, therefore, that there are no reliable palaeointensities for the moon, to the opposite extreme that sufficient palaeointensity data are available to discuss trends in the strength of the ancient lunar field (Stephenson *et al.*, 1974, 1977; Runcorn, 1978).

To make sense out of this controversy, it is helpful to consider the magnitude of the problem. Suppose there is an initial TRM (even this is debated; see Brecher, 1976), then the factors that could subsequently alter this TRM on the moon are: (i) chemical effects, (ii) irradiation effects, (iii) reheating effects, (iv) diurnal temperature cycle (roughly from −170°C to +140°C), (v) viscous effects, and (vi) shock effects.

Chemical effects on older rocks from the earth are common and make palaeointensity determinations from such rocks very difficult. The near vacuum-like conditions on the moon are often argued to result in little subsequent alteration, unless reheating occurs. However, there are at least two possible exceptions to this. The first is that Fe-Ni alloys, found in some lunar samples, typically undergo phase changes below their Curie temperatures (Wasilewski, 1974). For this reason alone, it might be wise to avoid using samples containing much Ni for palaeointensity studies. The second concerns the possibility of the breakdown of antiferromagnetic troilite (FeS) to ferromagnetic Fe and non-magnetic S at temperatures of a few hundreds of degrees (Larson *et al.*, 1974).

Fuller (1974) has reviewed the evidence on irradiation effects and found that such effects appear to play a negligible role on changing the stable part of the remanent magnetization. Samples that have been reheated should probably be avoided unless the sample has been heated above its Curie temperature (for example, recrystallized breccias could be acceptable). This follows since magnetic mineralogical changes may occur on reheating and because reheating to temperatures below the Curie temperature assures one that even in the best circumstances two components of TRM are likely, greatly compounding the difficulties in demonstrating that the consistency check associated with palaeointensity techniques has been satisfied. Since the moon presently has a very weak field, the effects of the diurnal temperature cycling should be to demagnetize regions with blocking temperatures lying below 137°C.

Viscous effects are more difficult to treat. Both the decay of NRM and any VRM acquisition can usually (but not always) be removed by AF demagnetization in a peak alternating field of 10–20 mT or by thermal demagnetization to 300°C or so. However, the understanding of multidomain remanence and its relaxation time spectrum is poor (see Chapter 3). Since multidomain grains are common both in the mare basalts and in the recrystallized breccias (§11.4.1), the rocks most likely to acquire an *initial* TRM, the problem is complicated.

The potential effects of shock on the remanence are complex, but can be summarized as follows:

(i) shock may produce new magnetic carriers (Housely *et al.*, 1973; Fuller, 1977; Cisowski and Fuller, 1978);
(ii) shock remanence can be acquired and the amount acquired is correlated with the amount of low coercive force material within the rock (Fuller, 1974; Cisowski *et al.*, 1975, 1976, 1977; Cisowski and Fuller, 1978). Unfortunately, multidomain grains statistically have lower coercive forces than single domain grains and they are common in those rocks thought most likely also to carry a TRM (igneous rocks and crystallized breccias, see §11.4.1);
(iii) the magnetic stability, as measured by AF demagnetization, for example, can be altered by shock. Cisowski and Fuller (1976) have shown that an impact bringing a mare basalt to the surface in a low magnetic field environment would lead to an increase in the stability of the NRM;
(iv) the direction of the remanence can be changed (Cisowski *et al.*, 1975, 1976, 1977; Fuller, 1977; Cisowski and Fuller, 1978); and
(v) some magnetic fabric might be introduced (Brecher, 1976).

Thus, even if a pure TRM were acquired by a lunar rock, it would be demagnetized up to 137°C or so and its remanence would be likely effected by shock, and perhaps by other processes as well. The manifestation of these processes are to be expected in the palaeomagnetic data.

11.4.3 Palaeointensity Estimates

Part of the problem in assessing the reliability of lunar palaeointensity determinations is that different reliability criteria are used. Coe *et al.* (1978) have discussed statistical procedures for dealing with terrestrial samples and adaptation of these procedures to lunar work would be helpful. Differences in reliability criteria are illustrated by the examples in Fig. 11.5a, b showing two different views of the existing "reliable" lunar palaeointensity data (the data sets are not identical). The first (Stephenson *et al.*, 1977) argues for a linear decrease in the lunar palaeointensity between 4.0 and 3.1 Ga ago (Fig. 11.5a). However, a careful review of the data by Cisowski *et al.* (1977) suggests that no linear decrease is seen statistically (Fig. 11.5b).

One of the best lunar palaeointensity determinations has been obtained from a modified Thelliers' method by Collinson *et al.* (1973) (see Fig. 11.6). The ARM technique (§3.4.3) was also used on this rock and gave a similar estimate for the lunar field intensity (Stephenson *et al.*, 1974). The line is dotted at low temperatures in Fig. 11.6 because the first one or two values are suspect due to the lunar diurnal cycle. (Note that these points just happen to fall on the same straight line, one indication that consistency checks are not foolproof). Only four points are thus used for a consistency check, not an exceptional one by terrestrial standards. However, even without the bottom two points, the modified Thellier curve (Fig. 11.6) could still be judged acceptable and since the ARM method gave similar results (Stephenson *et al.*, 1974), it can well be asked as to why the estimate of about 10^{-4} T should be questioned. Pearce *et al.* (1976) have examined another piece of sample 62235 and found it contained Fe-Ni minerals (Kamacite and Taeinite), troilite (FeS) in addition to Fe^0, so that this rock is a poor sample to carry out the heating required in the modified Thellier technique unless far better atmospheric control was used than was available to Collinson *et al.* (1973). For example, troilite commonly breaks down at temperatures well below the upper two points on the palaeointensity curve shown in Fig. 11.6 (Larson *et al.*, 1974). Using both an improved vacuum compared with that used by Collinson *et al.* (1973) and an atmosphere of 10% H_2, 90% N_2 (Pearce *et al.*, 1976) obtained very complicated thermal curves for sample 62235. They interpreted their results as implying that strong interactions between magnetized regions occurred during cooling. Larson *et al.* (1974) and Taylor (1979) correctly point out that the fundamental problem in essentially all

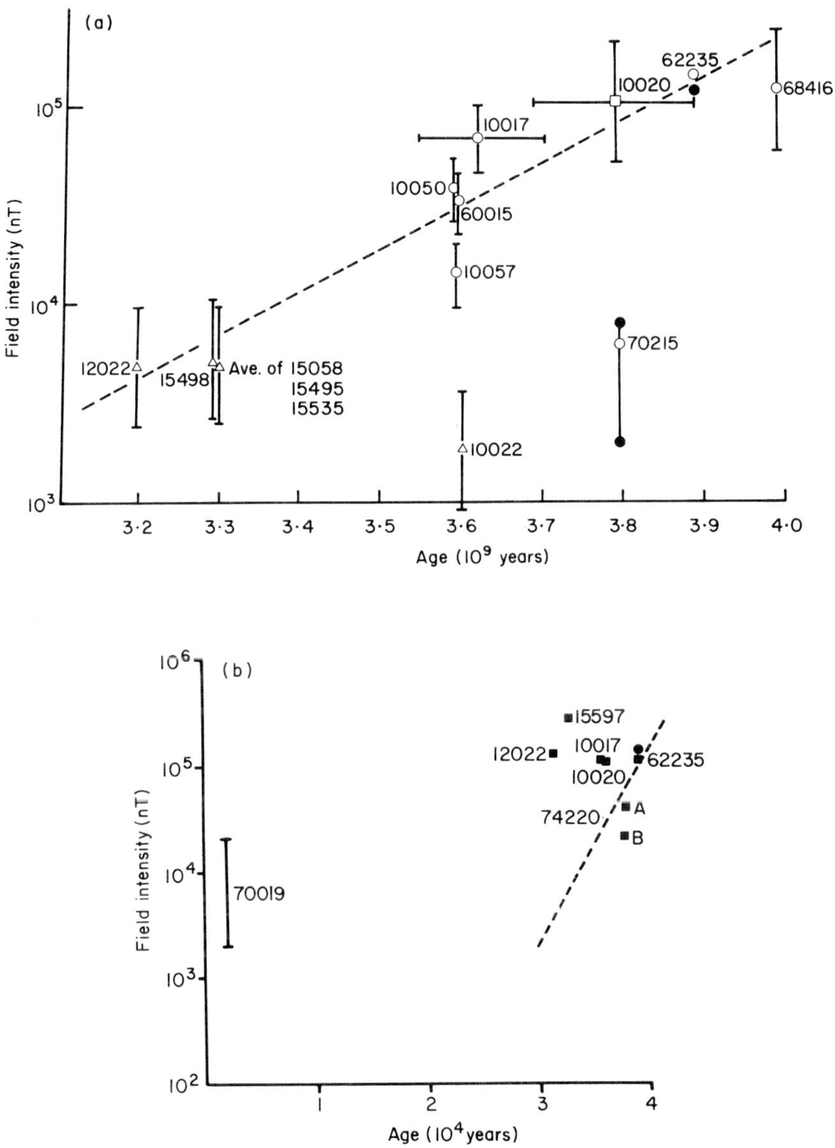

Fig. 11.5. Interpretation of lunar palaeointensities. Two views of "reliable" lunar palaeointensity results. (a) After Stephenson *et al.* (1977), (b) after Cisowski *et al.* (1977). The data shown in (a) have been used to argue for a linear decrease (dashed line) in the intensity of the lunar field between 4.0 and 3.1 Ga ago. This same proposed decrease is also shown in (b) by the dashed line. Note that different time scales have been used and different samples have been judged reliable. Sample numbers are indicated beside each point.

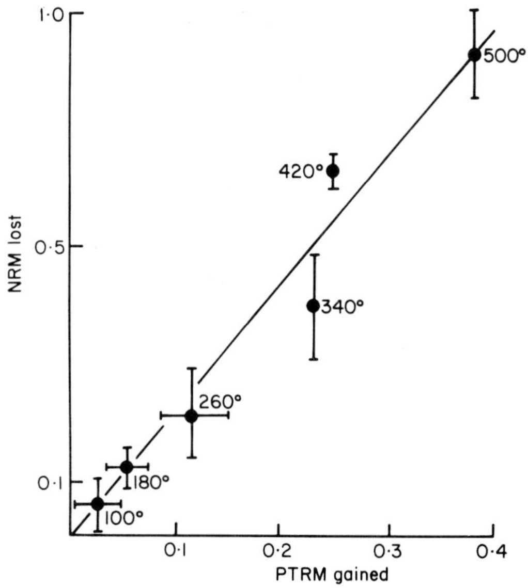

Fig. 11.6. Example of a lunar palaeointensity estimate. One of the best lunar palaeointensity estimates comes from the use of a modified Thellier method on sample 62235 (Collinson *et al.*, 1973). The temperatures used are given on the graph in °C. Note that the lowest one or two temperature values are suspect, because of the lunar diurnal temperature cycle. The estimated inducing field strength is 1.2×10^{-4} T.

viable palaeointensity techniques is to determine how to heat a sample without causing the breakdown of troilite (S buffering) or the oxidation of iron (oxygen buffering).

An additional problem with palaeointensity interpretation arises with the data of sample 70019, a rock reheated during a relatively "recent" crater impact (Cisowski *et al.*, 1977; see Fig. 11.5). Although it is common to suppose that high fields ($\approx 10^{-4}$ T) were present in the early history of the moon (first 1–2 Ga) and that low fields existed during the last 2 Ga, the data for sample 70019 suggest that a large magnetic field, between 3×10^{-6} T and 3×10^{-5} T, existed sometime during the last 200 Ma. These results have been supported by Sugiura *et al.* (1979a, b) who used a modified Thellier palaeointensity method on interstitial glass in this sample to obtain a value of 2.5×10^{-6} T. Magnetic anomaly fields of this magnitude are not expected at the Apollo 17 site, apparently demonstrating that there is, at least in one case, a mechanism for strongly intensifying the magnetic field locally, or affecting the remanence in some samples to produce an erroneous estimate of the inducing field.

The interpretation of the lunar palaeointensity data is controversial and is likely to remain so in the immediate future. At present most workers believe that the best estimate for the strength of the lunar magnetic field in the first 1–2 Ga is of the order of 10^{-4} T to 10^{-5} T (0.1 to 1 gauss). This follows because the handful of "best" lunar palaeointensity estimates give values in this range (Cisowski *et al.*, 1977; Sugiura and Strangway, 1980). Opponents question whether the "best" is good enough and whether there has not been some "reinforcement syndrome" in operation; that is, would a value radically different from 10^{-4} T be considered "poor" simply because one can always find something wrong with the details of a palaeointensity consistency check on any lunar sample and because the result does not agree with expectations? Also several samples have an NRM intensity value comparable to a TRM intensity acquired by the sample in a 10^{-4} T field (Fuller, 1974). Since TRM is one of the most efficient mechanisms known for acquiring a remanence, this result alone suggests a relatively high field. The counter argument acknowledges the fact that TRM is an extremely efficient mechanism for producing a remanence. However, opponents point out that there are mechanisms for magnifying a very weak remanence such as by chemical changes or shock processes. Therefore, even if there was an initially weak TRM, it could have been subsequently magnified and the estimates for the intensity of the lunar magnetic given above may be far too high. The estimates obtained from sample 70019 (Cisowski *et al.*, 1977; Sugiura *et al.*, 1969a, b) support the likelihood that either magnification of the remanence or local field magnification does happen sometimes. The electron mirroring data for Rima Sirsalis seem to require a relatively high intensity of NRM with a coherence in direction on a scale of 300 km or more. This result is interpreted to imply a strong inducing field with "moderate" scale lengths. Precise estimates of the strength of the inducing field cannot be established in this situation. This is probably the strongest argument for a moderately strong magnetic field early in the lunar history, but the argument is clearly indirect.

Since the field is very low today, the usual speculation that a field of 10^{-4} to 10^{-5} T was present on the moon sometime in its early history implies a decrease in field strength with time. The speculations of the Newcastle group (e.g. Stephenson *et al.*, 1977) that much of this decrease occurred between 4.0 and 3.0 Ga ago may be correct, even though this conclusion does not seem justified by the actual palaeointensity data (Fig. 11.5).

11.5 Origin of the Lunar Magnetic Field

Models to explain the alleged 10^{-4} T field early in the moon's history fall into two broad groups, those of internal and external origin. Several of the mechanisms proposed are outlined below.

(i) *Central dynamo*
This model was proposed by Runcorn (1978, 1979). Runcorn (1978) acknowledges problems in obtaining sufficient energy to derive a strong enough dynamo in a (maximum) 500 km radius core. Conventional earth dynamo sources seem inadequate (Chapter 7; and Runcorn, 1978). To avoid this problem, Runcorn suggests that heating by decay of super-heavy elements no longer present in the moon was the energy source (Runcorn, 1978). However, there are serious doubts regarding the existence of such elements in the abundance required for this mechanism to operate (e.g. Reedy, 1980). This raises serious questions about the existence of a central dynamo, even if a small lunar core is proved to exist. Also the properties of such a dynamo would probably be very different to those of the earth, particularly very short time constants such as a few tens of years for polarity reversals.

(ii) *Local Fe-FeS dynamos* (Murthy and Banerjee, 1973)
A dynamo mechanism is hypothesized to exist in a local region of Fe-FeS melt. There are serious problems associated both with the energy source and of maintaining a steady field in any given direction for even a short time (see Chapter 7). Also it is unlikely to explain the moderately large-scale anomaly observed for Rima Sirsalis. This model seems somewhat contrived and is not widely accepted.

(iii) *Thermoelectric processes*
A magnetic field is produced from thermoelectric currents that pass from one magma basin with a crust to another magma basin without a crust. The two basins thus have different temperatures, and the current circuit is completed above ground, passing through any ambient plasma present (Dyal *et al.*, 1977). Although this mechanism might explain some magnetic features, it also seems somewhat contrived and does not explain the Rima Sirsalis results.

(iv) *Primordial magnetization of the lunar core* (Runcorn and Urey, 1973; Strangway and Sharpe, 1974)
The crustal remanence is acquired in a magnetic field which originates from remanent magnetization in the lunar core or mantle. This primordial magnetization in the lunar interior is no longer present because of subsequent heating to temperatures above the Curie temperature. This mechanism is coupled to mechanism (vi) below and is discussed further there.

(v) *The magnification of pre-existing fields by shock, or the creation of fields by shock* (Srnka, 1977; Martelli and Newton, 1977; Srnka *et al.*, 1979)
Plasma forms on impact and strong magnetic fields associated with currents

in that plasma might produce the remanence. Alternatively, magnification of a pre-existing weak field might be intensified by the shock wave and produce the remanence. This might well be a very important mechanism. Because this mechanism could produce magnification of existing fields, a strong initial field is unnecessary. However, it is not clear how strong fields in a plasma could lead to the coherence of the remanence of rocks now in the lunar crust, such as seen in the Rima Sirsalis.

(vi) *Strong external field from the sun or the earth* (Runcorn and Urey, 1972; Strangway and Sharp, 1974; Banerjee and Mellema, 1976)
Strong fields might be associated with the T-Tauri phase of the sun (Banerjee and Mellema, 1976; see Chapter 12), or might be present when the moon was closer to the earth, but it is very doubtful that these strong fields would exist in the vicinity of the moon for one Ga or more. This mechanism can be combined with mechanism (iv) to bypass the time problem. However, a new time problem emerges. Can the supposed lunar core be kept cold enough for 1–2 Ga, and is it of sufficient size so that there is a large enough magnetic moment to produce a field of the order of 10^{-4} T at the lunar surface? Runcorn (1976) was an early advocate of this model but subsequently argued against it on the basis of this last argument. Nevertheless, the mechanism might be sufficient to produce some initial remanence which could subsequently be magnified.

The difficulty in choosing between the above models would be reduced if it was possible to distinguish experimentally between external and internal sources. Runcorn (1975) has developed a stimulating theorem to do this as illustrated in Fig. 11.7. This shows that little external field will be observed for the internal case, as opposed to the external case. Unfortunately, the oversimplified case theoretically treated by Runcorn (1975) is probably not applicable to the moon. The magnetization acquired from an "internal" source is likely to be very non-uniform. For example, even the largest possible lunar core is only 1/7 that for the earth. Therefore, the free decay time for the moon is 1/49 that for the earth (see §7.4) or roughly 200 years. Because this free decay time is small compared to the inferred life-time for the inducing field (as deduced from the ages of the mare), it is likely that rocks in close proximity would have substantially different directions and intensity, even if the field originated from a central dynamo. The only "internal" source that *might* be an exception to this is (iii). Similarly, the external field sources proposed would also vary considerably in intensity and with time.

In the limit that the remanence is random in direction, no external field would be measured today, regardless of whether the field which produced that remanence had an internal or external origin. Moreover, the magnetic

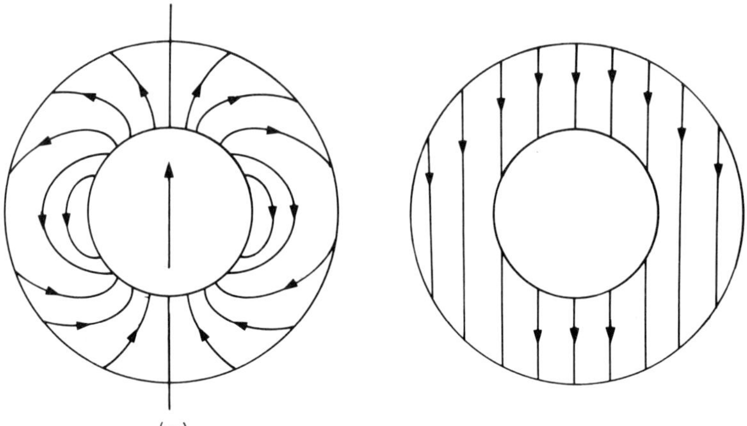

Fig. 11.7. Runcorn's theorem. The theorem states that if the susceptibility is high enough no magnetic field will exist outside of a spherical shell surrounding a dipole (or any other source). (a) illustrates the field lines for a spherical shell of intermediate susceptibility for which some field leakage occurs (see formal calculation by Stephenson (1976) or Merrill (1981)). (b) illustrates the case when the magnetization has been induced by a uniform field of external origin. (After Runcorn, 1975.)

anomaly measurements (§11.2.2) indicate that the remanence often does vary on a small scale over much of the moon (e.g. Hood *et al.*, 1979). Also magnetic anomalies are not associated with craters in maria (Lin, 1978; Hood *et al.*, 1979) casting serious doubts about the existence of large regions in maria with coherent magnetization. Thus, the theorem developed by Runcorn (1975) is probably not very useful in practice.

Serious problems thus exist for all the mechanisms proposed to explain the 10^{-4} T magnetic field speculated to be present during the first 1–2 Ga of lunar history. Perhaps the proper combination of mechanisms can remove some of these problems. Nevertheless, the most reasonable conclusion at present is that both the origin of the lunar remanent magnetization and the magnetic fields that produced that magnetization are still very poorly understood.

Chapter Twelve

Magnetic Fields of the Sun, Planets and Meteorites

12.1 Origin of the Solar System

Although it is generally agreed that major bodies of the solar system formed through the accretion of gas, dust and planetesimals 4.6×10^9 years ago, there is no general agreement over the detailed processes involved. The abundances of elements (excluding H, N and the noble gases) in the sun's photosphere are remarkably similar to those in type 1 carbonaceous chondrites. This observation has led to a widely-held belief that the elemental abundances in these carbonaceous chondrites represents the composition of the primitive solar nebula (see Taylor (1982) for a general overview of this subject).

Although details are at best sketchy, there appears to be a consensus that the sun formed at the centre of a rotating flattened disc of gas, dust and later planetesimals. In the outer, presumably colder, parts of this disc, substances such as water, methane and ammonia were in the form of solid crystals emersed in hydrogen and helium. The outer Jovian planets, Jupiter, Saturn, Uranus, Neptune and Pluto formed in this environment. Little is known about the small planet Pluto, but the other four appear to consist mainly of hydrogen and helium probably with central rocky cores whose sizes and compositions are very model dependent and controversial (Fig. 12.1).

In contrast, the inner terrestrial planets, Mercury, Venus, Earth and Mars

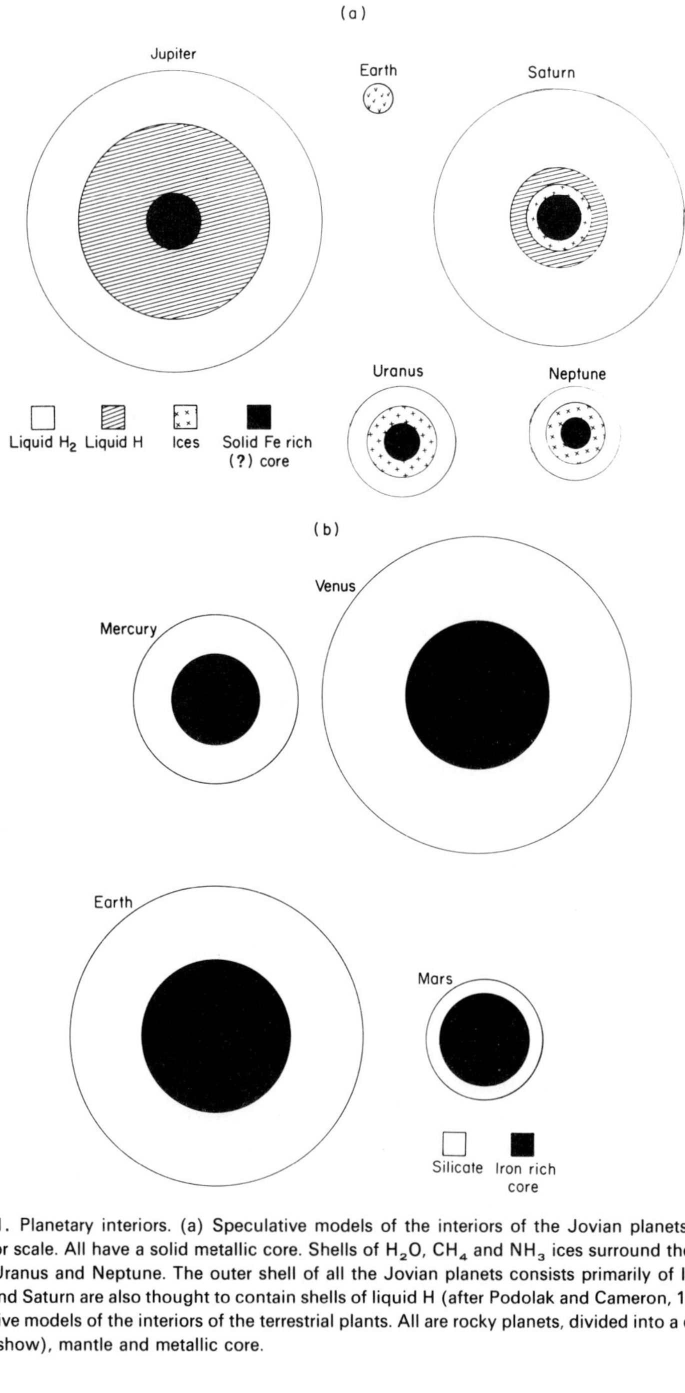

Fig. 12.1. Planetary interiors. (a) Speculative models of the interiors of the Jovian planets. Earth is shown for scale. All have a solid metallic core. Shells of H_2O, CH_4 and NH_3 ices surround the cores of Saturn, Uranus and Neptune. The outer shell of all the Jovian planets consists primarily of liquid H_2. Jupiter and Saturn are also thought to contain shells of liquid H (after Podolak and Cameron, 1974). (b) Speculative models of the interiors of the terrestrial plants. All are rocky planets, divided into a crust (too small to show), mantle and metallic core.

lost most of their hydrogen and helium during formation and probably possess iron-rich cores whose sizes, apart from the Earth, are the subject of considerable debate. Possible estimates for these core sizes that are consistent with the mass and moment of inertia data are given in Table 12.1 and Fig. 12.1.

One problem that faces all attempts to explain the origin of the solar system is why the Sun contains such a small fraction of the solar system's angular momentum. Indeed most calculations suggest that the Sun's angular momentum should be at least an order of magnitude larger, unless some process transferred angular momentum outward to the planets during its formation (e.g. see Alfven, 1954; Alfven and Arrhenius, 1975). Although the manner in which this arose is highly speculative, many workers have suggested that the sun went through a T-Tauri phase of evolution and it was during this phase that the transfer of angular momentum occurred. The T-Tauri phase is named after a class of T-Tauri stars, the best known member being T-Tauri, a star that lies in the constellation of Taurus, the bull. Stars in this class are pre-main sequence objects and are believed to be very young on the solar system time-scale ($\sim 10^6$ years?). They are unusually bright, have very strong solar winds and also are associated with very strong magnetic fields, typically several orders of magnitude larger than the present field of the sun. Transfer of angular momentum by the supersonic solar wind material and/or by the strong magnetic field during a T-Tauri phase have been suggested as possible mechanisms to resolve the angular momentum problems in the solar system (Alfven, 1954; Stacey, 1976; Alfven and Arrhenius, 1976). A rapidly rotating magnetic field that originates at a star transfers angular momentum to any conducting body via the magnetic induction equation. Just as a moving electrical conducting body tends to move the magnetic field in front of it, a moving magnetic field tends to move any conductor in front of it (see §7.4). The T Tauri stage is often used to explain the observed lack of H and He in the inner portion of the solar nebula. The strong solar wind is invoked to sweep away these elements from the inner portion of the young solar system.

The T-Tauri phase is believed to be very short, but if it occurred during the solar system formation and if the planets have already formed, as is sometimes supposed, then all of these bodies were probably exposed to very intense magnetic fields early in their histories. Indeed it is possible that these short-lived intense fields were the initial seed fields for dynamos that now appear to be operating in some planets.

12.2 The Sun

12.2.1 General Properties

The sun is a star which lies at present on the main sequences of the Hertzsprung-Russell (temperature-luminosity) diagram and which has a mass of 2.0×10^{30} kg and a radius of 7.0×10^5 km. Excellent general discussion of the sun can be found in the books by Brandt and Hodge (1964) and Akasofu and Chapman (1972). According to most theoretical models, the sun's luminosity has increased roughly 30% since the sun's birth 4.6 Ga ago. Thermonuclear reactions produced heat in the sun's core that is radiated out to a less massive (1% of the sun's mass) outer conducting shell. This outer shell has a radial dimension that is believed to be roughly 1/5 of the sun's radius. Evidence of convection in this shell is seen in granulation (typical dimensions of the order of 1000 km) and supergranulation (typical scale near 30,000 km). Lifetimes of the convection cells associated with the granulation are of the order of minutes and with the supergranulation of the order of an earth day (Gilman, 1974). Large giant convection cells with radial dimensions as large as the conducting shell are also believed to exist and to have lifetimes of the order of a month or more, even though very convincing observational evidence for their existence is still lacking (Gilman, 1974, 1977). There does not appear to be any significant axisymmetric meridional circulation (Gilman, 1974, 1977). The sun's rotation exerts an important influence on the convection and the convection is clearly turbulent.

The sun's rotation varies in location and time (see reviews by Gilman (1974) and Howard (1975)). Estimates of the sun's rotation have come from measurements of the Doppler shift of spectral lines, observations of sunspots, filaments, corona streamers, etc. Equatorial rotation rates are near 25 earth days while estimates for the polar regions are usually between 35 and 40 days. The decrease in rotation rate poleward goes roughly as $\sin^2 \lambda$, where λ is solar latitude. In the lower latitudes, where estimates are thought to be more reliable, the spectroscopic measurements indicate a rotation rate which is generally one day slower than that obtained from observations of sunspots, filaments, etc. The latter are associated with magnetic field features. Speculations on these differences include the possibility that the magnetic fields are connected to the sun's interior regions, for example through the frozen-in-field theorem operating in the conducting plasma (see §7.4). Dicke (1970) has produced evidence on the amount of the sun's flattening, which he interprets to demonstrate that the sun's core is rotating nearly five times as fast as the outer regions. However, it is not clear whether such a rapid rotation of the sun's interior would be mechanically stable. Also, even if the Dicke model were initially true, the frozen-in-field effect would eventually remove this large velocity gradient if the magnetic field is connected to the sun's interior. Alternatively, hydromagnetic wave mechanisms, similar to those discussed in §9.1, may lead to preferential "drift" of the solar field. The sun's rotation varies in time with periods sometimes as low as a couple of weeks or so, and with magnitudes of the order of 5% to 10% of the mean velocity (Gilman, 1974; Howard, 1975). There is some evidence that this variation may have been larger at times in the past (Eddy *et al.*, 1977). The sun's rotation axis makes a 7° angle with the normal to the ecliptic.

Magnetic fields of varying strengths are found in the photosphere. First, there is a general solar field with an intensity near 10^{-4} T; secondly, there are very intense local magnetic fields associated with sun spots, flares, coronal streamers, etc.

Only a little discussion will be given here concerning the magnetic fields of stars other than the sun. Some star types appear to possess fields from 10^2 to 10^8 times the general field of the sun (e.g. Preston, 1967; Landstreet and Angle, 1974; Angle, 1975) and fields up to 10^{12} times the sun's general field may exist in some neutron stars (Woltjer, 1975). Moffat (1978) suggests that many of these fields originated during the formation of the stars through compression of a galactic field, estimated to be 10^{-9} to 10^{-10} T. Moffat shows that the free-decay times in such bodies may easily be of the order of the lifetime of the star (§7.4.3). However, stars with oscillating magnetic fields that originate in their cores may have substantially shorter free decay.

12.2.2 Solar Magnetic Field

Information on the sun's magnetic field comes from several sources. The most recent measurements come from space missions which directly measure the interplanetary magnetic field (IMF) that originates at the sun. Probably the most valuable indirect measurements involve utilizing the Zeeman effect. Elemental spectral lines, such as those of hydrogen, are split in the presence of the solar magnetic field and the spacing between those lines is linearly proportional to the intensity of the magnetic field component that is directed along the line of observation. Other indirect measurements also provide information on the solar magnetic field, such as the direction solar plumes are emitted, since statistically such plumes will parallel magnetic field lines.

The magnetic field of the sun has a strong dipole component of the order of 10^{-4} T at the sun's photosphere. On this field is superimposed a complicated fine structure of fields some of which often exceed 10^{-2} T. To a first order approximation the sun appears to have a dipolar magnetic field at high latitudes, but at lower latitudes it appears to possess a more complicated sectorial pattern as shown in Fig. 12.2 (Wilcox and Ness, 1965; Rosenberg and Coleman, 1969; also see review by Akasofu and Chapman, 1972). In the equatorial regions the magnetic field enters the sun in two sectors and emerges from two other sectors. The nodal surfaces which separate the four sectors of opposite polarity are twisted in a spiral-like fashion away from the sun, due to the sun's rotation (Fig. 12.2). Although the sectorial pattern for the sun extends to solar latitudes sometimes exceeding 50° at the sun's surface, it has become restricted to a relatively narrow zone of a few degrees about the equator, at a distance of two solar radii from the sun's surface (Schatten, 1972; Smith *et al.*, 1978).

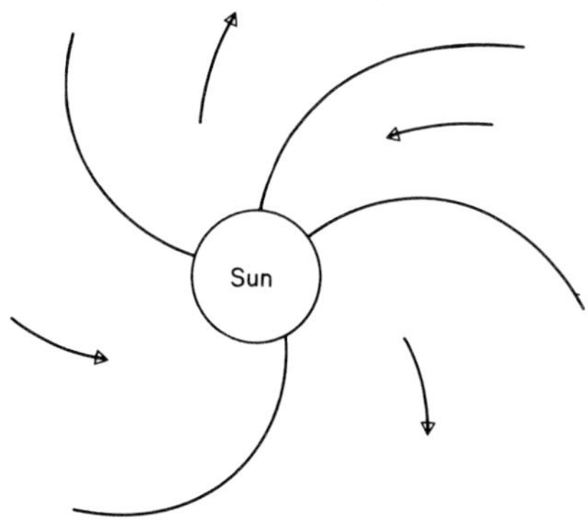

Fig. 12.2. Solar magnetic field sectors. Magnetic field entering and emerging from the sun in an equatorial cross-section exhibits four sectors of alternating polarity separated by nodal surfaces. This sectorial division extends up to mid-latitudes above which a more dipolar field occurs.

The temperatures at the photosphere are around 6000 K while in the outer corona they approach 10^6 K. Parker (1958) pioneered work showing that the heating of the corona gas is sufficient to overcome the sun's gravitational attraction and to produce supersonic outward transport (expansion) of mainly H^+ and $^4He^{2+}$, although some heavier elements are also present. The supersonic velocities are reached at a distance of roughly two or three solar radii, beyond which the rapidly outward moving plasma is referred to as the *solar wind*. This partially electrically conducting solar wind carries the sun's magnetic field with it to distances that reach beyond Saturn and probably beyond Pluto (see the frozen-in-field theorem in §7.4). However the work of Alfven and Arrhenius (1976) suggests that strict application of the frozen-in-field theorem to the solar wind can lead to misunderstandings in some situations.

Figure 12.3 shows the interplanetary magnetic field (IMF) in a plane taken perpendicular to the plane of the ecliptic (Smith *et al.*, 1978). The field is presently outward in the northern hemisphere and inward in the southern hemisphere. There is an electric current sheet associated with the magnetic field that lies close to the plane of the ecliptic (as can be determined from the curl of H). The current sheet is warped due to the sectorial division of the sun's magnetic field in the equatorial regions. The degree of warping of this current sheet is only a few degrees, since the sectorial division of the field is confined to very low latitudes at two or three solar radii, the distance at which the solar wind begins. This warped sheet will exhibit two maxima and minima, as illustrated in Fig. 12.3. The nodal surfaces shown in Fig. 12.3 correspond to the points A through I at which this warped sheet crosses the ecliptic (Smith *et al.*, 1978).

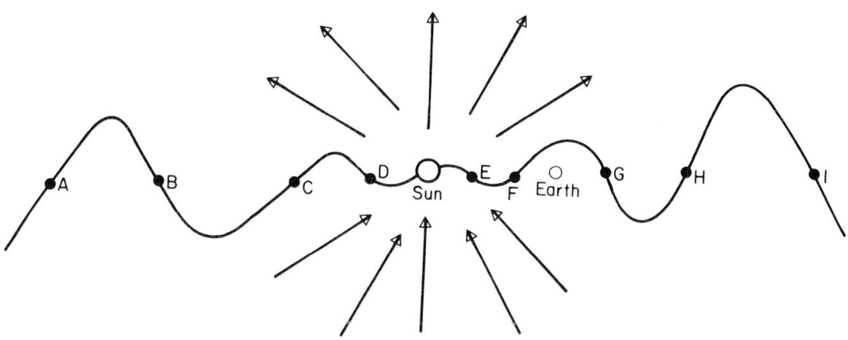

Fig. 12.3. The interplanetary field (IMF). Arrows indicate the direction of the *apparent* dipole magnetic field of the sun. Letters A to I indicate the locations of the nodal surfaces (after Smith *et al.*, 1978).

The strength of the IMF in the neighbourhood of the earth varies considerably but is around 5 nT. The component of the IMF seen by the earth's magnetosphere, which is perpendicular to the ecliptic, changes sign when the earth passes through a nodal surface in its orbit about the sun. This leads to the possibility that "reconnection" of field lines from the IMF with the earth's field lines occurs in the northern hemisphere when the IMF has a downward component (Fig. 12.4) but not in the southern hemisphere. This means that charged particles can more easily enter into the earth's ionosphere when the earth is in one sector of the IMF (with downward component) than in the opposite sector, because charged particles preferentially travel along field lines. Geomagnetic activity varies significantly depending on which sector the earth is in, as first implied by the work of Wilcox and Ness (1965). At present, considerable controversy exists as to whether the earth's climate is significantly affected by this variation in the incoming particle flux. Recent data showing apparent correlations between the size and magnitude of low pressure areas in the troposphere and geomagnetic activity are intriguing (e.g. see Bucha, 1980).

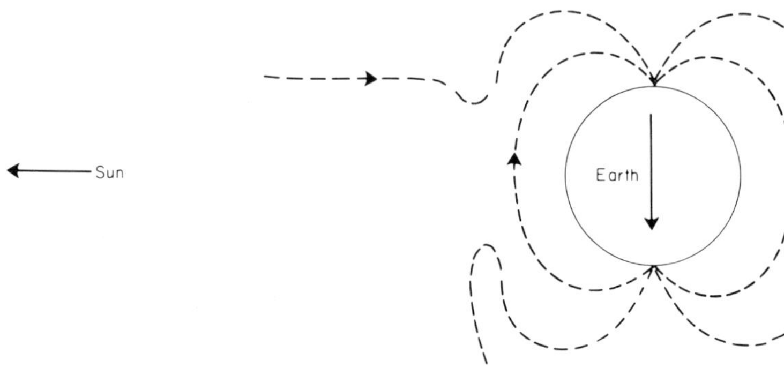

Fig. 12.4. The speculated "reconnection" of the IMF (possessing a downward field component) with the earth's magnetic field. In this case, reconnection occurs only in the northern hemisphere. Conversely (not shown), when the IMF has an upward component, reconnection occurs only in the southern hemisphere.

In addition to the sun's general field, there is the wide range of local magnetic fields. Only one example will be considered here, those fields associated with sunspots. These are darker regions on the sun's surface with diameters that typically range between several thousand to several tens of hundreds of kilometres. Although individual sunspots appear to have a lifetime of only a few weeks, they seem to occur in zones that vary latitudinally throughout the solar cycle. The spots first appear at latitudes near $\pm 30°$, reach sunspot maximum near $\pm 15°$ latitude and die off at low latitudes, $\sim 5°$. In addition, there is a general drift of sunspots towards the equator. This phenomenon, well described by the Maunder butterfly diagram (Fig. 12.5), is cyclic with a period of roughly 11 years. However, fluctuations occur in this periodicity, including a period of 70 years from 1645 to 1715 (the "Maunder minimum") during which there does not appear to have been any sunspot activity (Eddy, 1977; Eddy *et al.*, 1977).

Perhaps the most interesting aspect of sunspots is that they are very magnetic, with intensities of the order of 10^{-1} T. Although sunspot groups are often complex (Brandt and Hodge, 1964), about 90% of the time sunspots occur in pairs and the spots in any given pair exhibit opposite magnetic polarities. The sense of this polarity reverses every 11 years, coinciding with the transition of the sunspot zone from low latitudes to high latitudes. That is, if the most westerly sunspots statistically have normal polarity (downward inclination) in the northern hemisphere during a given 11 year cycle, then they will have a statistical reversed polarity preference during the next 11 year cycle. This adds to an overall sunspot period of 22 years (Fig. 12.5). Also, the most westerly member of sunspot groups in the southern hemisphere will have, statistically, the opposite polarity to that

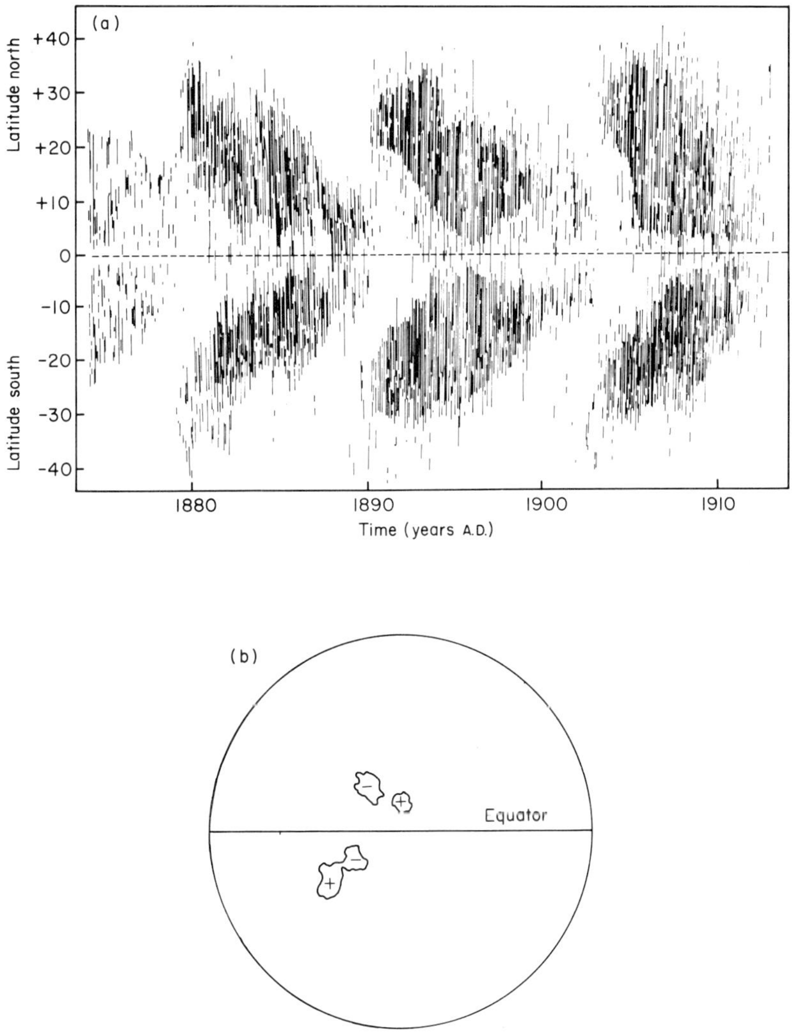

Fig. 12.5. Sunspot cycles. (a) The Maunder butterfly diagram shows the location of sunspot formation versus time in years. The beginning of a cycle is associated with sunspot formation at intermediate latitude ($\approx 30°$) and it ends 11 years later with the formation of sunspots near the equator. (b) The polarity of the first (left) sunspot in a pair depends on the hemisphere in which the sunspots form. The order of the polarity changes every 11 years when sunspots begin a new 11-year cycle. Combining this change in polarity with an 11-year cycle shown in (a) produces a 22-year cycle.

observed in the northern hemisphere. The sun's main dipole field also reverses polarity along with the change in polarity of the sunspots, a strong indication that the two phenomena are closely linked. Eddy (1977) has shown that this 11 year period is not invariant.

The origin of the magnetic fields in sunspots was first explained in magnetohydrodynamic terms by Cowling (1934). Subsequently, there have been numerous contributions to our understanding of the problem, the work of Babcock (1961) and Parker (1975) being especially stimulating. Because of the differential rotation between equatorial and polar latitudes the general poloidal magnetic field (the general dipole field) is rapidly wound up to produce a strong toroidal magnetic field (see §8.2). Thus the resulting intense toroidal field is *antisymmetric* with respect to the solar equator (§7.5.3) and it is this strong toroidal field that is believed to be associated with sunspots. The basic idea (Cowling, 1934; Babcock, 1961; Parker, 1975) is that an instability results in an outward flow of plasma which carries the toroidal field with it. This is illustrated in Fig. 12.6, which also shows why sunspots usually occur in pairs. One sunspot is characterized by the region in which the "toroidal" magnetic field leaves the sun's "surface", while the adjacent

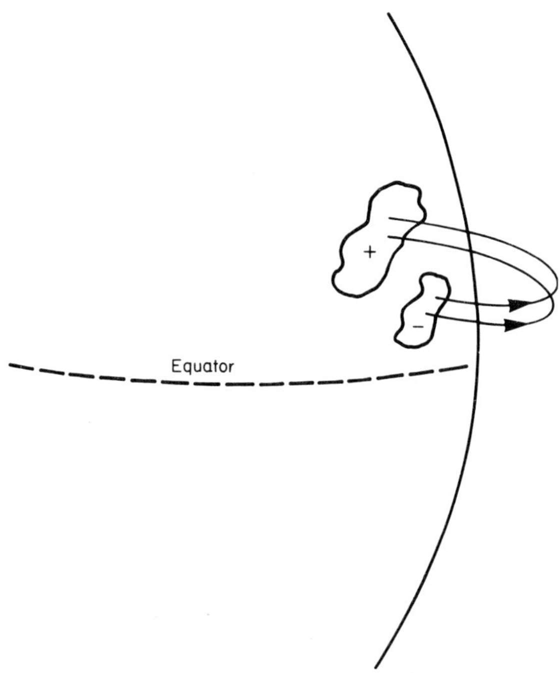

Fig. 12.6. Magnetic field lines emerge from one member of a sunspot pair and enter the sun through the second member.

sunspot is characterized by the region where the "toroidal" field re-enters the "surface".

The theoretical stability of sunspots is a perplexing problem. Parker (1975) points out that it is difficult to maintain sunspots with fields of the order of 10^{-1} T for even short times, unless possibly they originate near the bottom of the outer conducting layer of the sun. His theory suggests that sunspots must extend to depths of the order of 10^5 km or lower, a depth comparable to the thickness estimated for the sun's conducting layer. Yet, as Parker points out, it is doubtful that dynamo action can exist below the conducting layer. Although this problem will not be considered further here magnetohydrodynamics can offer an explanation for the observed lower temperatures of sunspots, temperatures near 4600 K, roughly 4/5 of the temperature of the surrounding photosphere.

The international gas law gives:

$$\frac{P_0 V_0}{T_0} = \frac{\left(P_0 + \dfrac{B^2}{8\mu}\right) V_i}{T_i}$$

where P is pressure, V is volume, T is temperature and the subscripts, 0 and i, refer to conditions outside the sunspot and inside the sunspot, respectively. B and μ are the magnetic induction and permeability. Because the pressure term inside the sunspot must include that arising from the magnetic pressure (§9.1.3), the temperature inside the sunspot will be lower. Assuming quasi-equilibrium conditions of no buoyancy ($V_0 = V_i$), the fractional decrease in temperature can be obtained from the last equation in the form:

$$\frac{T_0 - T_i}{T_0} = \frac{B^2}{8\mu P_0}$$

Dircct substitution of measured and probable values of T and P into this equation give a value for B of the order of 10^{-1} T. Sunspots exhibit considerable structure and better models of their magnetic fields can be constructed by using a collection of magnetic flux tubes (Parker, 1979).

There does not appear anything analogous to sunspots in the earth's core. At first glance, there is some similarity in that the scatter of magnetic field *directions* (as opposed to VGP; see Chapter 6) increases towards low latitudes in both the sun and the earth. However, on close inspection the analogy breaks down probably reflecting the differences in boundary conditions. The 11 year sunspot cycle is the same as the observed period of reversals for the sun's general poloidal field. If similar scaling is applied to the earth, then very large changes in the earth's magnetic field should occur in a few hundred thousand years in the Cenozoic era. Yet no large fields comparable to those in sunspots appear to be present at low latitudes on the earth with periods exceeding 1000 years. Also, unlike the westward drift of the terrestrial non-dipole field, sunspots appear at mid-latitudes and drift towards the equator. The lack of any good analogue is not surprising when one considers that the magnetic fields associated with sunspots probably involve gas instabilities interacting with a strong solar toroidal magnetic field.

Like the earth, the origin of the sun's magnetic field is still conjectural. However, it is generally agreed that the sun's outermost part is turbulent and that large velocity shear is present there. Therefore, an $\alpha\omega$-type dynamo (§7.5.3) is clearly the model-type that appears most acceptable at present (e.g. Gubbins, 1974; Gilman, 1977; Krause, 1977; Watanabe, 1981). Pressure gradients manifest themselves in outward gas expansion, which in turn will produce the necessary helicity for the α-effect to work (§7.5.2 and §8.3.1). Moreover, Steenbeck and Krause (1969a, b) (see review of solar dynamics by Gubbins, 1974) have pointed out that certain oscillatory $\alpha\omega$ dynamos not only reverse polarity, but exhibit some of the properties of sunspots.

It is clear from the above that the sun possesses a toroidal magnetic field, as manifested in sunspots, that is roughly three orders of magnitude larger than the poloidal field. Also, reversal of the poloidal field is accompanied by reversal of the toroidal magnetic field. The data still appear inadequate to resolve whether reversal of one of these precedes the other by a few months or whether both reverse together. Although these constraints are satisfied by many $\alpha\omega$ dynamo models, ultimately the successful model needs also to explain the origin of the sectorial fields, the apparent regularity of the reversals during this century, as well as certain anomalous behaviour, such as the apparent lack of sunspots during the Maunder minimum.

12.3 Meteorite Magnetism

Meteorites are iron or stony bodies that have fallen to earth from elliptical orbits suggesting that they probably originated within our solar system in the asteroid belt between Mars and Jupiter. Some meteorites have radiometric ages indicating that they were formed in the earliest stages of the solar system 4.6 Ga ago. Therefore, the study of meteorites is central to our understanding of the chemistry and origin of the solar system. From a magnetic point of view the study of meteorites may shed light on the magnetic field strength early in the solar system. Alfven (1954) suggested that a very strong interplanetary magnetic field played an important role in the early solar system dynamics by serving as a means of transferring angular momentum from the early sun to the planetary disc (see also Sonett *et al.*, 1970 and

Alfven and Arrhenius, 1976). It has often been thought that this hypothesis could be tested by palaeomagnetic measurements on meteorites to yield palaeointensity values. The magnetic properties of meteorites have also been used in the classification of meteorites, although schemes such as these are more detailed than will be given here (see Herndon and Rowe, 1974; Gose and Butler, 1975; Larson *et al.*, 1973, 1974; Watson *et al.*, 1975; Herndon *et al.*, 1976; Stacey, 1976, 1977a; Brecher, 1977; Sugiura, 1978; Nagata, 1980).

Although the origin of meteorites is obscure and much debated, often the formation of meteorites is attributed to the pre-existing small planetary bodies, typically a few tens of kilometres in diameter. The lack of high pressure mineral phases in meteorites indicates that the size of these bodies was never more than a few hundred kilometres (Wood, 1978, 1979). The interiors of some of these bodies became sufficiently hot to metamorphose some primitive planetary material (producing ordinary chondrites) and causing melting elsewhere (Wood, 1979). Components of the melted material separated gravitationally forming iron-rich cores (the source of iron meteorites) and pure silicates (achondrites). Achondrites are stony meteorites which are similar to terrestrial rocks consisting of ultramafic plutonic rocks and basaltic volcanic rocks. Chondrites are also stony meteorites but they are dissimilar to any terrestrial rocks. They are named after chondrules, rounded or partially rounded polyhedric inclusions, which they contain. Iron meteorites consist mostly of kamacite (a body centred cubic mineral consisting mostly of iron and a few percent of nickel) and taenite (a face centred cubic iron mineral which usually contains 25% or more nickel). Stony-iron meteorites have compositions intermediate to the achondrites and irons. The majority of meteorite finds are olivine-rich chondrites and iron-nickel meteorites containing both kamacite and taenite.

Meteorites have long served as the most diverse collection of samples of the solar system available for detailed analysis in the laboratory (see review by Wasson, 1974). Because of this diversification, numerous classification schemes, far more complex and complete than given above, have come into vogue and often act as a formidable barrier to the uninitiated. Fortunately, recent work using oxygen isotopes has greatly simplified this complex picture and suggests that the solar nebula was not completely homogenized prior to condensation and accretion of the meteorite parent bodies and, presumably, the planets (Clayton *et al.*, 1973; Clayton and Mayeda, 1975, 1978). The key to this analysis involves plots of $^{17}O/^{16}O$ versus $^{18}O/^{16}O$ for samples from various meteorites. Samples that are related to one and another by fractionation processes that depend only on the masses of the isotopes have predictable $^{17}O/^{16}O$ trends that will be seen on such plots. In addition, variations in composition produced by mixing of phases will fall along a straight line between the initial unmixed phases. Therefore, one can determine from these plots an isotope "fingerprint" that remains with the material and cannot be eradicated, no matter how much chemical processing or mass fractionation occurs. An excellent example of how this is used in classification and in understanding the origin of stony meteorites has been given by Clayton and Mayeda (1978). Also, there is at least one "exotic" meteorite, the Allende meteorite, portions of which show such large isotopic anomalies of oxygen (and other elements) that most investigators believe that these isotopes come from sources external to our solar system, perhaps from a supernova (Papanastassiou and Wasserberg, 1978). It is important to point out that the meteorite from which the most numerous and "reliable" palaeointensity estimates come is the Allende meteorite.

Carbonaceous chondrites, so named because they contain several percent free carbon and carbon compounds, are a small but important subclass of meteorites. C1 carbonaceous chondrites, a subtype containing the largest amount of carbon of any carbonaceous chondrites, are well recognized as representing some of the most primitive material in our solar system (see reviews by Wasson, 1974, Stacey, 1976; Papanastassiou and Wasserberg, 1978; Wood, 1979). Carbonaceous chondrites contain very little elemental metal and it is argued that magnetite is the chief carrier of a natural remanent magnetization (see Nagata, 1979a, b), although Wasilewski (1974) has suggested that small-grained Ni_3Fe may contribute much more to the remanence than is usually thought. Carbonaceous chondrites carry a much higher remanence than in lunar samples and often this remanence indicates good magnetic stability (Banerjee and Hargraves, 1971, 1972; Butler, 1972; Brecher, 1972; see also the reviews by Stacey, 1976, 1977a and Nagata, 1979a, b).

The most common palaeointensity technique used in meteorite research is the modified Thellier technique (see §3.4.2). Thermal experiments indicate that usually only the outer skin (approximately the outer 0.05 mm) of meteorites is heated sufficiently high in passing through the earth's atmosphere to reset the remanence significantly (Nagata, 1979a). Nevertheless, because of various chemical alteration problems palaeointensity estimates usually come only from the temperature spectrum below 200°C. This provides only a limited temperature range in which palaeointensity "consistency checks" can be obtained (see discussion on palaeointensity methods, §3.4). Several workers have now obtained palaeointensity estimates in this way to obtain values near 10^{-4} T (see reviews by Gose and Butler, 1975; Stacey, 1976, 1977a; Brecher, 1977; Nagata, 1979a). These estimates are extremely important if confirmed, because they are roughly four or five orders of magnitude larger than the present interplanetary field.

More palaeointensity estimates have come from the Allende meteorite than any other meteorite (Brecher, 1972; Brecher and Arrhenius, 1974; Banerjee and Hargraves, 1972; Butler, 1972; Lanoix *et al.*, 1977, 1978a, b). As in the case above, almost all estimates come from the low-temperature portion of the blocking spectrum and suggest fields with strengths near 10^{-5} to 10^{-4} T. Lanoix *et al.* (1977, 1978a, b) and Sugiura *et al.* (1979) have obtained estimates from the temperature range above 300°C for individual chondrules from the Allende. These estimates are near 16×10^{-4} T and are far higher than the whole rock estimates. Because the chondrules are so magnetic, one wonders why the chondrule estimates are so much larger than the whole rock estimates. Sugiura *et al.* (1979) point out that this must mean the directions of the remanence vary nearly randomly between chondrules. They argue that this is then a conglomerate test (§3.3, Fig. 3.6) showing that the remanence was acquired prior to the inclusion of the nodules into the meteorite and that both the whole rock and chondrule values are acceptable.

Although the above interpretations might be correct, some words of caution are needed. The chondrule studies of Lanoix *et al.* (1977, 1978a, b) and Sugiura *et al.* (1979) were carried out using gas mixtures in order to control the oxygen fugacity precisely and thus to minimize chemical changes. Yet significant deviations from ideal Thellier behaviour were observed near 130°C and 300°C. Sugiura *et al.* (1979) argue that these changes have not affected any of the remanence with blocking temperatures above 300°C, an argument supported by remarkable straight line Thellier plots. It is difficult to believe that such changes are occurring in pure magnetite. Both this and the work of Larson *et al.* (1974) show that "chemical changes" (including possible changes in phase) occur over a wide range of temperatures for different meteorite samples. The origin of these changes needs to be assessed properly since it is known that on some occasions "ideal Thellier behaviour" (see §3.4.2) over a temperature range of even a few hundred degrees can give rise to erroneous palaeointensity estimates.

If one assumes that the estimates of high palaeofields are correct, what are the possible sources for those fields? The list of possibilities is not unlike that encountered for the moon. Some of the magnetization may have been caused during the T-Tauri stage in the sun, in which the field is thought to have been very high because of angular momentum considerations (§12.1). However, even advocates of such a model find it hard to explain how to keep the meteorite stationary with respect to the interplanetary field during cooling (Stacey, 1976).

Dynamo models encounter problems similar to those faced for the moon; where did the dynamo originate? Rowe (1975) has suggested that more local processes, such as shock, have significantly altered the remanence. One could easily extend this list of mechanisms since almost all those given to explain the observed lunar remanence (§11.5) might be considered as possibilities. Stacey (1976, p. 156) summarizes the state of affairs very well as follows: "We have no explanation encompassing all observations on the magnetism of meteorites and even if we are allowed a variety of explanations, it is hard to avoid implausible assumptions."

12.4 Magnetic Fields of the Planets

Because of uncertainties in determining the origin of the Earth's magnetic field, it is useful to augment the empirical data as much as possible and to see if there are any critical similarities or differences in the magnetic parameters of the planets that might lend insight into magnetic field mechanisms. Until recently, the information about the magnetic fields of planets came mostly from indirect measurements, pioneered by Burke and Franklin (1955) who discovered strong bursts of polarized emissions from Jupiter at long radio wavelengths. Decimetre and decametre emissions from Jupiter and Saturn still provide valuable information on their magnetic fields (e.g. Stannard, 1975; Brown, 1975). Studies of planetary magnetic fields have now been enhanced by direct measurements received from various planetary space probes. For brevity, these measurements will not be discussed in detail.

Table 12.1 summarizes many of the known physical properties of planets, including the still scant information on their magnetic fields. The magnetic data for Table 12.1 come from several sources; Mercury (Ness *et al.*, 1975; Ness, 1979; Whang, 1977; Jackson and Beard, 1977), Venus (Dolginov *et al.*, 1969; Russell, 1976), Mars (Dolginov *et al.*, 1976), Jupiter and Saturn (Smith *et al.*, 1974, 1976; Acuna and Ness, 1976; Russell, 1980), the Moon (Chapter 11). These data and others have been reviewed by Strangway (1977a, b) and Russell (1980). Russell (1980) points out that a great deal of caution should be exercised in using these data because of the large variability in their quality. In particular, the offset for Mercury is very crudely estimated from combining the dipole and quadrupole terms given by Whang (1977) (see discussion by Jackson and Beard, 1977). This may well be wrong since a more inclined dipole ($\sim 14.5°$ with respect to the rotation axis) might be a preferable model (Ness, 1979), and in this case no estimate of the "offset" can be obtained. Also, the magnetic data for Saturn concerning the dipole offset and dipole, quadrupole and octupole ratios are obtained solely from modelling and should probably be regarded as somewhat speculative. Russell (1980) has also given good reasons why the smaller values for the magnetic moments of the planets given in Table 12.1 should be preferred (when two values are given). The presence of a magnetic field on Uranus is suggested from the tentative observation of 0.5 MHz pulses from the planet (Brown, 1976). Torbett and Smoluchowski (1980) have suggested that a dynamo is

TABLE 12.1

Some physical properties of the planets in the solar system

	(M) Mass (10^{24} kg)	Planet radius (km)	Core radius (km)	Moment of inertia (C/Ma^2)	Spin period ($\times 10^5$ s)	Magnetic dipole moment	Dipole tilt to rotation axis	Dipole offset (fraction of radius)	Dipole: quadrupole: octupole	Equivalent equatorial surface field (mT)	Closest distance to magnetopause in terms of the planet's radius, R
Mercury	0.33	2439	~1800	—	Prograde 76	$3.1 \times 10^{-4} M_e$ or $6.4 \times 10^{-4} M_e$	2.3° or 14.5°	(0.2)	(1.0:0.4:0.3)	191 or 370	$1.4 \pm 0.2\ R_M$
Venus	4.87	6055	~3000	—	Retrograde 210	$< 5 \times 10^{-5} M_e$	—	—	—	<1	Intersects planet (?)
Earth	5.97	6371	3485	0.3308	Prograde 0.862	M_e	11.5°	0.07 present	1.0:0.14:0.09	31,000	$10.9\ R_E$
Mars	0.642	3398	~1700	0.375	Prograde 0.886	$3.0 \times 10^{-4} M_e$ or $< 2.6 \times 10^{-4} M_e$	(15°–20°)	—	—	64	Intersects planet (?)
Jupiter	1899	71,600	~52,000	0.264	Prograde 0.354	$1\ 8 \times 10^{4} M_e$	11°	0.1	1.0:0.25:0.20	402,000	$50\ R_J$
Saturn	568	60,000	~28,000	0.207	Prograde 0.368	$0\ 5 \times 10^{3} M_e$	1.5° ± 0.5°	< (0.05)	(1.0:0.14:0.07)	55,000	$23\ R_s$
Uranus	87.2	25,900	—	(0.26)	Prograde 0.389	—	—	—	—	—	—
Neptune	102	25,200	—	0.26	Prograde 0.54	—	—	—	—	—	—
Pluto	0.66	1100	—	—	Prograde 5.52	—	—	—	—	—	—
Earth's moon	0.07	1738	<500	0.391	Prograde 23.6	$< 1.6 \times 10^{-4} M_e$	—	—	—	<0.02	—

operating on Uranus, but this suggestion like the measurements of the field of Uranus should be regarded as being very tentative.

Some of the uncertainty of the modelling can be illustrated by considering Saturn's magnetic field. Table 12.2 gives two different spherical harmonic representations of Saturn's magnetic field using data from Voyager 1 and 2 (Connerney *et al.*, 1981). The use of spherical harmonic analyses assumes $\nabla \times \boldsymbol{B} = \boldsymbol{0}$ (§2.2.1), that is it assumes there are no current sources in the spherical shell enclosing the observations and the region of representation. If this is violated, the spherical harmonic "fit" is incorrect. Connerney *et al.* (1981) point out that there are current sources external to this planet, primarily due to an equatorial azimuthal ring current between 8 and 16 R_s, where Saturn's radius R_s is 60,000 km. Measurements used to derive the values in Table 12.2 were obtained when Voyager I and II spacecraft were within 8 R_s.

TABLE 12.2
Spherical harmonic analysis of Saturn's magnetic field (after Connerney *et al.*, 1981). Units in nT

	Unconstrained ($l_{max} = 2$)	Axisymmetric ($l_{max} = 3$)
g_1^0	21,439	21,535
g_1^1	−143	—
h_1^1	143	—
g_2^0	1882	1642
g_2^1	−515	—
g_2^2	500	—
h_2^1	−433	—
h_2^2	−36	—
g_3^0	—	2743

The unconstrained model of Table 12.2 is not preferred because even with eight undetermined parameters it does not appear to fit the data much better than the three-term axisymmetric model. The greater efficiency of the axisymmetric model is consistent with the suggestion of Stevenson (1980) that the differential rotation of an outer fluid electrically conducting shell of Saturn above the active dynamo region greatly attenuates the non-axisymmetric components of the dynamo field. This illustrates that magnetic fields which *appear* to be axisymmetric away from the source region, may not be so in the source region. Thus Cowling's Theorem is not violated (§8.2.3). The offset dipole obtained from the axisymmetric model in Table 12.2 is 0.05 R_s.

12.5 Dynamos in the Solar System

Russell (1978) has up-dated the so-called magnetic Bode's law for planetary magnetism shown in Fig. 12.7. This is a convenient way for exhibiting some of the magnetic data and illustrates why it is tempting to look for a simple explanation of the planetary magnetic field data in terms of simple scaling laws. Blackett (1947) believed that there was some fundamental law relating angular momentum, L, and magnetic moment, μ, in planetary bodies (§1.6). Brecher and Brecher (1978) have attempted to revive this approach by using Schuster's relationship that argues for a linear relationship between μ and L. This is an empirical scaling law that works best on a log–log plot but does not seem supported on more close inspection. In precessional dynamos the Poincare acceleration is given by $\boldsymbol{\omega} \times \boldsymbol{\Omega}$, where $\boldsymbol{\omega}$ is the rotational angular velocity and $\boldsymbol{\Omega}$ is the precessional angular velocity (Malkus, 1968). Dolginov

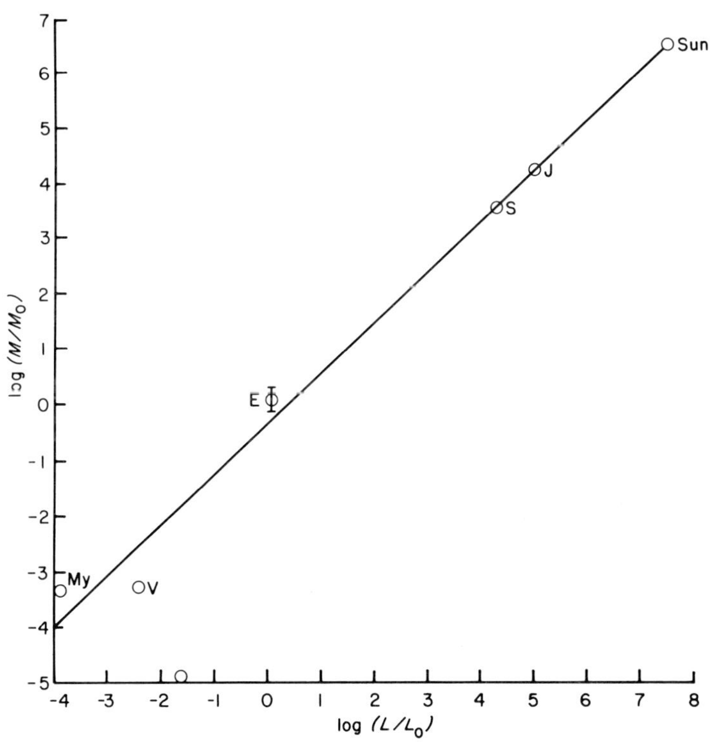

Fig. 12.7. The Magnetic Bode's Law. The log of magnetic moment M normalized to terrestrial moment M_o is plotted against the log of angular momentum L normalized to terrestrial angular momentum L_o. My, Mercury; V, Venus; Ms, Mars; E, Earth; J, Jupiter; S, Saturn. (After Russell, 1978.)

(1976, 1978) argues for a scaling law using the Poincare acceleration to define the proportionality between the magnetic field intensity and the Coriolis force. This scaling law requires a precessional driven dynamo and a dynamo in which magnetostrophic balance occurs (see §7.5.4 and §8.3.4). By contrast, Busse (1976) has developed a scaling law based on his own dynamo theory, a theory that requires geostrophic balance, as well as other assumptions (§8.3.4). Stevenson (1979) has developed a heuristic "scaling law" relating heat flux to magnetic moment. Although the number of "scaling laws" is on the increase, the conclusion of Russell and Goldstein (1976) that there is no completely satisfactory scaling law seems correct. This may simply reflect the fact that different processes (i.e. different dynamos) are acting in the solar system. In that case it would be expected that no single "scaling law" will ever be satisfactory based on physical principles.

The Sun, Earth, Jupiter and Saturn all have active dynamos today. Whether dynamos are active in other solar system bodies is arguable. For example, Stevenson (1983) argues that Mercury has a thin shell dynamo maintained by chemical convection and latent heat released by core growth, while Strangway (1977b) argues from thermal considerations that the field of Mercury is due to remanent magnetization. Russell (1978) also attributes the magnetic moment of Mars to remanence, possibly acquired at a time when Mars had an active dynamo.

There are only a few observations about dynamos and solar system magnetic fields that will probably not change with time. Although not rigorously proved, it appears that some necessary conditions for dynamos to exist in planets is that the planets have a sufficient fluid volume (e.g. a molten core), a sufficient energy source and sufficient rotation. Unfortunately, the word sufficient is difficult to define. For example, the lack of a strong magnetic field in Venus is sometimes attributed to its very slow rotation rate. However, Stevenson (1979) argues against this view since the Rossby number of Venus (Table 7.4) probably lies between that of the Earth and the Sun. Stevenson (1983) proposes that Venus has no dynamo because it has a completely fluid core which is stably stratified (i.e. there is no adequate energy source to power a dynamo).

Even though instantaneously axially symmetric magnetic fields might be unstable in planetary bodies (§8.2.3 and §9.2), providing no long-term asymmetries exist in the boundary conditions, the rotation requirement suggests that the magnetic fields of planets, when averaged over long periods, will be axially symmetric and central. The reason for this assertion is simple. Once variations in boundary conditions are disallowed, there is no longitudinal directional preference remaining in the problem. The data for the Earth indicate that the Earth's mean magnetic field is significantly more axially symmetric than the present day ("instantaneous") field. It is somewhat less clear whether the small dipolar offset from the planet's centre and the small angle the dipole makes with the rotation axis (near 10°; except for Saturn which has a dipole field aligned closely to the rotation axis—Table 12.1) is significant or simply the manifestation of some statistical fluctuation.

An important question is whether one dynamo model is applicable to the Sun, Earth, Saturn and Jupiter, or whether different types of dynamos are operating in these bodies. Several consequences follow from the assumption that only one dynamo type is necessary. For example, Busse's dynamo model (§8.3.4) would have to be rejected, since it does not allow for the very large toroidal magnetic field that probably is present in the Sun. Indeed, if forced to pick one dynamo type for all of these bodies, one would probably choose an $\alpha\omega$ dynamo based on the data for the Sun. Similarly, precessional energy could be eliminated from consideration, since significant differential precession of the Sun is highly unlikely (see arguments in §7.5.4).

Some insight can be gained as to whether one or more dynamo types are operating by comparing the properties of the magnetic fields of the Earth and the Sun. These are the two bodies that in most respects are most different of the four under consideration. Their magnetic fields are similar in that (to a first approximation) both have axially symmetric dipole fields which reverse polarity. On the other hand, the frequency of reversals of the Earth's magnetic field are highly irregular in time while those of the Sun appear highly regular, and the analogue to sunspot magnetic fields in the Earth is missing.

Mechanistically speaking, the magnetic field in both the Sun and Earth occurs in a spherical shell containing a conducting fluid that is undergoing convection, and where rotation plays an important role in influencing that convection. On the other hand, the dimensional scales for the Sun and Earth are different, the boundary conditions are different, the Sun's fluid is compressible while the Earth's is (essentially) not, and the Sun is partially ionized. The convection in the Sun is clearly turbulent and there clearly is a large velocity shear in rotation; the existence of these in the Earth is controversial.

Similar arguments to the above can be made for Earth–Saturn or Earth–Jupiter comparisons. However, it should be emphasized that the magnetic field data for Saturn and Jupiter are fewer and of poorer quality than for the Earth. Although these data are not sufficient to conclude that entirely different dynamo processes are presently acting in some solar system bodies, they are sufficient to cast serious doubt that any simple scaling law based on physical principles will give the relative magnetic properties (e.g. dipole strength) of solar system bodies.

Table 12.3 summarizes the present knowledge of dynamos in solar system objects. Probably only a few lessons can be learned about the necessary and sufficient conditions for dynamos from this summary, largely because it seems likely that more than one dynamo type is operating in our solar system. Dynamos seem to require electrically conducting rotating cores which are undergoing convection. The possibility of oscillatory dynamos involving no convection (waves only) cannot be eliminated, but the absence of dynamos in Venus, which probably has a large liquid metallic core, and Mars, which may have a liquid metallic core, suggests that convection is probably necessary in planetary objects. The minimum size required for the electrically conducting source region of the dynamo is not yet known. Limits

TABLE 12.3
Dynamos in the solar system

Sun	Active dynamo involving turbulence and large-scale velocity shear
Mercury	Source of magnetic field unknown with the most favoured mechanisms being an active dynamo or remanent magnetization. If a dynamo is presently active, it probably resides in a very thin liquid shell of a few hundred kilometres thickness at most. That is, conditions for an active dynamo are marginal considering the thermal history of the small planet
Venus	No dynamo, although it probably does have a moderate size liquid metallic core. This may be due to the low angular velocity of Venus, although Venus has a reasonably large Rossby number. Another explanation is that the Venusian core is stably stratified (Stevenson, 1983)
Earth	Active dynamo
Earth's Moon	Probably never possessed a dynamo, although the possibility of a dynamo operating in a small lunar core ($\sim$400 km radius) in the first billion years of the moon cannot be ruled out
Mars	No active dynamo, probably because it does not have a sufficiently large liquid electrically conducting core or if it does have a large liquid core, that core is thermally stratified
Meteorites/ asteroids	Small bodies possessing magnetic fields associated with remanent magnetizations acquired from unknown sources
Jupiter	Active dynamo driven by thermal convection. It is likely that the Jovian dynamo is operating in a much larger core (relative to the planet's radius) than that in the earth. This explains the large non-dipole component of the Jovian field compared to the terrestrial field. No clear-cut boundary between the Jovian "core" and "mantle" exists
Saturn	Active dynamo probably driven primarily by thermal convection. The very symmetric axial field either does not reflect the average situation or it is due to masking by an electrically conducting "mantle" of a less symmetric core-generated field. This mantle is rotating relative to the core, resulting in the selective screening of non-axially symmetric magnetic field components
Uranus, Neptune and Pluto	Too little observational evidence to make any convincing statements, although dynamos have been speculated to exist in Uranus and Neptune (e.g. see Stevenson, 1983)

may be better determined if the Earth's moon can definitely be shown not to have ever possessed a dynamo (or vice-versa) and if the ambiguity in origin of the field of Mercury can be resolved. Such limits inevitably involve a trade-off between size and the magnitude of the energy source: the smaller the size the larger the energy source required. Further discussion of this and other necessary conditions can be found in §8.2.3.

Interestingly, magnetic A stars (stars with surface temperatures near 10^4 K and with strong magnetic fields as high as 10^3 to 10^4 T), appear to exhibit some unexpected properties. The axis of symmetry of their magnetic field tends to be perpendicular to their spin axis, and more rapidly rotating A stars tend to have weaker magnetic fields than more slowly rotating ones (Parker, 1979). If a dynamo acts in such stars, it appears to be a very different type from those so-far considered. If no dynamo is responsible for the large fields in A stars, then what mechanism is responsible and could such an unidentified mechanism play a role in the generation of magnetic fields in our solar system?

Appendix A

SI and Gaussian CGS Units and Conversion Factors

Problems arise in the conversion of SI and CGS units in magnetism because there are two equations commonly used relating the magnetic induction $\boldsymbol{B}$, and magnetic field strength $\boldsymbol{H}$ and magnetization ($\boldsymbol{M}$ in SI, $\boldsymbol{J}$ in CGS). These are

$$\boldsymbol{B} = \mu_0(\boldsymbol{H} + \boldsymbol{M}) \tag{A.1}$$

and

$$\boldsymbol{B} = \mu_0 \boldsymbol{H} + \boldsymbol{J} \tag{A.2}$$

where μ_0 is the permeability of free space. To distinguish between these two forms of the equation $\boldsymbol{M}$ is usually referred to as the *magnetization* and $\boldsymbol{J}$ as the *magnetic polarization*.

The Kyoto General Assembly of IAGA in 1973 adopted the form of (A.1) for use in geomagnetism (Anonymous, 1974). The main reason for choosing this form is that unit conversions from CGS to SI are then simply a matter of powers of ten. Table A.1 gives SI and CGS units for the common magnetic terms used and provides conversion factors. More detailed comparisons including a resumé of the fundamental equations in electromagnetism and their form in the two systems have been given by Payne (1981).

TABLE A.1

Common magnetic terms in SI and CGS units with conversion factors

Magnetic term	Symbol	SI unit	CGS unit	Conversion factor
Magnetic induction	$\boldsymbol{B}$	Tesla(T) $= \text{kg A s}^{-2}$	Gauss	$1\text{T} = 10^4$ Gauss
Magnetic field strength	$\boldsymbol{H}$	A m^{-1}	Oersted (oe)	$1\text{ A m}^{-1} = 4\pi \times 10^{-3}$ oe
Magnetization	$\boldsymbol{M}$	A m^{-1}	emu cm^{-3} = Gauss	$1\text{ A m}^{-1} = 10^{-3}\text{ emu cm}^{-3}$
Magnetic polarization	$\boldsymbol{J}$	T	emu cm^{-3} = Gauss	$1\text{ T} = \frac{10^4}{4\pi}\text{ emu em}^{-3}$
Magnetic dipole moment	$\boldsymbol{m}$	A m^2	emu = Gauss cm^3	$1\text{ A m}^2 = 10^3$ emu
Magnetic pole strength	p	A m	emu = Gauss cm^2	1 A m = 10 emu
Magnetic flux	ϕ	Weber (Wb) $= \text{kg m}^2\text{ A s}^{-2}$	Maxwells	$1\text{ Wb} = 10^8$ Maxwells
Magnetic scalar potential	ψ	A	emu = oe cm	$1\text{ A} = 4\pi \times 10^{-1}$ emu
Magnetic vector potential	$\boldsymbol{A}$	$\text{Wb m}^{-1} = \text{kg m A s}^{-2}$	emu = Gauss cm	$1\text{ Wb m}^{-1} = 10^6$ emu
Permeability of free space	μ_o	Henry (H) $\text{m}^{-1} = \text{kg m A}^2\text{ s}^{-2}$	1	$4\pi \times 10^{-7}\text{ H m}^{-1} = 1$ cgs
Permittivity of free space	ε_o	Farad (F) $\text{m}^{-1} = \text{A s}^{24}\text{ kg m}^{-3}$	1	$8.85 \times 10^{-12}\text{ F m}^{-1} = 1$ cgs
Volume magnetic susceptibility	χ_m	Dimensionless	Dimensionless	4π (SI) = 1 (cgs)
Demagnetizing factor	N	Dimensionless	Dimensionless	1 (SI) = 4π (cgs)

Appendix B

Introduction to Universality Theory

There are many systems described by non-linear mathematics in which the variation of some critical parameter can lead to drastic changes in appearance of the system. Certain universal numbers exist which determine quantitatively the transition from smooth to turbulent, or erratic, behaviour of these systems (Feigenbaum, 1980). In spite of the apparent stochastic nature of these systems, they can often be described (and evolutionary behaviour predicted) by very simple mathematical equations. Apart from providing insight into the physics governed by such equations, and besides requiring some radical changes in basic philosophical foundations of physics, the study of such mathematics suggests more efficient ways to carry out experiments. This appendix gives an introduction to this newly developed mathematics, as can be found in reviews such as May (1976), Ruelle (1980) and Feigenbaum (1980).

Because of the universality of some numbers in these systems, considerable insight into them can be gained by considering the following simple recursion formula:

$$X_{n+1} = F(X_n) = 4\lambda X_n(1 - X_n) \tag{B.1}$$

in which λ is a parameter independent of X_n. Suppose (B.1) is examined for X in the interval [0, 1] with $\lambda = \frac{1}{2}$. Given any initial values X_1, all successive values $X_2, X_3 \ldots$ can be determined. However, there are two solutions of (B.1) for which $X_n = X_{n+1}$. These are called fixed-point solutions and occur at 0 and $\frac{1}{2}$. The solution of $\frac{1}{2}$ is an attractor, in that successive reiterations of

(B.1) using any initial value X_1 (not equal to a fixed-point) will converge to $\frac{1}{2}$. That is:

$$\lim_{n \to \infty} |X_n - \tfrac{1}{2}| = 0$$

It is relatively easy to show that a fixed point is an attractor if, and only if, the absolute value of the slope F at the fixed point, $|\partial F(X_n)/\partial X_n|$ is less than one (May, 1976; Feigenbaum, 1980). Let $F^2(X_n) \equiv F(F(X_n))$.The derivative of $F^2(X_n)$ with respect to X_n at a fixed point, x_0 (i.e. $x_{n+1} = X_n = x_0$) is:

$$\frac{\partial F^2(X_n)}{\partial X_n} = \frac{\partial F(X_{n+1})}{\partial X_n} = \frac{\partial F(X_{n+1})}{\partial X_{n+1}} \frac{\partial X_{n+1}}{\partial X_n} = \frac{\partial F(X_{n+1})}{\partial X_{n+1}} \frac{\partial F(X_n)}{\partial X_n}$$
$$= F'(X_0)^2$$

where the prime indicates the derivative evaluated at the fixed point. Generalizing this, one obtains:

$$F^{N'}(X_0) = F'(X_0)^N \tag{B.2}$$

in which $F^{N'}(X_0)$ means the derivative of $F^N(X_n)$ with respect to X_n evaluated at the fixed point. F^N is referred to as superstable if its derivative vanishes (from (B.2) this implies $F'(X_0) = 0$). The fixed point is stable if $F'(X_0) < 1$ and unstable if $F'(X_0) > 1$.

For the parameter choice $\lambda = \frac{1}{2}$ there are two fixed points in the interval $[0, 1]$, but only one of these is an attractor ($X_0 = \frac{1}{2}$). Suppose the value of λ is increased, then as λ increases to the value of $\frac{3}{4}$, $F'(X_0)$ goes to -1. For higher values of λ, $|F^{N'}(X_0)| > 1$ and $X_0 = \frac{1}{2}$ is no longer stable; indeed there are no stable fixed points for $F(X_n)$. However, there are two stable fixed points for $F(F(X_n)) \equiv F^2(X_n)$. This is referred to as a bifurcation leading to a stable 2-cycle. The critical value of λ at which this occurs is labelled $\Lambda_1 = \frac{3}{4}$. If λ is further increased these two new fixed points become unstable at $\lambda = \Lambda_2$, a new bifurcation occurs (producing four new stable fixed points) and this process continues on ad infinitum. This process can perhaps best be visualized graphically as discussed in an excellent review by Feigenbaum (1980). In any case, the amount of increase of λ between successive critical points decreases as λ increases. In particular, Feigenbaum also proved that:

$$\lim_{n \to \infty} \frac{\Lambda_{n+1} - \Lambda_n}{\Lambda_n - \Lambda_{n-1}} = \delta$$

where Λ_n represents the critical value of λ separating the stable n cycle from the stable $n + 1$ cycle.

δ is a universal number (4.669201609 . . .) which is surprisingly independent of the details (physics) of the governing equations (e.g. eqn. B.1), within very broad limits. For example, not only does the value for δ given above apply to the system described by (B.1), but it is identical to that applicable to the classic Duffing's harmonic oscillator (a damped harmonic oscillator driven by a sinusoidal varying force) and the onset of turbulence in some (and likely all) thermally-driven convection systems (Feigenbaum, 1980). There are other universal numbers, such as those which deal with the scaling between stable fixed points with increased bifurcation (the spacing between fixed points decreased). Finally, there is the limiting value, Λ_∞, found to be 0.8925 ... (applicable to eqn. B.1; this is not a universal number). For any value of λ greater than Λ_∞, the "simple" bifurcation process is not operating and one has passed into the chaotic regime. Mathematically, the chaotic regime is characterized by the following: (i) there is an infinite number of narrow n-cycle intervals which attract all x except for a set of that measure zero; (ii) there are aperiodic regions between periodic regions which have a finite measure; and (iii) these regions are never continuous but they are dense. All normal statistical analyses would describe the fixed points as being distributed randomly in either the chaotic or the very high order n-cycle regime, although in fact they are completely deterministic, since one generates them through successive reiterations of equations such as (B.1). The phrase *pseudo-random* is often used to describe such statistical behaviour.

Since geomagnetism is a subject which often deals with non-linear theory, any development such as above should be followed with interest. There are three examples of possible applicability which need explicit mention. First, there is the possible insight into dynamo theories such as discussed in §10.2. Secondly, the development of non-linear theory has shown conclusively that the ergodic hypothesis is not generally satisfied. This means, among other things, that the space-time averaging attempts used in this book regarding the palaeomagnetic data cannot be rigorously justified. Nevertheless it has been shown (§6.2.1) that such procedures are necessary to obtain a coherent global picture of the magnetic field in the past. Finally, if the speculations given in §10.2.2 are accepted and if a polarity transition is found to be (say) quadrupolar, the ratio of quadrupole to higher-order terms should be much smaller than the dipole to higher-order terms (measured in terms of spectral power) than observed during mean field times. If the scaling law between different convective regimes can be determined, then this might be the best way in which to predict the magnetic field evolution. Indeed, from a predictive point of view, and contrary to that carried out up to the present, it is far more critical to obtain the scaling law (as can be obtained from ratios of a few spectra in different regimes) than to carry out detailed spectral analysis in one given regime.

References

Abrahamsen, N. and Knudsen, K. L. (1979). Indication of a geomagnetic low-inclination excursion in supposed middle Weichselian interstadial marine clay at Rubjerg, Denmark. *Earth Planet. Sci. Lett.* **18**, 238.

Acheson, D. J. and Hide, R. (1973). Hydromagnetics of rotating fluids. *Rep. Prog. Phys.* **36**, 159.

Acuna, M. H. and Ness, N. F. (1976). The complex magnetic field of Jupiter. *Science* **81**, 2917.

Adam, N. V., Baranova, T. N., Benkova, N. P. and Cherevko, T. N. (1970a). Geomagnetic field variation according to magnetic declination data for 1550–1960. *Geomag. Aeron.* **10**, 1068 (English Translation, 854).

Adam, N. V., Baranova, T. N., Benkova, N. P. and Cherevko, T. N. (1970b). Spherical harmonic analysis of delination and secular geomagnetic variation 1550–1960. *Earth Planet. Sci. Lett.* **9**, 61.

Adam, N. V., Benkova, N. P., Khramov, A. N. and Cherevko, T. N. (1975). Spherical harmonic analysis of the geomagnetic field. *Stud. Geophys. Geod.* **19**, 141.

Ade-Hall, J. M., Palmer, H. C. and Hubbard, T. P. (1971). The magnetic and opaque petrological response of basalts to regional hydrothermal alteration. *Geophys. J. Roy. Astron. Soc.* **24**, 137.

Ahrens, T. J. (1979). Equations of state of iron sulfide and constraints on the sulfur content of the earth. *J. Geophys. Res.* **84**, 985.

Aitken, M. J. (1970). Dating by archaeomagnetic and thermoluminescent methods. *Phil. Trans. Roy. Soc. London* **A269**, 77.

Aitken, M. J. and Hawley, J. N. (1966). Magnetic dating. III. Further archaeomagnetic measurements in Britain. *Archaeometry* **9**, 187.

Aitken, M. J. and Hawley, J. N. (1967). Archaeomagnetic measurements in Britain, IV. *Archaeometry* **10**, 129.

Airken, M. J. and Weaver, G. H. (1962). Magnetic dating: some archaeomagnetic measurements in Britain. *Archaeometry* **5**, 5.

Aitken, M. J., Hawley, J. N. and Weaver, G. H. (1963). Magnetic dating: further archaeomagnetic measurements in Britain. *Archaeometry* **6**, 76.

Akaike, H. (1960a). Fitting autoregressive models for prediction. *Ann. Inst. Statist. Math.* **21**, 243.

Akaike, H. (1969b). Power spectrum estimation through autoregressive model fitting. *Ann. Inst. Statist. Math.* **21**, 407.

Akaike, H. (1970). Statistical predictor identification. *Ann. Inst. Statist. Math.* **22**, 203.

Akasofu, S. I. (1979a). Interplanetary energy flux associated with magnetospheric substorms. *Planet. Space Sci.* **27**, 425.

Akasofu, S. I. (1979b). Radial deformation of the solar current sheet as a cause of geomagnetic storms. *Planet. Space Sci.* **27**, 1055.

Akasofu, S. I. (1979c). A search for the interplanetary quantity controlling the development of geomagnetic storms. *Quart. J. Roy. Astron. Soc.* **20**, 119.

Akasofu, S. I. and Chapman, S. (1967). Geomagnetic storms and auroras. In *Physics of Geomagnetic Phenomena* (S. Matsushita and W. H. Campbell, eds), Vol. 2 (6). p. 1113. Academic Press, London and New York.

Akasofu, S. I. and Chapman, S. (1972). *Solar-terrestrial Physics*. Clarendon Press, Oxford.

Aki, K. and Richards, P. G. (1980). *Quantitative Seismology*, Vol. 2, Freeman, San Francisco.

Alfven, H. (1940). *Cosmical Electrodynamics*. Oxford University Press, Oxford.

Alfven, H. (1954). *On the Origin of the Solar System*. Clarendon Press, Oxford.

Alfven, H. and Arrhenius, G. (1975). *Structure and Evolutionary History of the Solar System*. D. Reidel Pub. Company, Boston, USA. (Geophys. and Astrophys. Monograph Series.)

Alfven, H. and Arrhenius, G. (1976). Evolution of the solar system. *NASA SP-345*.

Allan, J. L. and Shive, P. N. (1974). Mossbauer effect observations of the "x-phase" in the ilmenite-hematite series. *J. Geomag. Geoelect.* **26**, 329.

Alldredge, L. R. (1976). Effects of solar activity on geomagnetic component annual means. *J. Geophys. Res.* **81**, 2990.

Alldredge, L. R. (1977). Geomagnetic variations with periods from 13 to 30 years. *J. Geomag. Geoelect.* **29**, 123.

Alldredge, L. R. and Hurwitz, L. (1964). Radial dipoles as the sources of the earth's main magnetic field. *J. Geophys. Res.* **69**, 2631.

Alldredge, L. R. and Stearns, C. O. (1969). Dipole model of the sources of the earth's magnetic field and secular change. *J. Geophys. Res.* **74**, 6583.

Alldredge, L. R. and Stearns, C. O. (1974). A discussion of sources and description of the earth's magnetic field and its secular variation. *J. Geomag. Geoelect.* **26**, 393.

Anders, E. (1971). Meteorites and the early solar system. *Ann. Rev. Astron. Astrophys.* **9**, 1.

Anderson, K. A. and Wilhelms, D. E. (1979). Correlations of lunar farside magnetised regions with ringed impact basins. *Earth Planet. Sci. Lett.* **46**, 107.

Anderson, K. A., Lin, R. P., McCoy, J. E. and McGuire, R. E. (1976). Measurement of lunar and planetary magnetic fields by reflection of low energy electrons. *Sp. Sci. Instr.* **1**, 439.

Anderson, K. A., Lin, R. P. and McGuire, R. E. (1977). Linear magnetisation feature associated with Rima Sirsalis. *Earth Planet. Sci. Lett.* **34**, 141.

Angle, J. R. P. (1975). Strong magnetic fields in white dwarfs. *Ann. N.Y. Acad. Sci.* **257**, 80.

Anonymous (1974). Adoption of SI units in geomagnetism. (Trans. 2nd Gen. Ass. IAGA Kyoto, 1973). *IAGA Bull.* **35**, 148.

Appleton, E. V. and Barnett, M. (1925). On some direct evidence for downward atmospheric reflection of electric rays. *Proc. Roy. Soc. London* **A109**, 621.

Ashcroft, N. W. and Merman, N. D. (1976). *Solid State Physics*. Holt, Rinehart and Winston, New York.

Atwater, T. M. and Mudie, J. D. (1973). Detailed mean bottom geophysical study of the Gorda rise. *J. Geophys. Res.* **78**, 8865.

Axford, W. I. (1969). Magnetospheric convection. *Rev. Geophys. Space Phys.* **7**, 421.

Axford, W. I., Petschek, H. E. and Siscoe, G. L. (1965). Tail of the magnetosphere. *J. Geophys. Res.* **70**, 1231.

Baag, C. and Helsley, C. E. (1974). Geomagnetic secular variation model E. *J. Geophys. Res.* **79**, 4918.

Babcock, H. W. (1961). Topology of the Sun's magnetic field and the 22-year cycle. *Astrophys. J.* **133**, 572.

Backus, G. E. (1957). The axisymmetric self-excited fluid dynamo. *Astrophys. J.* **125**, 500.

Backus, G. E. (1958). A class of self sustaining dissipated spherical dynamos. *Ann. Phys., N.Y.* **4**, 372.

Backus, G. E. (1968). Kinematics of geomagnetic secular variation in a perfectly conducting core. *Phil. Trans. Roy. Soc. London* **A263**, 239.

Backus, G. E. (1975). Gross thermodynamics of heat engines in deep interior of earth. *Proc. Nat. Acad. Sci. USA* **72**, 1555.

Backus, G. E. (1983). Application of mantle filter theory to the magnetic jerk of 1969. *Geophys. J. Roy. Astron. Soc.* (in press).

Backus, G. E. and Chandrasekar, S. (1956). On Cowling's theorem on the impossibility of self-maintained axisymmetric homogeneous dynamos. *Proc. Nat. Acad. Sci. USA* **42**, 108.

Backus, G. E. and Gilbert, F. (1967). Numerical application of a formalism for geophysical inverse problems. *Geophys. J. Roy. Astron. Soc.* **13**, 247.

Backus, G. E. and Gilbert, F. (1968). The resolving power of gross earth data. *Geophys. J. Roy. Astron. Soc.* **16**, 169.

Backus, G. E. and Gilbert, F. (1970). Uniqueness in the inversion of inaccurate gross earth data. *Phil. Trans. Roy. Soc. London* **A266**, 123.

Bailey, M. E. and Dunlop, D. J. (1977). On the use of anhysteretic remanent magnetisation in paleointensity determination. *Phys. Earth Planet. Int.* **13**, 360.

Banerjee, S. K. (1977). On the origin of stable remanence in pseudo-single domain grains. *J. Geomag. Geoelect.* **29**, 319.

Banerjee, S. K. and Hargraves, R. B. (1971). Natural remanent magnetisation of carbonaceous chondrites. *Earth Planet. Sci. Lett.* **10**, 392.

Banerjee, S. K. and Hargraves, R. B. (1972). Natural remanent magnetisations of carbonaceous chondrites and the magnetic field in the early solar system. *Earth Planet. Sci. Lett.* **17**, 110.

Banerjee, S. K. and Mellema, J. P. (1976). Early lunar magnetism. *Proc. 7th Lunar Sci. Conf.* 3209.

Banks, R. J. (1969). Geomagnetic variations and the electrical conductivity of the upper mantle. *Geophys. J. Roy. Astron. Soc.* **17**, 457.

Barbetti, M. (1977). Measurements of recent geomagnetic secular variation in southeast Australia and the question of dipole wobble. *Earth Planet. Sci. Lett.* **36**, 207.

Barbetti, M. (1980). Geomagnetic strength over the last 50 000 years and changes in atmospheric ^{14}C concentration: emerging trends. *Radiocarbon* **22**, 192.

Barbetti, M. and Flude, K. (1979a). Palaeomagnetic field strengths from sediments baked by lava flows of the Chaine des Puys, France. *Nature* **278**, 153.

Barbetti, M. and Flude, K. (1979b). Geomagnetic variation during the Late Pleistocene period and changes in the radiocarbon time-scale. *Nature* **279**, 202.

Barbetti, M. and McElhinny, M. W. (1972). Evidence for a geomagnetic excursion 30 000 yr B.P. *Nature* **239**, 327.

Barbetti, M. F. and McElhinny, M. W. (1976). The Lake Mungo Geomagnetic Excursion. *Phil. Trans. Roy. Soc. London* **A281**, 515.

Barbetti, M. F., McElhinny, M. W., Edwards, D. J. and Schmidt, P. W. (1977). Weathering processes in baked sediments and their effects on archaeomagnetic field-intensity measurements. *Phys. Earth Planet. Int.* **13**, 346.

Barnes, A., Casser, P., Mihalou, J. D. and Eviatur, A. (1971). Permanent lunar surface magnetism and its deflection of the solar wind. *Science* **171**, 716.

Barnett, S. J. (1933). Gyromagnetic effects; history, theory and experiments. *Physica* **13**, 241.

Barraclough, D. R. (1974). Spherical harmonic analysis of the geomagnetic field for eight epochs between 1600 and 1910. *Geophys. J. Roy. Astron. Soc.* **36**, 497.

Barton, C. E. (1978). *Magnetic Studies of some Australian Lake Sediments.* Ph.D. thesis, Res. Sch. Earth Sci., Australian National University, Canberra.

Barton, C. E. (1982). Spectral analysis of palaeomagnetic time series and the geomagnetic spectrum. *Phil. Trans. Roy. Soc.* **A306**, 203.

Barton, C. E. (1983). Analysis of palaeomagnetic time series—techniques and applications. *Geophys. Surv.* **5**, 335.

Barton, C. E. and Burden, F. R. (1979). Modifications to the Mackereth cover. *Limnol. oceanogr.* **24**, 977.

Barton, C. E. and McElhinny, M. W. (1981). A 10 000 yr geomagnetic secular variation record from three Australian maars. *Geophys. J. Roy. Astron. Soc.* **67**, 465.

Barton, C. E. and McElhinny, M. W. (1982). Time series analysis of the 10 000 yr geomagnetic secular variation record from SE Australia. *Geophys. J. Roy. Astron. Soc.* **68**, 709.

Barton, C. E., Merrill, R. T. and Barbetti, M. (1979). Intensity of the earth's magnetic field over the last 10 000 years. *Phys. Earth Planet. Int.* **20**, 96.

Barton, C. E., McElhinny, M. W. and Edwards, D. J. (1980). Laboratory studies of depositional DRM. *Geophys. J. Roy. Astron. Soc.* **61**, 355.

Bauer, L. A. (1894). An extension of the Gaussian potential theory of terrestrial magnetism. *Proc. Amer. Assn. Adv. Sci.* **43**, 55.

Bauer, L. A. (1895). On the distribution and the secular variation of terrestrial magnetism, No. III. *Amer. J. Sci.* **50**, 314.

Benjamin, P. (1895). *The Intellectual Rise in Electricity.* Longmans Green, London.

Benkova, N. P. and Cherevko, T. N. (1972). Analytical representation of magnetic inclincations. *Geomag. Aeron.* **12**, 632.

Benkova, N. P., Adam, N. V. and Cherevko, T. N. (1970). Application of spherical harmonic analysis to magnetic declination data. *Geomag. Aeron.* **10**, 673.

Benkova, N. P., Kruglyakov, A. A., Khramov, A. N. and Cherevko, T. N. (1971). Spherical harmonic analysis of paleomagnetic data. *Geomag. Aeron.* **11**, 319.

Benkova, N. P., Khramov, A. N., Cherevko, T. N. and Adam, N. V. (1973). Spherical harmonic analysis of the paleomagnetic field. *Earth Planet. Sci. Lett* **18**, 1412.

Benkova, N. P., Kolomiytseva, G. I. and Cherevko, T. N. (1974). Analytical model of the geomagnetic field and its secular variations over a period of 400 years (1550–1950). *Geomag. Aeron.* **14**, 751. (English translation.)

Benton, E. R. (1979). On fluid circulation around null-flux curves at Earth's core—mantle boundary. *Geophys. Astrophys. Fluid Dyn.* **11**, 323.

Benton, E. R., Muth, L. A. and Stix, M. (1979). Magnetic contour maps at the core–mantle boundary. *J. Geomag. Geoelect.* **31**, 615.

Benton, E. R., Estes, R. H., Langel, R. A. and Muth, L. A. (1982). Sensitivity of selected geomagnetic properties to truncation level of spherical harmonic expansions. *Geophys. Res. Lett.* **9**, 254.

Bergh, H. W. (1970). Paleomagnetism of the Stillwater complex, Montana. In *Paleogeophysics* (S. K. Runcorn, ed.), p. 143. Academic Press, London and New York.

Bhargava, B. N. and Yacob, A. (1969). Solar cycle response in the horizontal force of the earth's magnetic field. *J. Geomag. Geoelect.* **21**, 385.

Biermann, L. (1951). Kometenschwerfe und solare Korpuskularstrahlung. *Zeits. Astrophys.* **29**, 274.

Birch, F. (1964). Density and composition of mantle and core. *J. Geophys. Res.* **69**, 4377.

Birch, F. (1972). The melting relations of iron and temperature in the earth's core. *Geophys. J. Roy. Astron. Soc.* **29**, 373.

Bird, M. K. and Beard, D. B. (1972a). Self-consistent description of the magnetotail current system. *J. Geophys. Res.* **77**, 4864.

Bird, M. K. and Beard, D. B. (1972b). The self-consistent geomagnetic tail under static conditions. *Planet. Space Sci.* **20**, 2057.

Blackett, P. M. S. (1947). The magnetic field of massive rotating bodies. *Nature* **159**, 658.

Blackett, P. M. S. (1952). A negative experiment relating to magnetism and the earth's rotation. *Phil. Trans. Roy. Soc. London* **A245**, 309.

Blackett, P. M. S. (1961). Comparison of ancient climates with the ancient latitudes deduced from rock magnetic measurements. *Proc. Roy. Soc. London* **A263**, 1.

Blackshear, W. T. and Gapcynski, J. P. (1977). An improved value of the lunar moment of inertia. *J. Geophys. Res.* **82**, 1647.

Blakely, R. J. (1976). An age-dependent, two-layer model for marine magnetic anomalies. In *The Geophysics of the Pacific Ocean Basin and its Margin* (G. H. Sutton, M. H. Manghnani, R. Moberly and E. U. McAgee, eds). *Amer. Geophys. Union Monograph* **19**, 227.

Blakely, R. J. and Cande, S. C. (1979). Marine magnetic anomalies. *Rev. Geophys. Space Phys.* **17**, 204.

Blakely, R. J. and Cox, A. (1972). Evidence for short geomagnetic polarity intervals in the early Cenozoic. *J. Geophys. Res.* **77**, 7065.

Bloomfield, P. (1976). *Fourier Analysis of Time Series: An Introduction.* Wiley, New York.

Bochev, A. (1965). The earth's magnetic field represented as dipoles. *Comptes rendus de l'academie bulgare des sciences* **18**(4), 319.

Bochev, A. (1969). Two and three dipoles approximating the earth's main magnetic field. *Pageoph.* **74**, 29.

Bochev, A. (1975). Presenting the earth's magnetic field as a field of six optimal dipoles. *Comptes rendus de l'academie bulgare des sciences* **28**(4), 469.

Bogue, S. W. and Coe, R. S. (1981). Thellier paleointensity results from a R–N transition zone on Kauai, Hawaii. *EOS* **62**, 853.

Bogue, S. W. and Coe, R. S. (1982). Successive paleomagnetic reversal records from Kauai. *Nature* **295**, 399.

Bonhommet, N. and Babkine, J. (1967). Sur la présence d'aimantations inversées dans la Chaine des Puys. *C.R. Acad. Sci. Paris* **B264**, 92.

Bonhommet, N. and Zähringer, J. (1969). Paleomagnetism and potassium-argon age determinations of the Laschamp geomagnetic polarity event. *Earth Planet. Sci. Lett.* **6**, 43.

Booker, J. R. (1969). Geomagnetic data and core motions. *Proc. Roy. Soc. London* **A309**, 27.

Bott, M. H. P. (1967). Solution to the linear inverse problem in magnetic interpretation with application to ocean magnetic anomalies. *Geophys. J. Roy. Astron. Soc.* **13**, 313.

Braginskiy, S. I. (1963). Structure of the F layer and reasons for convection in the earth's core. *Dokl. Akad. Nauk SSR* **149**, 8.

Braginskiy, S. I. (1964a). Magnetohydrodynamics of the earth's core. *Geomag. Aeron.* **4**, 698.

Braginskiy, S. I. (1964b). Kinematic models of earth's hydromagnetic dynamo. *Geomag. Aeron.* **4**, 732.

Braginskiy, S. I. (1965). Theory of the hydromagnetic dynamo. *Soviet Phys. J.E.T.P.* **20**, 1462.

Braginskiy, S. I. (1967). Magnetic waves in the earth's core. *Geomag. Aeron.* **7**, 851.

Braginskiy, S. I. (1972). Spherical harmonic analyses of the main geomagnetic field in 1550–1800. *Geomag. Aeron.* **12**, 464. (English translation.)

Braginskiy, S. I. (1976). On the nearly axially-symmetrical model of the hydromagnetic dynamo of the earth. *Phys. Earth Planet. Int.* **11**, 191.

Braginskiy, S. I. and Kulanin, N. V. (1971). Spherical analysis of the geomagnetic field from angular data and the extrapolated g_1^0 value. III. *Geomag. Aeron.* **11**, 931 (pp. 786–788 in English translation).

Brandt, S. O. and Hodge, P. (1964). *Solar System Astrophysics*. McGraw-Hill, New York.

Brecher, A. (1972). Memory of early magnetic fields in carbonaceous chondrites. In *On the Origin of the Solar System* (H. Reeves, ed.), p. 120. Centre Nationale de le Recherche Scientifique, Paris.

Brecher, A. (1976). Textural remanence: a new model of lunar rock magnetism. *Earth Planet. Sci. Lett.* **29**, 131.

Brecher, A. (1977). Meteorite magnetism: implications for parent bodies of origin. In *Comets, Asteroids, Meteorites: Interactions, Evolution and Origins* (A. H. Delsemore, ed.), p. 415. University of Toledo Press, Toledo, Ohio.

Brecher, A. and Arrhenius, G. (1974). The paleomagnetic record in carbonaceous chondrites: natural remanence and magnetic properties. *J. Geophys. Res.* **79**, 2081.

Brecher, A. and Brecher, K. (1978). On the observed relation between magnetic fields and rotation in astronomical objects. In *Conference on Origins of Planetary Magnetism*, Vol. 7. Lunar and Planetary Institute, Houston, Texas.

Breit, G. and Tuve, M. A. (1925). A test of the existence of the conducting layer. *Nature* **116**, 357.

Breit, G. and Tuve, M. A. (1926). A test of the existence of the conducting layer. *Phys. Rev.* **28**, 554.

Brett, R. (1973). A lunar core of Fe–Ni–S. *Geochim. Cosmochim. Acta.* **37**, 165.

Brett, R. (1976). The current status of speculations on the composition of the core of the earth. *Rev. Geophys. Space Phys.* **14**, 375.

Briden, J. C. (1966). Variation of the intensity of the paleomagnetic field through geological time. *Nature* **212**, 246.

Briden, J. C. (1968). Paleoclimatic evidence of a geocentric axial dipole field. In *History of the Earth's Crust* (R. A. Phinney, ed.), p. 178. Princeton University Press, Princeton, N.J.

Briden, J. C. and Irving, E. (1964). Paleoclimatic spectra of sedimentary palaeoclimatic indicators. In *Problems in Palaeoclimatology* (A. E. M. Nairn, ed.), p. 199. Wiley Interscience, New York.

Brock, A. (1971). An experimental study of palaeosecular variation. *Geophys. J. Roy. Astron. Soc.* **24**, 303.

Brown, J. M. and McQueen, R. G. (1980). Melting of iron under core conditions. *Geophys. Res. Lett.* **7**, 533.

Brown, L. W. (1975). Saturn rocks emission near 1 MHz. *Astrophys. J.* **198**, L89.

Brown, L. W. (1976). Possible radio emission from uranus at 0.5 MHz. *Astrophys. J.* **207**, L209.

Brown, W. F. (1959). Relaxation behaviour of fine magnetic particles. *J. Appl. Phys. Suppl.* **30**, 1305.

Brunhes, B. (1906). Recherches sur le direction d'aimantation des roches volcaniques. *J. Phys.* **5**, 705.

Bucha, V. (1967). Intensity of the earth's magnetic field during archaeological times in Czechoslovakia. *Archaeometry* **10**, 12.

Bucha, V. (1969). Changes in the earth's magnetic moment and radiocarbon dating. *Nature* **224**, 681.

Bucha, V. (1980). Mechanism of the relations between the changes of the geomagnetic field, solar corpuscular radiation, atmospheric circulation and climate. *J. Geomag. Geoelect.* **32**, 217.

Bucha, V. and Neustupny, E. (1967). Changes of the earth's magnetic field and radiocarbon dating. *Nature* **215**, 261.

Buddington, A. F. and Lindsley, A. F. (1964). Iron-titanium oxide minerals and synthetic equivalents. *J. Petrol.* **5**, 310.

Bullard, E. C. (1949a). The magnetic field within the earth. *Proc. Roy. Soc. London* **A197**, 433.
Bullard, E. C. (1949b). Electromagnetic induction in a rotating sphere. *Proc. Roy. Soc. London* **A199**, 413.
Bullard, E. C. and Gellman, H. (1954). Homogeneous dynamos and terrestrial magnetism. *Phil. Trans. Roy. Soc. London* **A247**, 213.
Bullard, E. C. and Gubbins, D. (1977). Generation of magnetic fields by fluid motions of global scale. *Geophys. Astrophy. Fluid Dynamics* **8**, 43.
Bullard, E. C., Freedman, C., Gellman, H. and Nixon, J. (1950). The westward drift of the earth's magnetic field. *Phil. Trans. Roy. Soc. London* **A243**, 67.
Burakov, K. S., Gurariy, G. Z., Khramov, A. N., Petrova, G. N., Rassanova, G. V. and Rodinov, G. P. (1976). Some peculiarities of the virtual geomagnetic pole position during reversals. *J. Geomag. Geoelect.* **28**, 295.
Burch, J. L. (1974). Observation of interactions between interplanetary and geomagnetic fields. *Rev. Geophys. Space Phys.* **12**, 363.
Burg, J. P. (1967). Maximum entropy spectral analysis. *37th An. Int. Meeting Soc. Explor. Geophys.*, Oklahoma, USA.
Burg, J. P. (1972). The relationship between maximum entropy spectra and maximum likelihood spectra. *Geophys.* **37**, 375.
Burke, B. F. and Franklin, K. L. (1955). Radio emission from Jupiter. *Nature* **175**, 1074.
Burlatskaya, S. P. (1972). Changes in geomagnetic field intensity in the last 8500 years, according to global archeomagnetic data. *Geomag. Aeron* **10**, 544.
Burlatskaya, S. P., Nechaeva, T. B. and Petrova, G. N. (1965). The westward drift of the secular variation of magnetic inclination and variations of the Earth's magnetic moment according to archeomagnetic data. *Izv. Akad. Nauk SSSR. Fizika Zemli* **6**, 31.
Burlatskaya, S. P., Nechaeva, T. B. and Petrova, G. N. (1969). Some archaeomagnetic data indicative of the westward drift of the geomagnetic field. *Archeometry* **11**, 115.
Burlatskaya, S. P., Kruchik, Y. A., Malkovskiy, Z., Nechaeva, T. B. and Petrova, G. N. (1970a). Comparison of secular variation in inclination in Poland and the Ukraine. *Geomag. Aeron.* **10**, 243.
Burlatskaya, S. P., Nachasova, I. E., Nachaeva, T. B., Rusakov, O. M., Zagniy, G. F., Tarhov, E. N. and Tchelidze, Z. A. (1970b). Archeomagnetic research in the USSR: recent results and spectral analysis. *Archeometry* **12**, 73.
Burlatskaya, S. P., Nachasova, I. E. and Burakov, K. S. (1975). New determinations of the parameters of the ancient geomagnetic field for Mongolia, Soviet Central Asia and Abhazia. *Geomag. Aeron.* **16**, 447.
Busse, F. H. (1970). Thermal instabilities in rapidly rotating systems. *J. Fluid Mech.* **44**, 441.
Busse, F. H. (1975). A model for the geodynamo. *Geophys. J. Roy. Astron. Soc.* **42**, 437.
Busse, F. H. (1976). Generation of planetary magnetism by convection. *Phys. Earth Planet. Int.* **12**, 350.
Busse, F. H. (1977a). An example of non-linear dynamo action. *J. Geophys.* **43**, 441.
Busse, F. H. (1977b). Mathematical problems of dynamo theory. In *Application of Bifurcation Theory* (P. H. Rabinowitz, ed.), p. 175. Academic Press, London and New York.
Busse, F. H. (1978). Magnetohydrodynamics of the earth's dynamo. *Ann. Rev. Fluid Mech.* **10**, 435.
Busse, F. H. and Cuong, P. G. (1977). Convection in rapidly rotating spherical fluid shells. *Geophys. Astrophys. Fluid Dyn.* **8**, 17.
Butler, R. F. (1972). Natural remanent magnetisation and thermomagnetic properties of the Allende Meteorite. *Earth Planet Sci. Lett.* **17**, 120.
Butler, R. F. (1982). Magnetic mineralogy of continental deposits, San Juan Basin, New Mexico, and Clark's Fork Basin, Wyoming. *J. Geophys. Res.* **87**, 7843.

Butler, R. F. and Banerjee, S. F. (1975). Theoretical single-domain grain size range in magnetite and titanomagnetite. *J. Geophys. Res.* **80**, 4049.

Butler, R. F., Lindsay, E. H., Jacobs, L. L. and Johnson, N. M. (1977). Magnetostratigraphy of the Cretaceous–Tertiary boundary in the San Juan Basin, New Mexico. *Nature* **267**, 318.

Cain, J. C. (1971). Geomagnetic models from satellite surveys. *Rev. Geophys. Space Phys.* **9**, 259.

Cain, J. C. (1975). Structure and secular change of the geomagnetic field. *Rev. Geophys. Space Phys.* **13**, 203.

Cain, J. C., Davis, W. M. and Regan, R. D. (1974). An $n = 22$ model of the geomagnetic field. *EOS* **56**, 1108.

Carmichael, C. M. (1961). The magnetic properties of ilmenite-hematite crystals. *Proc. Roy. Soc. London* **A263**, 508.

Carmichael, C. M. (1967). An outline of the intensity of the paleomagnetic field of the earth. *Earth Planet. Sci. Lett.* **3**, 35.

Carmichael, I. S. E. and Nicholls, J. (1967). Iron–titanium oxides and oxygen fugacities in volcanic rocks. *J. Geophys. Res.* **72**, 4665.

Champion, D. E. (1980). Holocene geomagnetic secular variation in the western United States: implications for the global geomagnetic field. *Rept. Open file series US Geol. Surv.* **80–824**, 314.

Chandrasekhar, S. (1961). *Hydrodynamic and Hydromagnetic Stability*. Clarendon Press, Oxford.

Chapman, S. (1942). Notes on isomagnetic charts; VI. Earth–air electric currents, and the mutual consistency of the H- and D-isomagnetic charts; VII. Mathematical notes on isotopic charts and their singular points; VIII. The mutual consistency of the declination and horizontal-intensity isoporic charts. *Terr. Mag. Atmosph. Elect.* **47**, 1, 115.

Chapman, S. (1951). Some phenomena of the upper atmosphere. *Proc. Phys. Soc. London* **B64**, 833.

Chapman, S. and Bartels, J. (1940, 1962). *Geomagnetism*, Vols. 1 and 2 (1940); 2nd edition (1962). Oxford University Press, Oxford.

Chapman, S. and Ferraro, V. C. A. (1931). A new theory of magnetic storms. *Terr. Magn. Atmosph. Elec.* **36**, 77 and 171.

Chapman, S. and Ferraro, V. C. A. (1932). A new theory of magnetic storms. *Terr. Magn. Atmosph. Elec.* **37**, 147 and 421.

Chapman, S. and Ferraro, V. C. A. (1933). A new theory of magnetic storms. *Terr. Magn. Atmosph. Elec.* **38**, 79.

Chatterjee, J. S. (1956). The crust as the possible seat of the earth's magnetism. *J. Atmosph. Terr. Phys.* **8**, 233.

Chen, Po-Fang (1981). Geomagnetic variations with periods from 5 to 30 years at Hong Kong. *J. Geomag. Geoelect.* **33**, 189.

Chevallier, R. (1925). L'aimantation des laves de l'Etna et l'orientation du champ terrestre en Sicile du XIIe and XVIIe siecle. *Ann. Phys.* **4**, 5.

Chikazumi, S. (1964). *Physics of Magnetism*. Wiley, New York.

Childers, D. C. (ed.) (1978). *Modern Spectrum Analyses*. IEEE Press, New York.

Childress, S. (1969). A class of solutions of the magnetohydrodynamic dynamo problem. In *The Application of Modern Physics to the Earth and Planetary Interiors* (S. K. Runcorn, ed.), p. 629. Wiley, New York.

Childress, S. (1970). New solutions of the kinematic dynamo problem. *J. Math. Phys.* **11**, 3063.

Childress, S. (1978). *Lectures on Geomagnetic Dynamo Theory*. In Woods Hole Ocean Inst. Tech. Report. WHO1-78-6 Part 1, 1.

Cisowski, S. M. and Fuller, M. (1978). The effect of shock on the magnetism of terrestrial rocks. *J. Geophys. Res.* **83**, 3441.

Cisowski, S. M., Dunn, J. R., Fuller, M., Rose, M. F. and Wu, Y. M. (1976). Magnetic effects of shock and their implications for lunar magnetism (II). *Proc. 7th Lunar Sci. Conf.* 3299.

Cisowski, S. M., Hale, C. and Fuller, M. (1977). On the intensity of ancient lunar fields. *Proc. 8th Lunar Sci. Conf.* 725.

Clark, H. C. and Kennett, J. P. (1973). Paleomagnetic excursion recorded in latest Pleistocene deep-sea sediments, Gulf of Mexico. *Earth Planet. Sci. Lett.* **19**, 267.

Clark, R. M. (1975). A calibration curve for radiocarbon dates. *Antiquity* **49**, 251 (and comments by Suess, and reply, *Antiquity* **50** 61).

Clark, R. M. and Thompson, R. (1979). A new statistical approach to the alignment of time series. *Geophys. J. Roy. Astron. Soc.* **58**, 593.

Clayton, R. N. and Mayeda, T. K. (1975). Genetic relations between the Moon and meteorites. *Proc. 6th Lunar Sci. Conf.* 1761.

Clayton, R. N. and Mayeda, T. K. (1978). Genetic relations between iron and stony meteorites. *Earth Planet. Sci. Lett.* **40**, 168.

Clayton, R. N., Grossman, L. and Mayeda, T. K. (1973). A component of primitive nuclear composition in carbonaceous meteorites. *Science* **182**, 485.

Coe, R. S. (1967a). The determination of paleointensities of the earth's magnetic field with emphasis on mechanisms which could cause nonideal behaviour in Thelliers method. *J. Geomag. Geoelect.* **19**, 157.

Coe, R. S. (1967b). Paleointensities of the earth's magnetic field determined from Tertiary and Quaternary rocks. *J. Geophys. Res.* **72**, 3247.

Coe, R. S. (1977). Source models to account for the Lake Mungo paleomagnetic excursion and their implications. *Nature* **269**, 49.

Coe, R. S. (1979). The effect of shape anisotropy on TRM direction. *Geophys. J. Roy. Astron. Soc.* **56**, 369.

Coe, R. S., Gromme, C. S. and Mankinen, E. A. (1978). Geomagnetic paleointensities from radiocarbon dated lava flows on Hawaii and the question of the Pacific non-dipole low. *J. Geophys. Res.* **83**, 1740.

Cole, K. D. (1967). On the Dst main phase and certain associated phenomena. In *Physics of Geomagnetic Phenomena* (S. Matsushita and W. H. Campbell, eds), Vol. 2 (6), p. 1300. Academic Press, London and New York.

Coleman, P. J. Jr., Lichtenstein, B. R., Russell, C. T., Sharp, L. R. and Schubert, G. (1972). Magnetic fields near the Moon. *Proc. 3rd Lunar Sci. Conf.* 2271.

Collinson, D. W. (1965). Origin of remanent magnetisation in certain red sandstones. *Geophys. J. Roy. Astron. Soc.* **9**, 203.

Collinson, D. W., Stephenson, A. and Runcorn, S. K. (1973). Magnetic properties of Apollo 15 and 16 rocks. *Proc. 4th Lunar Sci. Conf.* 2963.

Connerney, J., Acuna, M. and Ness, N. (1891). Saturn's ring current and inner magnetosphere. *Nature* **292**, 724.

Cook, P. J. and McElhinny, M. W. (1979). A re-evaluation of the spatial and temporal distribution of sedimentary phosphate deposits in the light of plate tectonics. *Econ. Geol.* **74**, 315.

Cook, R. M. and Belshe, J. C. (1958). Archaeomagnetism: a preliminary report from Britain. *Antiquity* **32**, 167.

Cooley, J. W. and Tukey, J. W. (1965). An algorithm for the machine calculation of complex fourier series. *Math. Comp.* **19**, 297.

Coupland, D. H. and Van der Voo, R. (1980). Long-term non-dipole components in the geomagnetic field during the last 130 Ma. *J. Geophys. Res.* **85**, 3529.

Courtillot, V. and Le Mouel, J. L. (1976). On the long-period variation of the earth's magnetic field from 2 months to 20 years. *J. Geophys. Res.* **81**, 2941.

Cowling, T. G. (1934). The magnetic field of sunspots. *Monthly Notices Roy. Astron. Soc.* **94**, 39.

Cowling, T. G. (1957). *Magnetohydrodynamics.* Wiley Interscience, New York.

Cox, A. (1962). Analysis of the present geomagnetic field for comparison with paleomagnetic results. *J. Geomag. Geoelect.* **13**, 101.

Cox, A. (1964). Angular dispersion due to random magnetisation. *Geophys. J. Roy. Astron. Soc.* **8**, 345.

Cox, A. (1968). Lengths of geomagnetic polarity intervals. *J. Geophys. Res.* **73**, 3247.

Cox, A. (1969a). Confidence limits for the precision parameter κ. *Geophys. J. Roy. Astron. Soc.* **18**, 545.

Cox, A. (1969b). Geomagnetic reversals. *Science* **163**, 237.

Cox, A. (1970). Latitude dependence of the angular dispersion of the geomagnetic field. *Geophys. J. Roy. Astron. Soc.* **20**, 253.

Cox, A. (1975). The frequency of geomagnetic reversals and the symmetry of the non-dipole field. *Rev. Geophys. Space Phys.* **13**, 35.

Cox, A. (1977). Erratum: Confidence limits for the precision parameter κ. *Geophys. J. Roy. Astron. Soc.* **48**, 136.

Cox, A. (1981). A stochastic approach towards understanding the frequency and polarity bias of geomagnetic reversals. *Phys. Earth Planet. Int.* **24**, 178.

Cox, A. (1982). Magnetostratigraphic time scale. In *A Geologic Time Scale* (W. B. Harland *et al.*, eds), p. 63. Cambridge University Press, Cambridge.

Cox, A. and Doell, R. R. (1960). Review of paleomagnetism. *Geol. Soc. Amer. Bull.* **71**, 645.

Cox, A. and Doell, R. R. (1964). Long period variations of the geomagnetic field. *Bull. Seismol. Soc. Amer.* **54**, 2243.

Cox, A., Doell, R. R. and Dalrymple, G. B. (1963a). Geomagnetic polarity epochs and Pleistocene geochronometry. *Nature* **198**, 1049.

Cox, A., Doell, R. R. and Dalrymple, G. B. (1963b). Geomagnetic polarity epochs: Sierra Nevada II. *Science* **142**, 382.

Cox, A., Doell, R. R. and Dalrymple, G. B. (1964a). Geomagnetic polarity epochs. *Science* **143**, 351.

Cox, A., Doell, R. R. and Dalrymple, G. B. (1964b). Reversals of the earth's magnetic field. *Science* **144**, 1537.

Creer, K. M. (1962). The dispersion of the geomagnetic field due to secular variation and its determination for remote times from paleomagnetic data. *J. Geophys. Res.* **67**, 3461.

Creer, K. M. (1974). Geomagnetic variations for the interval 7000–25 000 yr B.P. as recorded in a core of sediment from station 1474 of the Black Sea cruise of "Atlantis II". *Earth Planet. Sci. Lett.* **23**, 34.

Creer, K. M. (1977). Geomagnetic secular variation during the last 25 000 years: an interpretation of the data obtained from rapidly deposited sediments. *Geophys. J. Roy. Astron. Soc.* **48**, 91.

Creer, K. M. and Ispir, Y. (1970). An interpretation of the behaviour of the geomagnetic field during polarity transitions. *Phys. Earth Planet. Int.* **2**, 283.

Creer, K. M. and Kopper, J. S. (1976). Secular oscillations of the geomagnetic field recorded by sediments deposited in caves in the Mediterranean region. *Geophys. J. Roy. Astron. Soc.* **45**, 35.

Creer, K. M., Irving, E. and Runcorn, S. K. (1954). The direction of the geomagnetic field in remote epochs in Great Britain. *J. Geomag. Geoelect.* **6**, 163.

Creer, K. M., Irving, E. and Nairn, A. E. M. (1959). Palaeomagnetism of the Great Whin Sill. *Geophys. J. Roy. Astron. Soc.* **2**, 306.

Creer, K. M., Thompson, R., Molyneux, L. and Mackereth, F. H. (1972). Geomagnetic secular variation recorded in the stable magnetic remanence of recent sediments. *Earth Planet. Sci. Lett.* **14**, 115.

Creer, K. M., Georgi, D. T. and Lowrie, W. (1973). On the representation of the Quaternary and late Tertiary geomagnetic field in terms of dipoles and quadrupoles. *Geophys. J. Roy. Astron. Soc.* **33**, 323.

Creer, K. M., Gross, D. L. and Lineback, J. A. (1976a). Origin of regional geomagnetic variations recorded by Wisconsinan and Holocene sediments from Lake Michigan, USA and Lake Windermere, England. *Geol. Soc. Amer. Bull.* **87**, 531.

Creer, K. M., Anderson, T. W. and Lewis, C. F. M. (1976b). Late Quaternary geomagnetic stratigraphy recorded in Lake Erie sediments. *Earth Planet. Lett.* **31**, 37.

Creer, K. M., Hogg, E., Malkowski, Z., Mojski, J. E., Niedziolka-Krol, E., Readman, P. W. and Tucholka, P. (1979). Palaeomagnetism of Holocene lake sediments from north Poland. *Geophys. J. Roy. Astron. Soc.* **59**, 287.

Cullity, B. D. (1972). *Introduction to Magnetic Materials.* Addison-Wesley Pub. Co., Menlo Park, California.

Currie, R. G. (1966). The geomagnetic spectrum 40 days to 55 years. *J. Geophys. Res.* **71**, 4579.

Currie, R. G. (1967). Magnetic shielding properties of the earth's mantle. *J. Geophys. Res.* **72**, 2623.

Currie, R. G. (1968). Geomagnetic spectrum of internal origin and lower mantle conductivity. *J. Geophys. Res.* **73**, 2779.

Currie, R. G. (1973). Geomagnetic line spectra—2 to 70 years. *Astrophys. Space Sci.* **21**, 425.

Currie, R. G. (1974). Harmonics of the geomagnetic annual variation. *J. Geomag. Geoelect.* **25**, 319.

Currie, R. G. (1975). Long period magnetic activity—2 to 100 years. *Astrophys. Space Sci.* **23**, 1314.

Dagley, P. and Lawley, E. (1974). Paleomagnetic evidence for the transitional behaviour of the geomagnetic field. *Geophys. J. Roy. Astron. Soc.* **36**, 577.

Dagley, P. and Wilson, R. L. (1971). Geomagnetic field reversals—a link between strength and orientation of a dipole source. *Nature Phys. Sci.* **232**, 16.

Dainty, A. M., Toksov, M. N. and Stein, S. (1976). Seismic investigations of the lunar interior. *Proc. 7th Lunar Sci. Conf.* 3057.

Damon, P. E. (1970). Climatic vs. magnetic perturbation of the carbon-14 reservoir. In *Radiocarbon Variations and Absolute Chronology* (I. U. Olsson, ed.), p. 571. Alinqvist and Wiksell, Stockholm.

David, P. (1904). Sur la stabilité de la direction d'aimantation dans quelques roches volcaniques. *C.R. Acad. Sci. Paris* **138**, 41.

Day, R. (1977). TRM and its variation with grain size. *J. Geomag. Geoelect.* **29**, 233.

Day, R., Fuller, M. D. and Schmidt, V. A. (1976). Magnetic hysteresis properties of synthetic titanomagnetites. *J. Geophys. Res.* **81**, 873.

Day, R., Fuller, M. D. and Schmidt, V. A. (1977). Hysteresis properties of titanomagnetites: grain size and composition dependence. *Phys. Earth Planet. Int.* **13**, 260.

Denham, C. R. (1974). Counterclockwise motion of paleomagnetic directions 24 000 years ago at Lake Mono, California. *J. Geomag. Geoelect.* **26**, 487.

Denham, C. R. (1975). Spectral analysis of paleomagnetic time series. *J. Geophys. Res.* **80**, 1897.

Denham, C. R. and Cox, A. V. (1971). Evidence that the Laschamp Polarity Event did not occur 13 300–30 400 years ago. *Earth Planet. Sci. Lett.* **13**, 181.

Dicke, R. H. (1970). Internal rotation of the sun. *Annual Rev. of Astronomy and Astrophysics* **8**, 297.

Dickson, G. O., Pitman III, W. C. and Heirtzler, J. R. (1968). Magnetic anomalies in the South Atlantic and ocean floor spreading. *J. Geophys. Res.* **73**, 2087.

Dodson, R. E. (1979). Counterclockwise precession of the geomagnetic field vector and westward drift of the non-dipole field. *J. Geophys. Res.* **84**, 637.

Dodson, R. E. Fuller, M. D. and Kean, W. F. (1977). Paleomagnetic records of secular variation from Lake Michigan sediment cores. *Earth Planet. Sci. Lett.* **34**, 387.

Dodson, R. E., Dunn, J. R., Fuller, M. D., Williams, I., Ito, H., Schmidt, V. A. and Yee-Ming Wu (1978). Paleomagnetic record of a late Tertiary field reversal. *Geophys. J. Roy. Astron. Soc.* **53**, 373.

Doell, R. R. (1970). Paleomagnetic secular variation study of lavas from the Massif Central, France. *Earth Planet. Sci. Lett.* **8**, 352.

Doell, R. R. (1972). Paleomagnetic study of Icelandic lava flows. *Geophys. J. Roy. Astron. Soc.* **26**, 459.

Doell, R. R. and Cox, A. V. (1963). The accuracy of the paleomagnetic method as evaluted from historic Hawaiian lava flows. *J. Geophys. Res.* **68**, 1997.

Doell, R. R. and Cox, A. V. (1965). Paleomagnetism of Hawaiian lava flows. *J. Geophys. Res.* **70**, 3377.

Doell, R. R. and Cox, A. V. (1972). The Pacific geomagnetic secular variation anomaly and the question of lateral uniformity in the lower mantle. In *The Nature of the Solid Earth* (E. C. Robertson, ed.), p. 245. McGraw-Hill, New York.

Doell, R. R., Gromme, C. S., Thorpe, A. N. and Senftle, F. E. (1970). Magnetic studies of Apollo 11 lunar samples. *Proc. Apollo 11 Lunar Sci. Conf.* 2097.

Dolginov, Sh. Sh. (1976). On the question of the energy of the precessional dynamo. In *Solar Wind Interaction with the Planets Mercury, Venus and Mars* (N. F. Ness, ed.). *NASA Spec. Publ.* **SP-397**, 167.

Dolginov, Sh. Sh. (1978). On the magnetic field of Mars: Mars 5 evidence. *Geophys. Res. Lett.* **5**, 93.

Dolginov, Sh. Sh., Yeroshenko, E. G., Zhuzgov, L. N. and Pushkov, N. V. (1961). Investigation of the magnetic field of the Moon. *Geomag. Aeron.* **1**, 18.

Dolginov, Sh. Sh., Yeroshenko, E. G., Zhuzgov, L. N. and Pushkov, N. V. (1966). Measurement of the magnetic field in the vicinity of the Moon by the artificial satellite Luna 10. *Dokl. Akad. Nauk SSSR* **170**, 574.

Dolginov, Sh. Sh., Yeroshenko, E. G. and Davis, L. (1969). On the nature of the magnetic field near Venus. *Kosm. Issled.* **7**, 747.

Dolginov, Sh. Sh., Yeroshenko, E. G. and Zhuzgov, L. N. (1976). The magnetic field of Mars according to the data from the Mars 3 and Mars 5. *J. Geophys. Res.* **81**, 3353.

Domen, H. (1977). A single heating method of paleomagnetic field intensity determination applied to old roof tiles and rocks. *Phys. Earth Planet. Int.* **13**, 315.

Drewry, G. E., Ramsay, A. T. S. and Smith, A. G. (1974). Climatically controlled sediments, the geomagnetic field, and trade wind belts in Phanerozoic time. *J. Geol.* **82**, 531.

DuBois, R. L. (1974). Secular variation in southwestern United States as suggested by archeomagnetic studies. In *Proceedings of the Takesi Nagata Conference* (R. M. Fisher, M. D. Fuller, V. A. Schmidt and P. J. Wasilewski, eds), p. 133. University of Pittsburgh Press, Pittsburgh.

Dungey, J. W. (1961). Interplanetary magnetic field and the auroral zones. *Phys. Rev. Lett.* **6**, 47.

Dungey, J. W. (1963). The structure of the exosphere, or adventures in velocity space. In *Geophysics, the Earth's Environment* (C. DeWitt, J. Hieblot and A. Lebeau, eds), p. 526. Gordon and Breach, New York.

Dunlop, D. J. (1970). Magnetic properties of fine particle hematite. *Ann. Geophys.* **27**, 269.

Dunlop, D. J. (1973). Thermoremanent magnetisation in submicroscopic magnetite. *J. Geophys. Res.* **78**, 7602.

Dunlop, D. J. (1977). The huntng of the Psark, *J. Geomag. Geoelect.* **29**, 293.

Dunlop, D. J. (1979). On the use of Zijderveld vector diagrams in multicomponent paleomagnetic studies. *Phys. Earth Planet. Int.* **20**, 12.

Dunlop, D. J. and Waddington, C. J. (1975). The field dependence of thermoremanent magnetisation in igneous rocks. *Earth Planet. Sci. Lett.* **25**, 11.

Dunlop, D. J. and West, G. F. (1969). An experimental evaluation of single-domain theories. *Rev. Geophys. Space Phys.* **7**, 709.

Dunlop, D. J., Stacey, F. D. and Gillingham, D. E. W. (1974). The origin of thermoremanent magnetisation: contribution to pseudo-single-domain magnetic moments. *Earth Planet. Sci. Lett.* **21**, 288.

Dunn, J. R., Fuller, M. D., Ito, H. and Schmidt, V. A. (1971). Paleomagnetic study of a reversal of the earth's magnetic field. *Science* **172**, 840.

Dyal, P. and Parkin, C. W. (1971). The Apollo 12 magnetometer experiment: internal lunar properties from transient and steady magnetic field measurements. *Proc. 2nd Lunar Sci. Conf.* 2391.

Dyal, P., Parkin, C. W. and Sonett, C. P. (1970). Apollo 12 magnetometer: measurement of a steady magnetic field on the surface of the Moon. *Science* **169**, 762.

Dyal, P., Parkin, C. W. and Daily, W. D. (1974). Magnetism and the interior of the Moon. *Rev. Geophys. Space Phys.* **12**, 568.

Dyal, P., Parkin, C. W. and Daily, W. D. (1975). Lunar electrical conductivity and magnetic permeability. *Proc. 6th Lunar Sci. Conf.* 2909.

Dyal, P., Parkin, C. W. and Daily, W. D. (1977). Global lunar crust: electrical conductivity and thermoelectric origin of remanent magnetism. *Proc. 8th Lunar Sci. Conf.* 767.

Dzaloshinski, I. (1958). A thermodynamic theory of "weak" ferromagnetism of antiferromagnetics. *J. Phys. Chem. Solids* **4**, 241.

Eddy, J. A. (1977). The case of the missing sunspots. *Scientific American* **236**, 80.

Eddy, J. A., Gilman, P. A. and Trotter, D. E. (1977). Anomalous solar rotation in the early 17th century. *Science* **198**, 829.

Elsasser, W. M. (1946a). Induction effects in terrestrial magnetism. 1. Theory. *Phys. Rev.* **69**, 106.

Elsasser, W. M. (1946b). Induction effects in terrestrial magnetism. 2. The secular variations. *Phys. Rev.* **70**, 202.

Elsasser, W. M. (1947). Induction effects in terrestrial magnetism. 3. Electric modes. *Phys. Rev.* **72**, 821.

Elsasser, W. M., Ney, E. P. and Wenckler, J. R. (1956). Cosmic ray intensity and geomagnetism. *Nature* **178**, 1226.

Elston, D. P. and Purucker, M. E. (1979). Detrital magnetisation in red beds of the Moenkopi formation. *J. Geophys. Res.* **84**, 1653.

Evans, M. E. (1976). Test of the dipolar nature of the geomagnetic field throughout Phanerozoic time. *Nature* **262**, 676.

Evans, M. E. and McElhinny, M. W. (1966). The paleomagnetism of the Modipe gabbro. *J. Geophys. Res.* **71**, 6053.

Evans, M. E. and McElhinny, M. W. (1969). An investigation of the origin of stable remanence in magnetite-bearing igneous rocks. *J. Geomag. Geoelect.* **21**, 757.

Evans, M. E. and Wayman, M. C. (1974). An investigation of small magnetic particles by means of electron microscopy. *Earth Planet. Sci. Lett.* **9**, 365.

Feigenbaum, M. J. (1980). Universal behaviour in non-linear systems. *Los Alamos Science Summer 1980* **4**.

Fisher, R. A. (1953). Dispersion on a sphere. *Proc. Roy. Soc. London* **A217**, 295.

Folgerhaiter, G. (1899). Sur les variations seculaires de l'inclinaison magnetique dans antiquite. *J. Phys.* **8**, 5.

Foster, J. H. and Opdyke, N. D. (1970). Upper Miocene to Recent magnetic stratigraphy in deep-sea sediments. *J. Geophys. Res.* **75**, 4465.

Foster, T. D. (1971). Intermittent convection. *Geophys. Fluid Dyn.* **2**, 201.

Freed, W. K. (1977). The virtual geomagnetic pole path during the Brunhes-Matuyama polarity change when viewed from equatorial latitudes. *EOS* **58**, 6, 380.

Freed, W. K. and Healy, N. (1974). Excursions of the Pleistocene geomagnetic field recorded in Gulf of Mexico sediments. *Earth Planet. Sci. Lett.* **24**, 99.

Fuller, M. D. (1970). Geophysical aspects of paleomagnetism. *Crit. Rev. Solid state Phys.* **1**, 137.

Fuller, M. D. (1974). Lunar magnetism. *Rev. Geophys. Space Phys.* **12**, 23.

Fuller, M. D. (1977). Review of effects of shock (60 kbars; 6×10^9 Pa) on magnetism of lunar samples. *Phil. Trans. Roy. Soc. London* **A285**, 409.

Fuller, M. D., Williams, I. and Hoffman, K. A. (1979). Paleomagnetic records of geomagnetic field reversals and the morphology of the transitional fields. *Rev. Geophys. Space Phys.* **17**, 179.

Galliher, S. C. and Mayhew, M. A. (1982). On the possibility of detecting large-scale crustal remanent magnetisation with MAGSAT vector magnetic anomaly data. *Geophys. Res. Lett.* **9**, 325.

Games, K. P. (1977). The magnitude of the paleomagnetic field: a new non-thermal, non-detrital method using sun-dried bricks. *Geophys. J. Roy. Astron. Soc.* **48**, 315.

Games, K. P. (1980). The magnitude of the archaeomagnetic field in Egypt between 3000 and 0 B.C. *Geophys. J. Roy. Astron. Soc.* **63**, 45.

Gans, R. F. (1972). Viscosity of the earth's core. *J. Geophys. Res.* **77**, 360.

Gardiner, R. B. and Stacey, F. D. (1971). Electrical resistivity of the core. *Phys. Earth Planet. Int.* **4**, 406.

Gauss, C. F. (1838). Allgemeine Theorie des Erdmagnetismus. In *Resultate magn. verein* (reprinted in Werke, Band 5, 121).

Georgi, D. T. (1974). Spherical harmonic analysis of paleomagnetic inclination data. *Geophys. J. Roy. Astron. Soc.* **39**, 71.

Gibson, R. D. and Roberts, P. H. (1969). The Bullard-Gellman dynamo. In *Application of Modern Physics to the Earth and Planetary Interiors* (S. K. Runcorn, ed.), p. 577. Wiley Interscience, New York.

Gilman, P. A. (1974). Solar rotation. *Ann. Rev. Astron. & Astrophys.* **12**, 47.

Gilman, P. A. (1977). Nonlinear dynamics of Boussinesq convection in a deep rotating spherical shell I. *Geophys. Astrophys. Fluid Dyn.* **8**, 93.

Goins, N. R., Dainty, A. M. and Tokooz, M. N. (1977). The deep seismic structure of the Moon. *Proc. 8th Lunar Sci. Conf.* 471.

Gold, T. (1959). Motions in the magnetosphere of the earth. *J. Geophys. Res.* **64**, 1219.

Goldstein, B. E., Phillips, R. J. and Russell, C. T. (1976). Magnetic evidence concerning a lunar core. *Proc. 7th Lunar Sci. Conf.* 3321.

Gose, W. A. and Butler, R. F. (1975). Magnetism of the Moon and meteorites. *Rev. Geophys. Space Phys.* **13**, 189.

Gose, W. A. and Morris, R. V. (1977). Depositional history of the Apollo 16 deep drill core. *Proc. 8th Lunar Sci. Conf.* 2909.

Gose, W. A., Strangway, D. W. and Pearce, G. W. (1978). Origin of magnetisation in lunar breccias: an example of thermal overprinting. *Earth Planet. Sci. Lett.* **38**, 373.

Graham, J. W. (1949). The stability and significance of magnetism in sedimentary rocks. *J. Geophys. Res.* **54**, 131.

Graham, K. W. T. and Hales, A. L. (1957). Palaeomagnetic measurements on some Karroo dolerites. *Phil. Mag. Adv. Phys.* **6**, 149.

Graham, K. W. T., Helsley, C. E. and Hales, A. L. (1964). Determination of the relative positions of continents from paleomagnetic data. *J. Geophys. Res.* **69**, 3895.

Griffiths, D. H. (1953). Remanent magnetism of varved clays from Sweden. *Nature* **172**, 539.

Gromme, C. S., Wright, T. L. and Peck, D. L. (1969). Magnetic properties and oxidation of iron-titanium oxide minerals in Alae and Makaopuhi lava lakes, Hawaii. *J. Geophys. Res.* **74**, 5277.

Gubbins, D. (1974). Theories of the geomagnetic and solar dynamos. *Rev. Geophys. Space Phys.* **12**, 137.

Gubbins, D. (1975a). Numerical solutions of the hydromagnetic dynamo problem. *Geophys. J. Roy. Astron. Soc.* **42**, 295.

Gubbins, D. (1975b). Can the earth's magnetic field be sustained by core oscillations? *Geophys. Res. Lett.* **2**, 409.

Gubbins, D. (1976). Observational constraints on the generation process of the earth's magnetic field. *Geophys. J. Roy. Astron. Soc.* **47**, 19.

Gubbins, D. (1977). Energetics of the earth's core. *J. Geophys.* **43**, 453.

Gubbins, D., Thomson, G. J. and Whaler, K. A. (1982). Stable regions in the earth's liquid core. *Geophys. J. Roy. Astron. Soc.* **68**, 241.

Guerin, G. and Valladas, G. (1980). Thermoluminescence dating of volcanic plagioclases. *Nature* **286**, 697.

Haddon, R. A. W. and Cleary, J. R. (1974). Evidence for scattering of seismic PKP waves near the mantle–core boundary. *Phys. Earth Planet. Int.* **8**, 211.

Häge, H. and Muller, G. (1979). Changes in dimensions, stresses and gravitational energy of the earth due to crystallisation at the inner core boundary under isochemical conditions. *Geophys. J. Roy. Astron. Soc.* **58**, 495.

Hager, B. H. and O'Connell, R. J. (1978). Subduction zone dip angles and flow driven by plate motion. *Tectonophysics* **50**, 111.

Haggerty, S. E. (1970). Magnetic minerals in pelagic sediments. *Carnegie Inst. Wash. Yearbook* **68**, 332.

Haggerty, S. E. (1976). Chapters 4 and 8. In *Oxide Minerals* (Douglas Rumble III, ed.). Southern Printing Co., Blacksburg, Virginia.

Haggerty, S. E. (1978). Mineralogical constraints on Curie isotherms in deep crustal magnetic anomalies. *Geophys. Res. Lett* **5**, 105.

Haigh, G. (1958). The process of magnetisation by chemical change. *Phil. Mag.* **3**, 267.

Hall, C. M. and York, D. (1978). K–Ar and $^{40}Ar/^{39}Ar$ age of the Laschamp geomagnetic polarity reversal. *Nature* **274**, 462.

Hall, H. and Murthy, V. R. (1971). The early chemical history of the Earth: some critical elemental fractionations. *Earth Planet. Sci. Lett.* **11**, 239.

Halls, H. C. (1978). The use of converging remagnetisation circles in paleomagnetism. *Phys. Earth Planet. Int.* **16**, 1.

Hamano, Y. (1974). The melting of pure iron at high pressure. In *Rock Magnetism and Paleogeophysics* 2, p. 86. Rock Magnetism and Paleogeophysics Research Group, Japan.

Harrison, C. G. A. (1968). Formation of magnetic anomaly patterns by dyke injection. *J. Geophys. Res.* **73**, 2137.

Harrison, C. G. A. (1969). What is the true rate of reversals of the earth's magnetic field? *Earth Planet. Sci. Lett.* **6**, 186.

Harrison, C. G. A. (1980). Secular variation and excursions of the earth's magnetic field. *J. Geophys. Res.* **85**, 3511.

Harrison, C. G. A. and Carle, H. M. (1981). Intermediate wavelength magnetic anomalies over ocean basins. *J. Geophys. Res.* **86**, 11585.

Harrison, C. G. A. and Ramirez, E. (1975). Areal coverage of spurious reversals of the earth's magnetic field. *J. Geomag. Geoelect.* **27**, 139.

Harrison, C. G. A. and Somayajulu, B. L. K. (1966). Behaviour of the earth's magnetic field during a reversal. *Nature* **212**, 1193.

Head, J. W. (1976). Lunar volcanism in space and time. *Rev. Geophys. Space Phys.* **14**, 265.

Heirtzler, J. R., Dickson, G. O., Herron, E. M., Pittman II, W. C. and Le Pichon, X. (1968). Marine magnetic anomalies, geomagnetic field reversals and motions of the ocean floor and continents. *J. Geophys. Res.* **73**, 2119.

Heller, F. (1980). Self-reversal of natural remanent magnetisation in the Olby-Laschamp lavas. *Nature* **284**, 334.

Helsley, C. E. and Steiner, M. B. (1969). Evidence for long intervals of normal polarity during the Cretaceous period. *Earth Planet. Sci. Lett.* **5**, 325.

Henshaw, P. C., Jr. and Merrill, R. T. (1980). Magnetic and chemical changes in marine sediments. *Rev. Geophys. Space Phys.* **18**, 483.

Herndon, J. M. and Rowe, M. W. (1974). Magnetism in meteorites. *Meteoritics* **9**, 289.

Herndon, J. M., Rowe, M. W., Larson, E. E. and Watson, D. E. (1976). Thermomagnetic analysis of meteorites, 3. C3 and C4 chondrites. *Earth Planet. Sci. Lett.* **29**, 283.

Hess, H. H. (1960). Evolution of ocean basins: Report to Office of Naval Research on research supported by ONR Contract Nonr 1858 (10).

Hess, H. H. (1962). History of Ocean basins. In *Petrologic Studies: A Volume to Honor A. F. Buddington* (A. E. J. Engel, H. L. James and B. F. Leonard, eds), p. 599. Geological Society of America.

Hibberd, F. H. (1979). The origin of the earth's magnetic field. *Proc. Roy. Soc. London* **A369**, 31.

Hide, R. (1966). Free hydromagnetic oscillations of the earth's core and the theory of geomagnetic secular variation. *Phil. Trans. Roy. Soc. London* **A259**, 615.

Hide, R. (1967). Motion of the Earth's core and mantle, and variations of the main geomagnetic field. *Science* **157**, 55.

Hide, R. (1970). On the earth's core–mantle interface. *Quart. J. Roy. Soc.* **96**, 579.

Hide, R. (1981). Self-exciting dynamos and geomagnetic polarity changes. *Nature* **293**, 728.

Hide, R. and Malin, S. R. C. (1970). Novel correlations between global features of the earth's gravitational and magnetic fields. *Nature* **225**, 605.

Hide, R. and Malin, S. R. C. (1971a). Novel correlation between global features of the Earth's gravitational and magnetic fields: further statistical considerations. *Nature Phys. Sci.* **230**, 63.

Hide, R. and Malin, S. R. C. (1971b). Effect of rotation in latitude on correlations between Earth's gravitational and magnetic fields. *Nature Phys. Sci.* **232**, 31.

Hide, R. and Palmer, T. (1982). Generalisation of Cowling's theorem. *Geophys. Astrophys. Fluid Dyn.* **19**, 301.

Hide, R. and Roberts, P. H. (1979). How strong is the magnetic field in the earth's liquid core? *Phys. Earth Planet. Int.* **20**, 124.

Hide, R. and Stewartson, K. (1972). Hydromagnetic oscillations of the earth's core. *Rev. Geophys. Space Phys.* **10**, 579.

Higgins, G. and Kennedy, G. C. (1971). The adiabatic gradient and the melting point gradient in the core of the earth. *J. Geophys. Res.* **76**, 1870.

Hill, T. W. (1974). Origin of the plasma sheet. *Rev. Geophys. Space Phys.* **12**, 379.

Hillhouse, J. and Cox, A. (1976). Brunhes-Matuyama polarity transition. *Earth Planet. Sci. Lett.* **29**, 51.

Hirooka, K. (1971). Archeomagnetic study for the past 2000 years in southwest Japan. *Mem. Fac. Sci. Kyoto Univ. Ser. Geol. & Min.* **38**, 167.

Hodges, R. R., Jr. and Hoffman, H. H. (1975). Implications of atmospheric ^{40}Ar escape on the interior structure of the moon. *Proc. 6th Lunar Sci. Conf.* 3039.

Hoffman, K. A. (1975). Cation diffusion processes and self reversal of thermoremanent magnetisation in the ilmenite-hematite solid solution series. *Geophys. J. Roy. Astron. Soc.* **41**, 65.

Hoffman, K. A. (1977). Polarity transition records and the geomagnetic dynamo. *Science* **196**, 1329.

Hoffman, K. A. (1979). Behaviour of the geodynamo during reversal: a phenomenological model. *Earth Planet. Sci. Lett.* **44**, 7.

Hoffman, K. A. (1981a). Quantitative description of the geomagnetic field during the Matuyama–Brunhes polarity transition. *Phys. Earth Planet. Int.* **24**, 229.

Hoffman, K. A. (1981b). Paleomagnetic excursions, aborted reversals and transitional fields. *Nature* **294**, 67.

Hoffman, K. A. and Day, R. (1978). Separation of multicomponent NRM: a general method. *Earth Planet. Sci. Lett.* **40**, 433.

Hoffman, K. A. and Fuller, M. (1978). Transitional field configurations and geomagnetic reversal. *Nature* **273**, 715.

Holcomb, R. T., Champion, D. E. and McWilliams, M. O. (1984). Recent geomagnetic secular variation in Hawaii from lava flows dated by ^{14}C. *Geol. Soc. Amer. Bull.* **95**.

Hood, L. L. (1981). The enigma of lunar magnetism. *EOS* **62**, 161.

Hood, L. L., Coleman, P. J. and Wichelms, D. E. (1979). The Moon: sources of crustal anomalies. *Science* **204**, 53.

Hospers, J. (1953–1954). Reversals of the main geomagnetic field I, II and III. *Proc. Kon. Nederl. Akad, Wet. B.* **56**, 467, 477, **57**, 112.

Hospers, J. (1954). Rock magnetism and polar wandering. *Nature* **173**, 1183.

Housley, R. M., Grant, R. W. and Patton, N. E. (1973). Origin and characteristics of excess Fe metal in lunar glass welded aggregates. *Proc. 4th Lunar Sci. Conf.* 2737.

Housley, R. M., Cirlin, E. H., Goldberg, I. B. and Crowe, H. (1976). Ferromagnetic resonance studies of lunar core stratigraphy. *Proc. 7th Lunar Sci. Conf.* 13.

Houtermans, J. (1966). On the quantitative relationships between geophysical parameters and the natural ^{14}C inventory. *Z. Phys.* **193**, 1.

Houtermans, J. C., Suess, H. E. and Oeschger, H. (1973). Reservoir models and production rate variations of natural radiocarbon. *J. Geophys. Res.* **78**, 1897.

Howard, R. (1975). The rotation of the sun. *Scientific American* **232**, 106.

Howe, H. C., Lin, R. P., McGuire, R. E. and Anderson, F. A. (1974). Energetic electron scattering from the lunar remanent magnetic field. *Geophys. Res. Lett.* **1**, 101.

Hubbard, N. J. and Minear, J. W. (1975). A physical and chemical model of early lunar history. *Proc. 6th Lunar Sci. Conf.* 1057.

Hurwitz, L. (1960). Eccentric dipoles and spherical harmonic analysis. *J. Geophys. Res.* **65**, 2555.

Huxtable, J. and Aitken, M. J. (1977). Thermoluminescent dating of Lake Mungo geomagnetic polarity excursion. *Nature* **265**, 40.

Huxtable, J., Aitken, M. J. and Bonhommet, N. (1978). Thermoluminescence dating of sediments baked by lava flows of the Chaine des Puys. *Nature* **275**, 207.

IGRF Evaluation Group of Japan (1975). Evaluation of IGRF models proposed for 1975. *J. Geomag. Geoelect.* **27**, 441.

Inglis, D. R. (1955). Theories of the earth's magnetism. *Rev. Mod. Phys.* **27**, 212.

Irving, E. (1956). Paleomagnetic and paleoclimatological aspects of polar wandering. *Geofis. Pura. Appl.* **33**, 23.

Irving, E. (1957). Rock magnetism: a new approach to some palaeogeographic problems. *Phil. Mag. Suppl. Adv. Phys.* **6**, 194.

Irving, E. (1964). *Palaeomagnetism and its Application to Geological and Geophysical Problems.* Wiley, New York.

Irving, E. (1970). The Mid-Atlantic ridge at 45°N. XIV. Oxidation and magnetic properties of basalts. *Can. J. Earth Sci.* **7**, 1528.

Irving, E. (1977). Drift of the major continental blocks since the Devonian. *Nature* **270**, 304.

Irving, E. (1979). Pole positions and continental drift since the Devonian. In *The Earth: Its Origin, Structure and Evolution* (M. W. McElhinny, ed.), p. 567. Academic Press, London and New York.

Irving, E. and Briden, J. C. (1962). Palaeolatitude of evaporite deposits. *Nature* **196**, 425.

Irving, E. and Brown, D. A. (1964). Abundance and diversity of the labyrinthodonts as a function of paleolatitude. *Amer. J. Sci.* **262**, 689.

Irving, E. and Gaskell, T. F. (1962). The palaeogeographic latitude of oil fields. *Geophys. J. Roy. Astron. Soc.* **7**, 54.

Irving, E. and Major, A. (1964). Post-depositional detrital remanent magnetisation in a synthetic sediment. *Sedimentology* **3**, 135.

Irving, E. and McGlynn, J. C. (1976). Proterozoic magnetostratigraphy and the tectonic evolution of Laurentia. *Phil. Trans. Roy. Soc. London* **A280**, 433.

Irving, E. and Naldrett, A. J. (1977). Paleomagnetism in Abitibi greenstone belt, and Abitibi and Metachewan diabase dykes: evidence of the Archean geomagnetic field. *J. Geol.* **85**, 157.

Irving, E. and Parry, L. G. (1963). The magnetism of some Permian rocks from New South Wales. *Geophys. J. Roy. Astron. Soc.* **7**, 395.

Irving, E. and Pullaiah, G. (1976). Reversals of the geomagnetic field, magnetostratigraphy, and relative magnitude of paleosecular variation in the Phanerozoic. *Earth Sci. Rev.* **12**, 35.

Irving, E. and Ward, M. A. (1964). A statistical model of the geomagnetic field. *Pure Appl. Geophys.* **57**, 47.

Irving, E., Park, J. K., Haggerty, S. E., Aumento, F. and Loncarevic, B. (1970). Magnetism and opaque mineralogy of basalts from the Mid-Atlantic Ridge at 45°N. *Nature* **228**, 974.

Ishikawa, Y. and Akimoto, S. (1958). Magnetic property and crystal chemistry of ilmenite ($FeTiO_3$) and hematite (Fe_2O_3) system, 2. Magnetic property. *J. Phys. Chem. Solids* **24**, 517.

Ishikawa, Y. and Syono, Y. (1963). Order-disorder transformation and reverse thermoremanent magnetism in the $FeTiO_3$–Fe_2O_3 system. *J. Phys. Chem. Solids* **4**, 517.

Jackson, J. D. (1962). *Classical Electrodynamics*. Wiley, New York.

Jackson, J. D. and Beard, D. B. (1977). The magnetic field of Mercury. *J. Geophys. Res.* **82**, 2828.

Jacobs, J. A. (1975). *The Earth's Core*. Academic Press, London and New York.

Jacobs, J. A. (1981). Heat flow and reversals of the earth's magnetic field. *J. Geomag. Geoelect.* **33**, 527.

James, R. W. (1971). More on secular variation. *Comments on earth sciences, Geophysics* **2**, 28.

James, R. W. (1974). The inability of latitude-dependent westward drift to account for zonal secular variation. *J. Geomag. Geoelect.* **26**, 359.

James, R. W. and Winch, D. E. (1967). The eccentric dipole. *Pure Appl. Geophys.* **66**, 77.

James, R. W., Roberts, P. H. and Winch, D. E. (1980). The Cowling anti-dynamo theorem. *Geophys. Astrophys. Fluid Dyn.* **15**, 149.

Jeanloz, R. (1979). Properties of iron at high pressures and the state of the core. *J. Geophys. Res.* **84**, 6059.

Johnson, E. A., Murphy, T. and Torrenson, O. W. (1948). Pre-history of the Earth's magnetic field. *Terr. Magn. Atmosph. Elect.* **53**, 349.

Johnson, H. P. (1979). Magnetisation of the oceanic crust. *Rev. Geophys. Space Phys.* **17**, 215.

Johnson, H. P. and Atwater, T. (1977). A magnetic study of basalts from the Mid-Atlantic Ridge at 37°N. *Geol. Soc. Amer. Bull.* **88**, 637.

Johnson, H. P. and Hall, H. M. (1978). A detailed rock magnetism and opaque mineralogy study of the basalts from the Nazca Plate. *Geophys. J. Roy. Astron. Soc.* **52**, 45.

Johnson, H. P. and Merrill, R. T. (1978). A direct test of the Vine-Mathews hypothesis. *Earth Planet. Sci. Lett.* **40**, 263.

Johnson, H. P., Kinoshita, H. and Merrill, R. T. (1975a). Rock magnetism and paleomagnetism of some north Pacific deep-sea sediment cores. *Geol. Soc. Amer. Bull.* **86**, 412.

Johnson, H. P., Lowrie, W. and Kent, D. V. (1975b). Stability of anhysteretic remanent magnetisation in fine and coarse magnetite and maghemite particles. *Geophys. J. Roy. Astron. Soc.* **41**, 7.

Johnson, M. J. S. and Strens, R. G. J. (1973). Electrical conductivity of molten Fe–Ni–S–C core mix. *Phys. Earth Planet. Int.* **7**, 217.

Jones, G. M. (1977). Thermal interaction of the core and the mantle and long term behaviour of the geomagnetic field. *J. Geophys. Res.* **82**, 1703.

Kanasewich, E. R. (1973). *Time Sequence—Analysis in Geophysics*. University of Alberta Press, Edmonton, Alberta.

Kaula, W. M. (1971). Dynamic aspects of lunar origin. *Rev. Geophys. Space Phys.* **9**, 217.

Kaula, W. M. (1979). Thermal evolution of Earth and Moon growing by planetesimal impacts. *J. Geophys. Res.* **84**, 999.

Kaula, W. M., Schubert, O., Lingenfelter, R. E., Sjogren, W. L. and Wollenhaupt, W. R. (1974). Apollo laser altimetry and inferences as to lunar structure. *Proc. 5th Lunar Sci. Conf.* 3049.

Kawai, N., Hirooka, K. and Sasajima, S. (1965). Counterclockwise rotation of the geomagnetic dipole axis revealed in the worldwide archaeosecular variations. *Proc. Japan Acad.* **41**, 398.

Keefer, C. M. and Shive, P. N. (1981). Curie temperature and lattice constant reference contours for synthetic titanomaghemites. *J. Geophys. Res.* **86**, 987.

Keeler, R. N., Matassov, G. M., Mitchell, A. C. and Nellis, W. J. (1978). The electrical and thermal conductivity of various terrestrial core materials to 1.8 Mb (abstract), Conf. on Origins of Planetary Magnetism. *Lunar Planet. Inst. Contr.* **348**, 55.

Kennedy, G. C. and Higgins, G. H. (1973). The core paradox. *J. Geophys. Res.* **78**, 900.

Kennett, J. P. (ed.) (1980). *Magnetic Stratigraphy of Sediments*. Dowden, Hitchingson and Ross, Stroudsburg, Pennsylvania.

Kent, D. V. and Lowrie, W. (1974). Origin of magnetic instability in sediment cores from the Central North Pacific. *J. Geophys. Res.* **79**, 2987.

Kent, D. V. and Opdyke, N. D. (1977). Palaeomagnetic field intensity variation recorded in a Brunhes Epoch deep-sea sediment core. *Nature* **266**, 256.

Khan, M. A. (1971). Correlation function in geophysics. *Nature* **230**, 57.

Khramov, A. N. (1960). *Palaeomagnetism and Stratigraphic Correlation*. (Gostoptechizdat, Leningrad), 1958. (English translation by A. J. Lojkine, Australian National University, Canberra.)

Khramov, A. N. and Rodionov, V. P. (1981). The geomagnetic field during Palaeozoic time. In *Global Reconstruction and the Geomagnetic Field During the Palaeozoic* (M. W. McElhinny, A. N. Khramov, M. Ozima and D. A. Valencio, eds), p. 99. Center for Academic Publications, Japan.

Kidd, R. G. W. (1977). The nature and shape of the sources of marine magnetic anomalies. *Earth Planet. Sci. Lett.* **33**, 310.

King, J. H. and Ness, N. F. (1977). Lunar magnetic permeability studies and magnetometer sensitivity. *Geophys. Res. Lett.* **4**, 129.

King, R. F. and Rees, A. I. (1966). Detrital magnetism in sediments: an examination of some theoretical models. *J. Geophys. Res.* **71**, 561.

Kirschvink, J. L. (1980). The least squares line and plane and the analysis of palaeomagnetic data. *Geophys. J. Roy. Astron. Soc.* **62**, 699.

Kittel, C. (1949). Physical theory of ferromagnetic domains. *Rev. Mod. Phys.* **21**, 541.

Knopoff, L. and MacDonald, G. (1958). The magnetic field and the central core of the Earth. *Geophys. J. Roy. Astron. Soc.* **1**, 216.

Kobayashi, K. (1961). An experimental demonstration of the production of chemical remanent magnetisation with Cu–Co alloy. *J. Geomag. Geoelect.* **12**, 148.

Kobayashi, K. (1968). Palaeomagnetic determination of the intensity of the geomagnetic field in the Precambrian period. *Phys. Earth. Planet. Int.* **1**, 387.

Kobayashi, K. and Nomura, N. (1974). Ferromagnetic minerals in the sediment cores collected from the Pacific basin. *J. Geophys.* **40**, 501.

Koci, A. and Sibrava, V. (1976). The Brunhes-Matuyama boundary at central European localities, Report No. 3 of project 72/1/24 in *Quaternary Glaciations* in the Northern Hemisphere, Prague.

Kolesova, V. I. and Kropachev, E. P. (1973). Spherical harmonic analysis of the geomagnetic field for the 1965 epoch up to $m = 23$ according to ground-based data results. *Geomag. Aeron.* **13**, 127.

Kono, M. (1972). Mathematical models of the earth's magnetic field. *Phys. Earth Planet. Int.* **5**, 140.

Kono, M. (1973). Spherical harmonic analysis of the geomagnetic field from inclination and declination data. *Rock Magn. Paleogeophys. Tokyo* **1**, 124.

Kono, M. (1978). Reliability of paleointensity methods using alternating field demagnetisation and anhysteretic remanence. *Geophys. J. Roy. Astron. Soc.* **54**, 241.

Kovacheva, M. (1980). Summarised results of the archeomagnetic investigations of the geomagnetic field variation for the last 8000 yr in southeastern Europe. *Geophys. J. Roy. Astron. Soc.* **61**, 57.

Kovacheva, M. and Veljovich, D. (1977). Geomagnetic field variations in southeastern Europe between 6500 and 100 years B.C. *Earth Planet. Sci. Lett.* **37**, 131.

Krause, F. (1977). Mean-field electrodynamics and dynamo theory of the earth's magnetic field. *J. Geophys.* **43**, 421.

Krause, F. and Steenbeck, M. (1967). Some simple models of magnetic field regeneration by non-mirror symmetric turbulence. *Z. Naturforsch.* **A22**, 671.

Kraut, E. A. and Kennedy, G. C. (1966). New melting law at high pressures. *Phys. Rev.* **151**, 668.

Kristjansson, L. and McDougall, I. (1982). Some aspects of the late Tertiary geomagnetic field in Iceland. *Geophys. J. Roy. Astron. Soc.* **68**, 273.

LaBrecque, J. L., Kent, D. V. and Cande, S. C. (1977). Revised magnetic polarity time scale for Late Cretaceous and Cenozoic time. *Geology* **5**, 330.

Lahiri, B. N. and Price, A. T. (1939). Electromagnetic induction in non-uniform conductors, and the determination of the conductivity of the earth from terrestrial magnetic variations. *Phil. Trans. Roy. Soc. London* **A237**, 509.

Lamb, H. (1883). On electrical motions in a spherical conductor. *Phil. Trans. Roy. Soc. London* **A174**, 519.

Lambeck, K. (1980). *The Earth's Variable Rotation: Geophysical Causes and Consequences.* Cambridge University Press, Cambridge.

Landstreet, J. D. and Angel, J. R. P. (1974). The polarisation spectrum and magnetic field strength of the White Dwarf Grw + 70° 8247, 1974. *Astrophys. J.* **196**, 819.

Langel, R. A. and Estes, R. H. (1982). A geomagnetic field spectrum. *Geophys. Res. Lett.* **9**, 250.

Lanoix, M., Strangway, D. W. and Pearce, G. W. (1977). Anomalous acquisition of thermoremanence at 130°C in iron and paleointensity of the Allende Meteorite. *Proc. 8th Lunar Sci. Conf.* 689.

Lanoix, M., Strangway, D. W. and Pearce, G. W. (1978a). The primordial magnetic field preserved in chondrules of the Allende meteorite. *Geophys. Res. Lett.* **5**, 73.

Lanoix, M., Strangway, D. W. and Pearce, G. W. (1978b). Paleointensity determinations from Allende chondrules. *Lunar Planet. Sci.* **IX**, 630.

Larmor, J. (1919a). Possible rotational origin of magnetic fields of sun and earth. *Elec. Rev.* **85**, 412.

Larmor, J. (1919b). How could a rotating body such as the sun become a magnet? *Rept. Brit. Assoc.* 159.

Larson, E. E. and Walker, T. R. (1975). Development of chemical remanent magnetisation during early stages of red bed formation in late Cenozoic sediments, Baja, California. *Geol. Soc. Amer. Bull.* **86**, 639.

Larson, E. E., Ozima, M., Nagata, T. and Strangway, D. (1969). Stability of remanent magnetisation of rocks. *Geophys. J. Roy. Astron. Soc.* **17**, 263.

Larson, E. E., Watson, D. E. and Jennings, W. (1971). Regional comparison of a Miocene geomagnetic transition in Oregon and Nevada. *Earth Planet. Sci. Lett.* **11**, 391.

Larson, E. E., Watson, C. E., Herndon, J. M. and Rowe, M. W. (1973). Partial AF demagnetisation studies of 40 meteorites. *J. Geomag. Geoelect.* **25**, 331.

Larson, E. E., Watson, D. E., Herndon, J. M. and Rowe, M. W. (1974). Thermomagnetic analysis of meteorites, 1. C1 chondrites. *Earth Planet. Sci. Lett.* **21**, 345.

Latham, G. V., Dorman, H. J., Horvath, P., Ibrahin, A. K., Koyama, J. and Nakamura, Y. (1978). Passive seismic experiment: a summary of current status. *Proc. 9th Lunar Planet. Sci. Conf.* 3609.

Lawley, E. (1970). The intensity of the geomagnetic field in Iceland during Neogene polarity transitions and systematic deviations. *Earth Planet. Sci. Lett.* **10**, 145.

Leaton, B. R. and Malin, S. R. C. (1967). Recent changes in the magnetic dipole moment of the earth. *Nature* **213**, 1110.

Lee, S. H. and McElhinny, M. W. (1984). Estimates of the time-variation of the non-dipole part of the time-averaged paleomagnetic field (in press).

Le Mouël, J. L. and Courtillot, V. (1982). On the outer layers of the core and geomagnetic secular variation. *J. Geophys. Res.* **87**, 4103.

Lepaluto, D. A. (1972). Melting of iron by significant structure theory. *Phys. Earth Planet. Int.* **6**, 175.

Le Pichon, X. (1968). Seafloor spreading and continental drift. *J. Geophys. Res.* **73**, 3661.

Le Pichon, X. and Heirtzler, J. R. (1968). Magnetic anomalies in the Indian Ocean and seafloor spreading. *J. Geophys. Res.* **73**, 2101.

Levi, S. (1975). Comparison of two methods of performing the Thellier experiment (or, How the Thellier experiment should not be done). *J. Geomag. Geoelect.* **27**, 245.

Levi, S. and Merrill, R. T. (1976). A comparison of ARM and TRM in magnetite. *Earth Planet. Sci. Lett.* **32**, 171.

Levi, S. and Merrill, R. T. (1978). Properties of single-domain, pseudo-single-domain, and multidomain magnetite. *J. Geophys. Res.* **83**, 309.

Levy, E. H. (1972a). Effectiveness of cyclonic convection for producing the geomagnetic field. *Astrophys. J.* **171**, 621.

Levy, E. H. (1972b). Kinematic reversal schemes for the geomagnetic dipole. *Astrophys. J.* **171**, 635.

Levy, E. H. (1972c). On the state of the geomagnetic field and its reversals. *Astrophys. J.* **175**, 573.

Liddicoat, J. C. and Coe, R. S. (1979). Mono Lake geomagnetic excursion. *J. Geophys. Res.* **84**, 261.

Liddicoat, J. C., Coe, R. S., Lambert, P. W. and Valastro, S. (1979). Palaeomagnetic record in late Pleistocene and Holocene dry lake deposits at Tlapacoya, Mexico. *Geophys. J. Roy. Astron. Soc.* **59**, 367.

Lilley, F. E. M. (1970a). On kinematic dynamos. *Proc. Roy. Soc. London* **A316,** 153.

Lilley, F. E. M. (1970b). Geomagnetic reversals and the position of the north magnetic pole. *Nature* **227**, 1336.

Lin, R. P. (1978). A search for impact crater-associated surface magnetic fields in mare regions. *Lunar Planet. Sci.* **IX**, 651.

Lin, R. P., Anderson, K. A., Russell, C. T., Boyce, J., Masursky, H., Wilhelms, D. W. and Stuart-Alexander, D. E. (1977). Correlations of lunar surface remanent magnetic fields to surface geologic age. *Lunar Sci.* **VIII** (Supp. A), 14.

Lindsley, D. H. (1976). Chapters 1 and 2. In *Oxide Minerals* (Douglas Rumble III, ed.). Southern Printing Co., Blacksburg, Virginia.

Lingenfelter, R. E. and Ramaty, R. (1970). Astrophysical and geophysical variations in C-14 production. In *Radiocarbon Variations and Absolute Chronology* (I. U. Olsson, ed.), p. 513. Alinqvist and Wiksell, Stockholm.

Liu, Lin-gun and Bassett, W. A. (1975). The melting of iron up to 200 Kbar. *J. Geophys. Res.* **80**, 3777.

Loper, D. E. (1978a). The gravitationally powered dynamo. *Geophys. J. Roy. Astron. Soc.* **54**, 389.

Loper, D. E. (1978b). Some thermal consequences of a gravitationally powered dynamo. *J. Geophys. Res.* **83**, 5961.

Loper, D. E. and Roberts, P. H. (1978). On the motion of an iron-alloy core containing a slurry. *Geophys. Astrophys. Fluid Dyn.* **9**, 289.

Loper, D. E. and Roberts, P. H. (1979). Are planetary dynamos driven by gravitational settling? *Phys. Earth Planet. Int.* **20**, 192.

Lowes, F. J. (1955). Secular variation and the non-dipole field. *Ann. Geophys.* **11**, 91.

Lowes, F. J. (1971). Significance of the correlation between spherical harmonic fields. *Nature* **230**, 61.

Lowes, F. J. (1974). Spatial power spectrum of the main geomagnetic field and extrapolation to the core. *Geophys. J. Roy. Astron. Soc.* **36**, 717.

Lowes, F. J. (1975). Vector errors in spherical harmonic analysis of scalar data. *Geophys. J. Roy. Astron. Soc.* **42**, 637.

Lowes, F. J. (1976). The effect of a field of external origin on spherical harmonic analysis using only internal coefficients. *J. Geomag. Geoelect.* **28**, 515.

Lowes, F. J. and Runcorn, S. K. (1951). The analysis of the geomagnetic secular variation. *Phil. Trans. Roy. Soc. London* **A243**, 525.

Lowes, F. J. and Wilkinson, I. (1963). Geomagnetic dynamo: a laboratory model. *Nature* **198**, 1158.

Lowes, F. J. and Wilkinson, I. (1967). A laboratory self-exciting dynamo. In *Magnetism and the Cosmos*—NATO Advanced Study Inst. on Planetary and Steller Magnetism, pp. 121–125, University of Newcastle Upon Tyne, 1965. Elsevier, Amsterdam.

Lowes, F. J. and Wilkinson, I. (1968). Geomagnetic dynamo: an improved laboratory model. *Nature* **219**, 717.

Lowrie, W. and Alvarez, W. (1981). One hundred million years of geomagnetic polarity history. *Geology* **9**, 392.

Lowrie, W. and Fuller, M. (1971). On the alternating field demagnetisation characteristics of multidomain thermoremanent magnetisation magnetite. *J. Geophys. Res.* **76**, 6339.

Lowrie, W. R., Lovlie, R. and Opdyke, N. D. (1973a). Magnetic properties of deep sea drilling project basalts from the North Pacific Ocean. *J. Geophys. Res.* **78**, 7647.

Lowrie, W. R., Lovlie, R. and Opdyke, N. D. (1973b). Magnetic properties of deep sea drilling project basalts from the Atlantic Ocean. *Earth Planet. Sci. Lett.* **17**, 338.

Lund, S. P. and Banerjee, S. K. (1979). Paleosecular variation from lake sediments. *Rev. Geophys. Space Phys.* **17**, 244.

MacDonald, K. and Holcombe, T. L. (1978). Inversion of magnetic anomalies and seafloor spreading in the Cayman Trough. *Earth Planet. Sci. Lett.* **40**, 407.

Mackereth, F. J. H. (1958). A portable core sampler for lake deposits. *Limnol. oceanogr.* **3**, 181.

Mackereth, F. J. H. (1971). On the variation in direction of the horizontal component of remanent magnetisation in lake sediments. *Earth Planet. Sci. Lett.* **12**, 332.

Malin, S. R. C. and Hodder, B. M. (1982). Was the 1970 geomagnetic jerk of internal or external origin? *Nature* **296**, 726.

Malkus, W. V. R. (1963). Precessional torques as the cause of geomagnetism. *J. Geophys. Res.* **68**, 2871.

Malkus, W. V. R. (1968). Precession as the cause of geomagnetism. *Science* **160**, 259.

Mankinen, E. A. and Dalrymple, G. B. (1979). Revised geomagnetic polarity time scale for the interval 0–5 my B.P. *J. Geophys. Res.* **84**, 615.

Marshall, M. and Cox, A. (1971). Magnetism of pillow basalts and their petrology. *Geol. Soc. Amer. Bull.* **82**, 6459.

Martelli, G. and Newton, G. (1977). Hypervelocity cratering and impact magnetisation of basalts. *Nature* **269**, 478.

Mason, R. G. and Raff, A. D. (1961). Magnetic survey off the west coast of North America, 32°N latitude to 42°N latitude. *Geol. Soc. Amer. Bull.* **72**, 1259.

Matsushita, S. (1967). Solar quiet and lunar daily variation fields. In *Physics of Geomagnetic Phenomena* (S. Matsushita and W. H. Campbell, eds), Vol. 1 (3), p. 302. Academic Press, London and New York.

Matthews, D. H. and Bath. J. (1967). Formation of magnetic anomaly pattern of Mid-Atlantic Ridge. *Geophys. J. Roy. Astron. Soc.* **13**, 349.

Mattis, D. G. (1965). *The Theory of Magnetism.* Harper & Row, New York.

Matuyama, M. (1929). On the direction of magnetisation of basalt in Japan, Tyosen and Manchuria. *Proc. Imp. Acad. Japan* **5**, 203.

Maxwell, A. E., Von Herzen, R. P., Hsu, K. T., Andrews, J. E., Saito, T., Percival, S. F., Mitlow, E. D. and Boyce, R. E. (1970). Deep-sea drilling in the South Atlantic. *Science* **168**, 1047.

May, R. M. (1976). Simple mathematical models with very complicated behaviour. *Nature* **261**, 459.

Mayhew, M. and Galliher, S. (1982). An equivalent layer magnetization model for the United States derived from MAGSAT data. *Geophys. Res. Lett.* **9**, 311.

McCammon, C. A., Ringwood, A. E. and Jackson, I. (1983). Thermodynamics of the sytem Fe–FeO–MgO at high pressure and temperature and a model for formation of the Earth's core. *Geophys. J. Roy. Astron. Soc.* **72**, 577.

McDonald, K. L. (1957). Penetration of a geomagnetic secular field through a mantle with variable conductivity. *J. Geophys. Res.* **62**, 117.

McDonald, K. L. and Gunst, R. H. (1968). Recent trends in the earth's magnetic field. *J. Geophys. Res.* **73**, 2057.

McDougall, I. (1979). The present status of the geomagnetic polarity time scale. In *The Earth: Its Origin, Structure and Evolution* (M. W. McElhinny, ed.), p. 543. Academic Press, London and New York.

McDougall, I. and Tarling, D. H. (1963). Dating of polarity zones in the Hawaiian Islands. *Nature* **200**, 54.

McDougall, I. and Tarling, D. H. (1964). Dating geomagnetic polarity zones. *Nature* **202**, 171.

McDougall, I., Watkins, N. D. and Kristjansson, L. (1976a). Geochronology and paleomagnetism of a Miocene-Pliocene lave sequence at Bessatadaá, eastern Iceland. *Amer. J. Sci.* **276**, 1078.

McDougall, I., Watkins, N. D., Walker, G. P. L. and Kristjansson, L. (1976b). Potassium-argon and paleomagnetic analysis of Icelandic lava flows: limits on the age of anomaly 5. *J. Geophys. Res.* **81**, 1055.

McDougall, I., Saemundsson, K., Johannesson, H., Watkins, N. D. and Kristjansson, L. (1977). Extension of the geomagnetic polarity time scale to 6.5 my: K–Ar dating, geological and paleomagnetic study of a 3500 m lave succession in western Iceland. *Geol. Soc. Amer. Bull.* **88**, 1.

McElhinny, M. W. (1971). Geomagnetic reversals during the Phanerozoic. *Science* **172**, 157.

McElhinny, W. M. (1973). *Palaeomagnetism and Plate Tectonics*. Cambridge University Press, Cambridge.

McElhinny, M. W. (1977). The magnetic properties of Indian Ocean basalts. In *Indian Ocean Geology and Biostratigraphy* (J. Heirtzler, J. Bolli, T. Davies, J. Saunders and J. Scater, eds), p. 301. A.G.U., Washington D.C.

McElhinny, M. W. and Brock, A. (1975). A new paleomagnetic result from East Africa and estimates of the Mesozoic paleoradius. *Earth Planet. Sci. Lett.* **27**, 321.

McElhinny, M. W. and Evans, M. E. (1968). An investigation of the strength of the geomagnetic field in the early Precambrian. *Phys. Earth Planet. Int.* **1**, 435.

McElhinny, M. W. and Merrill, R. T. (1975). Geomagnetic secular variation over the past 5 my. *Rev. Geophys. Space Phys.* **13**, 687.

McElhinny, M. W. and Senanayake, W. E. (1980). Paleomagnetic evidence for the existence of the geomagnetic field 3.5 Ga ago. *J. Geophys. Res.* **85**, 3523.

McElhinny, M. W. and Senanayake, W. E. (1982). Variations in the geomagnetic dipole 1: the past 50 000 years. *J. Geomag. Geoelect.* **34**, 39.

McElhinny, M. W., Taylor, S. R. and Stevenson, D. J. (1978). Limits to the expansion of Earth, Moon, Mars and Mercury and to changes in the gravitational constant. *Nature* **271**, 316.

McFadden, P. L. (1984). Statistical tools for the analysis of geomagnetic reversal sequences. *J. Geophys. Res.* **89**.

McFadden, P. L. and McElhinny, M. W. (1982). Variations in the geomagnetic dipole 2: Statistical analysis of VDM's for the past 5 million years. *J. Geomag. Geoelect.* **34**, 163.

McFadden, P. L. and McElhinny, M. W. (1984). A physical model for palaeosecular variation. *Geophys. J. Roy. Astron. Soc.* (in press).

McFadden, P. L. and Merrill, R. T. (1984). Lower mantle convection and geomagnetism. *J. Geophys. Res.*, **89**.

McKenzie, D. P. and Parker, R. L. (1967). The North Pacific: an example of tectonics on a sphere. *Nature* **216**, 1276.

McKenzie, D. P, and Richter, F. M. (1981). Parameterised thermal convection in a layered region and the thermal history of the earth. *J. Geophys. Res.* **86**, 11, 667.

McNish, A. G. (1940). Physical representations of the geomagnetic field. *Trans. Am. Geophys. Union* **21**, 287.

McQueen, R. G. and Marsh, S. P. (1966). Shockwave compression of iron–nickel alloys and the earth's core. *J. Geophys. Res.* **71**, 1751.

Mercanton, P. L. (1926). Inversion de l'inclinaison magnétique terrestre aux agês géologiques. *Terr. Magn. Atmosph. Elec.* **31**, 187.

Merrill, R. T. (1975). Magnetic effects associated with chemical changes in igneous rocks. *Geophys. Surv.* **2**, 227.

Merrill, R. T. (1977). The demagnetisation of multidomain grains. *J, Geomag. Geoelect.* **29**, 185.

Merrill, R. T. (1981). Towards a better theory of thermal remanent magnetisation. *J. Geophys. Res.* **86**, 937.

Merrill, R. T. and McElhinny, M. W. (1977). Anomalies in the time-averaged paleomagnetic field and their implications for the lower mantle. *Rev. Geophys. Space Phys.* **15**, 309.

Merrill, R. T., McElhinny, M. W. and Stevenson, D. J. (1979). Evidence for long-term asymmetrics in the earth's magnetic field and possible implications for dynamo theory. *Phys. Earth Planet. Int.* **20.** 75.

Merrill, R. T., McFadden, P. L. and McElhinny, M. W. (1984). Non-linear processes in the geodynamo: palaeomagnetic evidence. *Geophys. J. Roy. Astron. Soc.* (in press).

Metzger, A. E., Lin, R. P. and Russell, C. T. (1977). On a correlation between surface remanent magnetisation and chemistry for the lunar frontside and limbs. *Proc. 8th Lunar Sci. Conf.* 1187.

Mitchell, A. C. (1939). Chapter on the history of terrestrial magnetism. *Terr. Magn. Atmosph. Elec.* **44**, 77.

Moffat, H. K. (1961). The amplification of a weak applied magnetic field by turbulence in fluids of moderate conductivity. *J. Fluid Mech.* **11**, 625.

Moffat, H. K. (1970). Turbulent dynamo action at low magnetic Reynolds numbers. *J. Fluid Mech.* **41**, 435.

Moffat, H. K. (1978). *Magnetic Field Generation in Electrically Conducting Fluids*. Cambridge University Press, Cambridge.

Momose, K. (1963). Studies on the variation of the earth's magnetic field during Pleistocene time. *Bull. Earthquake Res. Inst. Tokyo Univ.* **41**, 487.

Morel, P. and Irving, E. (1978). Tentative paleocontinental maps for the early Phanerozoic and Proterozoic. *J. Geol.* **86**, 535.

Morgan, W. J. (1968). Rises, trenches, great faults and crustal blocks. *J. Geophys. Res.* **73**, 1959.

Morley, L. W. and Laurchelle, A. (1964). Paleomagnetism as a means of dating geological events. *Roy. Soc. Canad. Spec. Publ.* **8**, 512.

Morner, N. A. and Lanser, J. P. (1974). Gothenberg magnetic "flip". *Nature, Phys. Sci.* **251**, 408.

Morner, L. A. and Lanser, J. P. (1975). Paleomagnetism in deep-sea core A179–15. *Earth Planet. Sci. Lett.* **26**, 121.

Morner, N. A., Lanser, J. P. and Hospers, J. (1971). Late Weichselian paleomagnetic reversal. *Nature Phys. Sci.* **234**, 173.

Morris, R. V. (1976). Surface exposure indices of lunar soils: a comparative FMR study. *Proc. 7th Lunar Sci. Conf.* 315.

Morrish, P. H. (1965). *The Physical Principles of Magnetism*. Wiley, New York.

Morrison, L. V. (1979). Re determination of the decade fluctuations in the rotation of the earth in the period 1861–1978. *Geophys. J. Roy. Astron. Soc.* **58**, 349.

Morse, P. M. and Feshbach, H. (1953). *Methods of Theoretical Physics*, Vol. 1 and 2. McGraw-Hill, New York.

Murthy, V. R. and Banerjee, S. K. (1973). Lunar evolution: how well do we know it now? *The Moon* **7**, 149.

Nachasova, I. Ye (1972). Magnetic field in the Moscow area from 1480 to 1840. *Geomag. Aeron* **12**, 277.

Nagata, T. (1961). *Rock Magnetism*, revised edition. Maruzen Co., Tokyo.

Nagata, T. (1965). Main characteristics of recent geomagnetic secular variation. *J. Geomag. Geoelect.* **17**, 263.

Nagata, T. (1969). Lengths of geomagnetic polarity intervals. *J. Geomag. Geoelect.* **21**, 701.

Nagata, T. (1979a). Meteorite magnetism and the early solar system magnetic field. *Phys. Earth Planet. Int.* **20**, 324.

Nagata, T. (1979b). Magnetic classification of stony meteorites (IV). *Mon. Natl. Inst. Polar Res., Ser. C*, No. 10 (T. Nagata, ed.) 273, Natl. Inst. Polar Res., Tokyo.

Nagata, T. (1980). Magnetic classification of Antarctic Meteorites. *Proc. 11th Lunar Sci. Conf.* 1789.

Nagata, T., Ishikawa, Y., Kinoshita, H., Kuno, M., Syono, Y. and Fisher, R. M. (1970). Magnetic properties and natural remanent magnetisation of lunar materials. *Proc. Apollo 11 Lunar Sci. Conf.* 2325.

Naidu, P. S. (1971). Statistical structure of geomagnetic field reversals. *J. Geophys. Res.* **76**, 2649.

Naidu, P. S. (1974). Are geomagnetic field reversals independent? *J. Geomag. Geoelect.* **26**, 101.

Nakajimi, T., Yashkawa, K., Natsuhara, N., Kawai, H. and Hones, S. (1973). Very short geomagnetic excursion 18 000 yr B.P. *Nature* **244**, 8.

Nakamura, Y., Latham, G., Lovimleur, D., Ewing, M., Duennebier, F. K. and Dorman, J. (1974). Deep lunar interior inferred from recent seismic data. *Geophys. Res. Lett.* **1**, 137.

Nakamura, Y., Latham, G. V., Dorman, H. J. and Duennebier, F. K. (1976). Seismic structure of the Moon: a summary of current status. *Proc. 7th Lunar Sci. Conf.* 3113.

Needham, J. (1962). *Science and Civilisation in China*, Vol. 4. *Physics and Physical Technology*, part 1. *Physics*. Cambridge University Press, Cambridge.

Néel, L. (1949). Théorie du trainage magnétique des ferromagnétiques aux grains fins avec applications aux terres cuites. *Ann. Geophys.* **5**, 99.

Néel, L. (1955). Some theoretical aspects of rock magnetism. *Adv. Phys.* **4**, 191.

Ness, G., Levi, S. and Couch, R. (1980). Marine magnetic anomalies time scales for the Cenozoic and Late Cretaceous: A précis, critique and synthesis. *Rev. Geophys. Space Phys.* **18**, 753.

Ness, N. F. (1979). The magnetic fields of Mercury, Mars and the Moon. *Ann. Rev. Earth Planet. Sci.* **7**, 249.

Ness, N. F., Behannon, K. W., Lepping, R. P. and Whang, Y. C. (1975). The magnetic field of Mercury. *J. Geophys. Res.* **80**, 2708.

Niitsuma, N. (1971). Detailed study of the sediments recording the Matuyama-Brunhes geomagnetic reversal. *Tohoku Univ. Sci. Rpt 2nd Ser.* (Geology) **43**, 1.

Ninkovich, D., Opdyke, N. D., Heezen, B. C. and Foster, J. H. (1966). Paleomagnetic stratigraphy, rates of deposition and tephrachronology in North Pacific deep-sea sediments. *Earth Planet. Sci. Lett.* **1**, 476.

Nishitani, T. (1979). Grain size effects on the low-temperature oxidation of titanomagnetites. *Rock Magn. Paleogeophys.* Tokyo **6**, 128.

Nodia, M. Z. and Chelidze, Z. A. (1972). Position of the geomagnetic pole in the past. *Geomag. Aeron.* **12**, 518.

Noel, M. and Tarling, D. H. (1975). The Laschamp geomagnetic 'event'. *Nature* **253**, 705.

Noltimeier, H. C. and Colinvaux, P. A. (1976). Geomagnetic excursion from Imuruk Lake, Alaska. *Nature* **259**, 197.

Olson, P. L. (1977). *Internal Waves and Hydromagnetic Induction in the Earth's Core*, Ph.D. dissertation, University of California, Berkeley.

Olson, P. L. (1981a). Mantle convection with spherical effects. *J. Geophys. Res.* **86**, 4881.

Olson, P. L. (1981b). A simple model for the terrestrial dynamo. *J. Geophys. Res.* **86**, 10875.

Onwumechilli, A. (1967). Geomagnetic variations in the equatorial zone. In *Physics of Geomagnetic Phenomena* (S. Matsushita and W. H. Campbell, eds), Vol. 1 (3), p. 425. Academic Press, London and New York.

Opdyke, N. D. (1961). The palaeoclimatological significance of desert sandstone. In *Descriptive Palaeoclimatology* (A. E. M. Nairn, ed.), p. 45. Wiley Interscience, New York.

Opdyke, N. D. (1962). Palaeoclimatology and Continental drift. In *Continental Drift* (S. K. Runcorn, ed.), p. 41. Academic Press, London and New York.

Opdyke, N. D. (1972). Paleomagnetism of deep sea cores. *Rev. Geophys. Space Phys.* **10**, 213.

Opdyke, N. D. (1976). Discussion of paper by Morner and Lanser concerning the paleomagnetism of deep-sea core A179–15. *Earth Planet. Sci. Lett.* **29**, 238.

Opdyke, N. D. and Glass, B. P. (1969). The paleomagnetism of sediment cores from the Indian Ocean. *Deep-Sea Res.* **16**, 249.

Opdyke, N. D. and Henry, K. W. (1969). A test of the dipole hypothesis. *Earth Planet. Sci. Lett.* **6**, 139.

Opdyke, N. D. and Runcorn, S. K. (1956). New evidence for reversal of the geomagnetic field near the Pliocene–Pleistocene boundary. *Science* **123**, 1126.

Opdyke, N. D. and Runcorn, S. K. (1960). Wind direction in the western United States in the late Paleozoic. *Geol. Soc. Amer. Bull.* **71**, 959.

Opdyke, N. D., Ninkovich, D., Lowrie, W. and Hays, J. D. (1972). The paleomagnetism of two Aegean deep-sea cores. *Earth Planet, Sci. Lett.* **14**, 145.
Opdyke, N. D., Kent, D. V. and Lowrie, W. (1973). Details of magnetic polarity transitions recorded in a high deposition rate deep-sea core. *Earth Planet. Sci. Lett.* **20**, 315.
Opdyke, N. D., Burckle, L. H. and Todd, A. (1974). The extension of the magnetic time scale in sediments of the Central Pacific Ocean. *Earth Planet. Sci. Lett.* **22**, 300.
O'Reilly, W. and Banerjee, S. K. (1966). Oxidation of titanomagnetites and self-reversal. *Nature* **211**, 26.
Oversby, V. M. and Ringwood, A. E. (1972). Potassium distribution between metal and silicate and its bearing on the occurrence of potassium in the Earth's core. *Earth Planet. Sci. Lett.* **14**, 345.
Ozima, M. and Ozima, M. (1965). Origin of thermoremanent magnetisation. *J. Geophys. Res.* **70**, 1363.
Ozima, M. and Ozima, M. (1967). Self-reversal of remanent magnetisation in some dredged submarine basalts. *Earth Planet. Sci. Lett.* **3**, 213.
Ozima, M. and Ozima, M. (1971). Characteristic thermomagnetic curves in submarine basalts. *J. Geophys. Res.* **76**, 2051.
Ozima, M. and Sakamoto, N. (1971). Magnetic properties of synthesised titanomagnetite. *J. Geophys. Res.* **76**, 7035.
Papanastassiou, D. A. and Wasserberg, G. J. (1978). Sr isotopic anomalies in the Allende meteorite. *Geophys. Res. Lett.* **5**, 595.
Parker, E. N. (1955). Hydromagnetic dynamo models. *Astrophys. J.* **122**, 293.
Parker, E. N. (1958). Dynamics of the interplanetary gas and magnetic fields. *Astrophys. J.* **128**, 664.
Parker, E. N. (1967). Disturbance of the geomagnetic field by the solar wind. In *Physics of Geomagnetic Phenomena* (S. Matsushita and W. H. Campbell, eds), Vol. 2 (6), p. 1154. Academic Press, London and New York.
Parker, E. N. (1969). The occasional reversal of the geomagnetic field. *Astrophys. J.* **158**, 815.
Parker, E. N. (1975). The generation of magnetic fields in astrophysical bodies. X. Magnetic buoyancy and the solar dynamo. *Astrophys. J.* **198**, 205.
Parker, E. N. (1979). *Cosmical Magnetic Fields.* Clarendon Press, Oxford.
Parker, R. L. (1973). The rapid calculation of potential anomalies.*Geophys. J. Roy. Astron. Soc.* **31**, 447.
Parker, R. L. (1977). Understanding inverse theory. *Ann. Rev. Earth Planet. Sci.* **5**, 35.
Parker, R. L. and Huestis, S. P. (1974). The inversion of magnetic anomalies in the presence of topography. *J. Geophys. Res.* **79**, 1587.
Patton, B. J. and Fitch, T. L. (1962). An hysteretic remanent magnetisation in small steady fields. *J. Geophys. Res.* **67**, 507.
Payne, M. A. (1981). SI and Gaussian CGS units, conversions and equations for use in geomagnetism. *Phys. Earth Planet. Int.* **26**, 10.
Peale, S. J. (1981). Mercurian liquid core. *Icarus* **48**, 143.
Pearce, G. W., Hage, G. S., Strangway, D. W., Walker, B. M. and Taylor, L. A. (1976). Some complexities in the determination of lunar paleointensities. *Proc. 7th Lunar Sci. Conf.* 3271.
Peddie, N. W. (1979). Current loop models of the earth's magnetic field. *J. Geophys. Res.* **84**, 4517.
Perreault, P. and Akasofu, S. I. (1978). A study of geomagnetic storms. *Geophys. J. Roy. Astron. Soc.* **54**, 547.
Peterson, N. and Bliel, U. (1973). Self-reversal of remanent magnetisation in synthetic titanomagnetite. *Zeits. Geophys.* **39**, 965.
Petherbridge, J. (1977). A magnetic coupling occurring in partial self-reversal of magnetism and

its association with increased magnetic viscosity in basalts. *Geophys. J. Roy. Astron. Soc.* **50**, 395.

Phillips, J. D. (1977). Time variation and asymmetry in the statistics of geomagnetic reversal sequences. *J. Geophys. Res.* **82**, 835.

Phillips, J. D. and Cox, A. (1976). Spectral analysis of geomagnetic reversal time scales. *Geophys. J. Roy. Astron. Soc.* **43**, 19.

Phillips, J. D., Blakely, R. J. and Cox, A. (1975). Independence of geomagnetic polarity intervals. *Geophys. J. Roy. Astron. Soc.* **43**, 747.

Piddington, J. H. (1967). A hydromagnetic model of geomagnetic storms and auroras. In *Physics of Geomagnetic Phenomena* (S. Matsushita and W. H. Campbell, eds), Vol. 2 (6), p. 1203. Academic Press, London and New York.

Pitman, W. C., Herron, E. M. and Heirtzler, J. R. (1968). Magnetic anomalies in the Pacific and seafloor spreading. *J. Geophys. Res.* **73**, 2069.

Podolak, M. and Cameron, A. G. W. (1974). Models of giant plants. *Icarus* **22**, 123.

Preston, G. W. (1967). Studies of stellar magnetism—past, present and future. In *Proc. AAS–NASA Symposium. The Magnetic and Related Stars* (R. C. Cameron, ed.), p. 3. Mono Book Corp., Baltimore, Maryland.

Prevot, M. (1977). Large intensity changes of the nondipole field during a polarity transition. *Phys. Earth Planet. Int.* **13**, 342.

Price, A. T. (1967). Electromagnetic induction within the earth. In *Physics of Geomagnetic Phenomena* (Matsushita and Campbell, eds), Vol. 1 (2), p. 235. Academic Press, London and New York.

Pudovkin, I. M. and Kolesova, V. I. (1968). Dipole model of the main geomagnetic field (based on an analysis of the Z_{ao} field). *Geomag. Aeron.* (English trans.) **8** (2), 301.

Pudovkin, I. M., Kolesova, V. I. and Valuyeva, G. Ye. (1968). Spatial structure of the residual Z_{ao} field. *Geomag. Aeron.* (English trans.) **8** (4), 483.

Readman, P. W. and O'Reilly, W. (1970). The synthesis and inversion of non-stoichiometric titanomagnetites. *Phys. Earth Planet. Int.*, **4**, 121.

Readman, P. W. and O'Reilly,W. (1972). Magnetic properties of oxidised (cation deficient) titanomagnetites (Fe, Ti, $\square)_3O_4$. *J. Geomag. Geoelect.* **24**, 69.

Reedy, R. C. (1980). Silver isotope anomalies in iron meteorites: cosmic ray production and other possible sources. *Proc. 11th Lunar Planet. Sci. Conf.* 1169.

Rikitake, T. (1958). Oscillations of a system of disc dynamos. *Proc. Cambridge Philos. Soc.* **54**, 89.

Rikitake, T. (1966). *Electromagnetism and the Earth's Interior*. Elsevier, Amsterdam.

Rimbert, F. (1959). Contribution à l'étude de l'action de champs alternatifs sur les aimantations remanentes des roches, applications geophysiques. *Rev. Inst. Fr. Pet.* **14**, 17 and 123.

Ringwood, A. E. (1977). Composition of the core and implications for the origin of the earth. *Geochim. J.* **11**, 111.

Ringwood, A. E. (1978). Origin of the Moon. *Lunar & Planet. Sci.* **IX**, 961.

Ringwood, A. E. (1979). *Origin of the Earth and Moon*. Springer-Verlag, New York.

Rivin, Yu. R. (1974). 11-year periodicity in the horizontal component of the geomagnetic field. *Geomag. Aeron.* **14**, 97.

Roberts, G. O. (1969). Dynamo waves. In *The Application of Modern Physics to the Earth and Planetary Interiors* (S. K. Runcorn, ed.), p. 603. Wiley, New York.

Roberts, G. O. (1970). Spatially periodic dynamos. *Phil. Trans. Roy. Soc. London* **A266,** 535.

Roberts, G. O. (1972). Dynamo action of fluid motions with two dimensional periodicity. *Phil. Trans. Roy. Soc. London* **A271**, 411.

Roberts, N. and Shaw, J. (1982). Relationship between the magnitude and direction of the geomagnetic field over the late Tertiary in Iceland. *Geophys. J. Roy. Astron. Soc.*

Roberts, P. H. (1967). *An Introduction to Magnetohydrodynamics*. Longmans. London; Elsevier, New York.

Roberts, P. H. (1971). Dynamo theory. In *Mathematical Problems in the Geophysical Sciences*. (W. H. Reid, ed.), p. 129. American Mathematics Society, Providence, R.I.

Roberts, P. H. (1972). Electromagnetic core–mantle coupling. *J. Geomag. Geoelect.* **24**, 231.

Roberts, P. H. and Stix, M. (1972). α-effect dynamos by the Bullard-Gellman formalism. *Astron. Astrophys.* **18**, 453.

Roche, A. (1951). Sur les inversions de l'aimantation remanente des roches volcaniques dans les monts d'Auvergne. *C.R. Acad. Sci. Paris* **233**, 1132.

Roche, A. (1953). Sur l'origine des inversions l'aimentation constantees dans les roches d'Auvergne. *C.R. Acad. Sci. Paris* **236**, 107.

Roche, A. (1956). Sur la date de la dernière inversion due champ magnetique terrestre. *C.R. Acad. Sci. Paris* **243**, 812.

Roche, A. (1958). Sur les variation de direction du champ magnétique terrestre au cours du Quaternaire. *C.R. Acad. Sci. Paris* **246**, 3364.

Rochester, M. G. (1968). Perturbation in the earth's rotation and geomagnetic core–mantle coupling. *J. Geomag. Geoelect.* **20**, 387.

Rochester, M. G., Jacobs, J. A., Smylie, D. E. and Chong, K. F. (1975). Can precession power the geomagnetic dynamo? *Geophys. J. Roy. Astron. Soc.* **43**, 661.

Roden, R. B. (1963). Electromagnetic core–mantle coupling. *Geophys. J. Roy. Astron. Soc.* **7**, 363.

Rosenberg, R. L. and Coleman, P. J., Jr. (1969). Heliographic latitude dependence of the dominant polarity of the interplanetary magnetic field. *J. Geophys. Res.* **74**, 5611.

Rossi, B. (1964). *Cosmic Rays*. McGraw-Hill, New York.

Rowe, M. W. (1975). Constraints on magnetic field which magnetised the Farmington meteorite parent body. *Meteoritics* **10**, 23.

Roy, J. L. and Park, J. K. (1972). Red beds: DRM or CRM? *Earth Planet. Sci. Lett.* **17**, 211.

Ruelle, D. (1980). Strange attractions, mathematical intelligences. Translated from *La Recherche* **108**, 126.

Ruff, L. J. and Helmberger, D. V. (1982). The structure of the lowermost mantle determined by short-period P-wave amplitudes. *Geophys. J. Roy. Astron. Soc.* **68**, 95.

Runcorn, S. K. (1955). The electrical conductivity of the earth's mantle. *Trans. Am. Geophys. Union* **36**, 191.

Runcorn, S. K. (1956). Paleomagnetic comparisons between North America and Europe. *Proc. Geol. Assoc. Canada* **8**, 103.

Runcorn, S. K. (1959). On the theory of geomagnetic secular variation. *Ann. Geophys.* **15**, 87.

Runcorn, S. K. (1961). Climatic change through geological time in the light of paleomagnetic evidence for polar wandering and continental drift. *Quart. J. Roy. Met. Soc.* **87**, 282.

Runcorn, S. K. (1975). On the interpretation of lunar magnetism. *Phys. Earth Planet. Int.* **10**, 327.

Runcorn, S. K. (1976). Inferences concerning the early thermal history of the Moon. *Proc. 7th Lunar Sci. Conf.* 3321.

Runcorn, S. K. (1978). The ancient lunar core dynamo. *Science* **199**, 771.

Runcorn, S. K. (1979). An iron core in the Moon generating an early magnetic field? *Proc. 10th Lunar Planet. Sci. Conf.* 2325.

Runcorn, S. K. and Urey, H. C. (1973). A new theory of lunar magnetism. *Science* **180**, 636.

Runcorn, S. K., Collinson, D. W., O'Reilly, W., Stephenson, A., Battey, M. H., Manson, A. J. and Readman, P. W. (1970). Magnetic properties of Apollo 11 lunar samples. *Proc. Apollo 11 Lunar Sci. Conf.* 2369.

Rusakov, O. M. and Zagniy, G. F. (1973). Intensity of the geomagnetic field in the Ukraine and Moldavia during the past 6000 years. *Archeometry* **15**, 275.

Russell, C. T. (1976). The magnetic moment of Venus: Venera-4 measurements reinterpreted. *Geophys. Res. Lett.* **3**, 125.

Russell, C. T. (1978). Re-evaluating Bode's law of planetary magnetism. *Nature* **272**, 147.

Russell, C. T. (1980). Planetary magnetism. *Rev. Geophys. Space Phys.* **18**, 77.

Russell, C. T. and Goldstein, B. E. (1976). The geomagnetic dynamos of the Moon and Venus: comparisons with a recent scaling law. *Proc. 7th Lunar Sci. Conf.* 3343.

Russell, C. T., Coleman, P. J., Jr. Lichtenstein, B. R. and Schubert, G. (1974). The permanent and induced magnetic dipole moment of the Moon. *Proc. 5th Lunar Sci. Conf.* 2747.

Russell, C. T., Weiss, H., Coleman, P. J., Jr., Soderblom, L. A. and Stuart-Alexander, D. E. (1977). Geologic-magnetic correlations on the Moon: Apollo subsatellite results. *Proc. 5th Lunar Sci. Conf.* 1171.

Rutten, M. G. (1959). Paleomagnetic reconnaissance of mid-Italian volcanoes. *Geol. Mijnbouw.* **21**, 373.

Ruzmaikin, A. A. and Sokoloff, D. D. (1980). Helicity, linkage and dynamo action. *Geophys. Astrophys. Fluid Dyn.* **16**, 73.

Ryall, P. J. C. and Ade-Hall, J. M. (1975). Laboratory-induced self-reversal of thermoremanent magnetisation in pillow basalts. *Nature* **257**, 117.

Sabatier, P. C. (1977). On geophysical inverse problems and constraints. *J. Geophys.* **43**, 115.

Sato, M. and Wright, T. L. (1966). Oxygen fugacities directly measured in volcanic gases. *Science* **153**, 1103.

Schatten, K. H. (1972). Large scale properties of the interplanetary magnetic field. In *Solar Wind* (C. P. Sonnett *et al.*, eds), p. 469. NASA Sp-308.

Schmidt, A. (1939). Zur Frage der Hypothetischen die erdoberfläche durchdringenden Ströme. Mit einen zusatz von J. Bartels. *Gerlands Beitr. Geophys.* **55**, 292.

Schmidt, V. A. (1973). A multidomain model of thermoremanence. *Earth Planet. Sci. Lett.* **20**, 440.

Schmidt, V. A. (1976). The variation of the blocking temperature in models of thermal remanences (TRM). *Earth Planet. Sci. Lett.* **29**, 146.

Schmitz, D., Frayser, J. B. and Cain, J. C. (1982). Application of dipole modelling to magnetic anomalies. *Geophys. Res. Lett.* **9**, 307.

Schouten, H. and Denham, C. R. (1979). Modelling the oceanic magnetic source layer. In *Deep Sea Drilling Results in the Atlantic Ocean: Ocean Crust* (M. Talwani, C. G. Harrison and D. E. Hayes, eds), p. 151. Maurice Ewing Ser. Amer. Geophys. Union.

Schouten, H. and McCarry, K. (1972). Filtering marine magnetic anomalies. *J. Geophys. Res.* **77**, 7089.

Schubert, G., Stevenson, D. and Cassen, P. (1980). Whole planet cooling and radiogenic heat source contents of the Earth and Moon. *J. Geophys. Res.* **85**, 2531.

Schult, A. (1968). Self-reversal of magnetisation and chemical composition of titanomagnetites in basalts. *Earth Planet. Sci. Lett.* **4**, 440.

Schure, L. and Parker, R. L. (1981). An alternative explanation for intermediate wavelength magnetic anomalies. *J. Geophys. Res.* **86**, 11600.

Schuster, A. (1898). On the investigation of hidden periodicities with application to a supposed 26 day period of meteorological phenomena. *Terr. Magn.* **3**, 33.

Schwarz, E. J. and Symons, D. T. A. (1969). Geomagnetic intensity between 100 and 2500 million years ago. *Phys. Earth Planet. Int.* **2**, 11.

Senanayake, W. E. and McElhinny, M. W. (1981). Hysteresis and susceptibility characteristics of magnetite and titanomagnetites: interpretation of results from basaltic rocks. *Phys. Earth Planet. Int.* **26**, 47.

Senanayake, W. E. and McElhinny, M. W. (1983). A paleointensity method for use with highly oxidised basalts, and application to some Permian volcanics. *J. Geophys.* **53**, 85.

Senanayake, W. E., McElhinny, M. W. and McFadden, P. L. (1982). Comparisons between the Thelliers' and Shaws's paleointensity methods using basalts less than 5 million years old. *J. Geomag. Geoelect.* **34**, 141.

Sharp, L. R., Coleman, P. J., Jr., Lichtenstein, B. R., Russell, C. T. and Schubert, G. (1973). Orbital mapping of the lunar magnetic field. *The Moon* **7**, 322.

Shaw, J. (1974). A new method of determining the magnitude of the paleomagnetic field. *Geophys. J. Roy. Astron. Soc.* **39**, 133.

Shaw, J. (1975). Strong geomagnetic fields during a single Icelandic polarity transition. *Geophys. J. Roy. Astron. Soc.* **40**, 345.

Shaw, J. (1977). Further evidence for a strong intermediate state of the paleomagnetic field. *Geophys. J. Roy. Astron. Soc.* **48**, 263.

Skiles, D. D. (1970). A method of inferring the direction of drift of the geomagnetic field from paleomagnetic data. *J. Geomag. Geoelect.* **22**, 441.

Skiles, D. D. (1972a). The laws of reflection and refraction of incompressible magnetohydrodynamic waves at a fluid-solid interface. *Phys. Earth Planet. Int.* **5**, 90.

Skiles, D. D. (1972b). On the transmission of the energy in an incompressible magnetohydrodynamic wave into a conducting solid. *Phys. Earth Planet. Int.* **5**, 99.

Smith, A. G., Hurley, A. M. and Briden, J. C. (1981). *Phanerozoic Paleocontinental World Maps.* Cambridge University Press, Cambridge.

Smith, E. J., Davis, L., Jr., Jones, D. E., Coleman, P. J., Jr., Colburn, D. S., Dyal, P., Sonett, C. P. and Fransden, A. M. A. (1974). The planetary magnetic field and magnetosphere of Jupiter: Pioneer 10. *J. Geophys. Res.* **79**, 3501.

Smith, E. J., Davis, L., Jr., Jones, D. E., Coleman, P. J., Jr., Colburn, D. S., Dyal, P. and Sonett, C. P. (1975). Jupiter's magnetic field and magnetosphere and interaction with the solar wind: Pioneer II. *Science* **188**, 451.

Smith, E. J., Tsurutani, B. T. and Rosenberg, R. L. (1978). Observations of the interplanetary sector structure up to heliographic latitudes of 16°: Pioneer II. *J. Geophys. Res.* **83**, 717.

Smith, P. J. (1967a). The intensity of the Tertiary geomagnetic field. *Geophys. J. Roy. Astron. Soc.* **12**, 239.

Smith, P. J. (1967b). The intensity of the ancient geomagnetic field: a review and analysis. *Geophys. J. Roy. Astron. Soc.* **12**, 321.

Smith, P. J. (1967c). On the suitability of igneous rocks for ancient geomagnetic field intensity determination. *Earth Planet. Sci. Lett.* **2**, 99.

Smith, P. J. (1968). Pre-Gilbertian conceptions of terrestrial magnetism. *Tectonophysics* **6**, 499.

Smith, P. J. (1970a). Petrus Peregrinus Epistola—The beginning of experimental studies of magnetism in Europe. *Atlas (News Supp. to Earth Sci. Revs.)* **6**, A11.

Smith, P. J. (1970b). Do magnetic polarity-oxidation state correlations imply self-reversal? *Comm. Earth Sci. Geophys.* **1**, 74.

Smith, P. J. and Needham, J. (1967). Magnetic declination in medieval China. *Nature* **214**, 1213.

Smylie, D. E., Clark, G. K. C. and Ulrych, T. J. (1973). Analysis of irregularities in the earth's rotation. In *Methods in Computational Physics*, Vol. 13, p. 391. Academic Press, London and New York.

Soderblom, L. A., Arnold, J. R., Boyce, J. M. and Lin, R. P. (1977). Regional variations in the lunar maria: age, remanent magnetism, and chemistry. *Proc. 8th Lunar Sci. Conf.* 1191.

Solomon, S. C. (1974). Density within the Moon and implications for lunar composition. *The Moon* **9**, 147.

Solomon, S. C. and Chaiken, J. (1976). Thermal expansion and thermal stress in the Moon and terrestrial planets: clues to early thermal history. *Proc. 7th Lunar Sci. Conf.* 3229.

Solomon, S. C. and Head, J. W. (1979). Vertical movement in mare basins: relation to mare emplacement, basin tectonics, and lunar thermal history. *J. Geophys. Res.* **84**, 1667.

Sonett, C. P. and Wiskerchen, M. J. (1977). A new source of lunar electromagnetic induction: forcing by the diamagnetic cavity. *Geophys. Res. Lett.* **4**, 307.

Sonett, C. P., Colburn, D. S. and Currie, R. G. (1967). The intrinsic magnetic field of the Moon. *J. Geophys. Res.* **72**, 5503.

Sonett, C. P., Colburn, D. S., Schwartz, K. and Kiel, K. (1970). The melting of asteroidal-sized bodies by unipolar induction from a primordial T-Tauri sun. *Astrophys. Space Sci.* **7**, 1446.

Soward, A. M. (1975). Random waves and dynamo action. *J. Fluid Mech.* **69**, 145.

Soward, A. M. (1979). Some new results on spherical dynamos. *Phys. Earth Planet. Int.* **20**, 134.

Speiser, T. W. and Ness, N. F. (1967). The neutral sheet in the geomagnetic tail: its motion, equivalent currents, and field line convection through it. *J. Geophys. Res.* **72**, 131.

Spiess, F. N. and Mudie, J. D. (1970). Small scale topographic and magnetic features of the seafloor. In *The Sea* (A. E. Maxwell, ed.), Vol. 4, p. 205. Wiley, New York.

Srivastava, B. J. and Abbas, H. (1977). Geomagnetic secular variation in India: regional and local features. *J. Geomag. Geoelect.* **29**, 51.

Srnka, L. J. (1977). Spontaneous magnetic field generation in hypervelocity impact. *Proc. 8th Lunar Sci. Conf.* 785.

Srnka, L. J., Martelli, G., Newton, G., Cisowski, S. M., Fuller, M. D. and Scholl, R. B. (1979). Magnetic field and shock effects and remanent magnetisation in a hypervelocity impact experiment. *Earth Planet. Sci. Lett.* **42**, 127.

Stacey, F. D. (1963). The physical theory of rock magnetism. *Adv. Phys.* **12**, 45.

Stacey, F. D. (1972). Physical properties of the earth's core. *Geophys. Surv.* **1**, 99.

Stacey, F. D. (1976). Paleomagnetism of meteorites. *Rev. Earth Planet. Sci.* **4**, 147.

Stacey, F. D. (1977a). *Physics of the Earth*, 2nd edition. Wiley, New York.

Stacey, F. D. (1977b). Applications of thermodynamics to fundamental earth physics. *Geophys. Surv.* **3**, 175.

Stacey, F. D. (1980). The cooling earth: A reappraisal. *Phys. Earth Planet. Int.* **22**, 89.

Stacey, F. D. and Banerjee, S. K. (1974). *The Physical Principles of Rock Magnetism*. Elsevier, Amsterdam.

Stannard, D. (1975). Observations of the magnetic field structure of Jupiter. *Geophys. J. Roy. Astron. Soc.* **41**, 327.

Steenbeck, M. and Krause, F. (1969a). On the dynamo theory of stellar and planetary magnetic fields. 1. AC dynamos of solar type. *Astron. Nach.* **291**, 49.

Steenbeck, M. and Krause, F. (1969b). On the dynamo theory of stellar and planetary magnetic fields. II. DC dynamos of the planetary type. *Astron. Nach.* **291**, 271.

Steenbeck, M., Kirko, I. M., Gailitis, A., Klawina, A. P., Krause, F., Laumonis, I. J. and Lielausis, O. A. (1967). An experimental verification of the α-effect. *Monatsber, Deut. Akad. Wiss. Berlin* **9**, 716.

Stehli, F. G. (1968). A paleoclimatic test of the hypothesis of an axial dipolar magnetic field. In *History of the Earth's crust* (R. A. Phinney, ed.), p. 195. Princeton University Press, Princeton, N.J.

Steiger, R. H. and Jager, E. (1977). Subcommission on geochronology: Convention on the use of decay constants in geo- and cosmochronology. *Earth Planet. Sci. Lett.* **36**, 359.

Stephenson, A. (1975). The observed moment of a magnetised inclusion of high Curie point within a titanomagnetite particle of lower Curie point. *Geophys. J. Roy. Astron. Soc.* **40**, 29.

Stephenson, A. (1976). The residual permanent magnetic dipole moment of the Moon. *The Moon* **15**, 67.

Stephenson, A. and Collinson, D. W. (1974). Lunar magnetic field paleointensities determined by anhysteretic remanent magnetisation method. *Earth Planet. Sci. Lett.* **23**, 220.

Stephenson, A. Collinson, D. W. and Runcorn, S. K. (1974). Lunar magnetic field paleointensity determinations on Apollo 11, 16 and 17 rocks. *Proc. 5th Lunar Sci. Conf.* 2859.

Stephenson, A., Runcorn, S. K. and Collinson, D. W. (1977). Paleointensity estimates from lunar samples 10017 and 10020. *Proc. 8th Lunar Sci. Conf.* 679.

Stern, D. P. (1977). Large scale electric fields in the earth's magnetosphere. *Rev. Geophys. Space Phys.* **15**, 156.

Stevenson, D. (1974). Planetary magnetism. *Icarus* **22**, 403.

Stevenson, D. J. (1979). Turbulent thermal convection in the presence of rotation and a magnetic field: a heuristic theory. *Geophys. Astrophys. Fluid Dyn.* **12**, 139.

Stevenson, D. J. (1980). Saturn's luminosity and magnetism. *Science* **208**, 746.

Stevenson, D. J. (1981). Models of the earth's core. *Science* **214**, 611.

Stevenson, D. J. (1983). Planetary magnetic fields. *Rept. Prog. Phys.* **46**, 555.

Stevenson, D. J. and Turner, J. S. (1979). Fluid models of mantle convection. In *The Earth: Its Origin, Structure and Evolution* (M. W. McElhinny, ed.), p. 227. Academic Press, London and New York.

Stewart, R. M. (1973). Composition and temperature of the outer core. *J. Geophys. Res.* **78**, 2586.

Stewartson, K. (1967). Slow oscillations of a fluid in a rotating cavity in the presence of a toroidal magnetic field. *Proc. Roy. Soc. London* **A299**, 173.

Stober, J. C. and Thompson, R. (1977). Palaeomagnetic secular variation studies of Finnish lake sediment and the carriers of remanence. *Earth. Planet Sci. Lett.* **37**, 139.

Stoner, E. C. and Wohlfarth, E. P. (1948). A mechanism of magnetic hysteresis in heterogeneous alloys. *Phil. Trans. Roy. Soc. London* **A240**, 599.

Strangway, D. W. (1977a). The record of magnetic fields in the early solar system. *The Moon and Planets* **18**, 273.

Strangway, D. W. (1977b). The magnetic fields of the early planets. *Phys. Earth Planet. Int.* **5**, 121.

Strangway, D. W. and Sharpe, H. A. (1974). Lunar magnetism, an early cold Moon. *Nature* **249**, 227.

Strangway, D. W., Larson, E. E. and Goldstein, M. (1968). A possible cause of high magnetic stability in volcanic rocks. *J. Geophys. Res.* **73**, 3787.

Strangway, D. W., Larson, E. E. and Pearce, G. W. (1970). Magnetic studies of lunar samples—breccia and fines. *Proc. Apollo 11 Lunar Sci. Conf.* 2435.

Strangway, D. W., Gose, W., Pearce, G. and McConnell, R. K. (1973). Lunar magnetic anomalies and the Cayley formation. *Nature* **246**, 112.

Stupavsky, M., Gravenor, C. P. and Symons, D. T. A. (1979). Paleomagnetic stratigraphy of the Meadowcliffe till, Scarborough Bluffs, Ontario: A late Pleistocene excursion? *Geophys. Res. Lett.* **6**, 269.

Sugiura, N. (1978). Magnetic properties and remanent magnetisation of stony meteorites. *J. Geomag. Geoelect.* **29**, 519.

Sugiura, N. and Strangway, D. W. (1980). Comparisons of magnetic paleointensity methods using a lunar sample. *Proc. 11 Lunar Planet. Sci. Conf.* 1801.

Sugiura, N., Lanoix, M. and Strangway, D. W. (1979). Magnetic fields of the solar nebula as recorded in chrondules from the Allende meteorite. *Phys. Earth Planet. Int.* **20**, 342.

Sugiura, N., Wu, Y. M., Strangway, D. W., Pearce, G. W. and Taylor, L. A. (1979a). A new magnetic paleointensity value for a young lunar glass. *Proc. 10th Lunar Sci. Conf.* 2189.

Sugiura, N., Wu, Y. M., Strangway, D. W., Pearce, G. W. and Taylor, L. A. (1979b). Palaeointensity studies on 70019, a young glass sample for Apollo 17. In *Lunar and Planetary Science X*, pp. 1195–1196. Lunar and Planetary Institute, Houston, Texas.

Suzuki, Y. and Sato, R. (1970). Viscosity determination in the earth's outer core from ScS and SKS Phases. *J. Phys. Earth* **18**, 157.

Swingler, D. N. (1979). A comparison between Burg's maximum entropy method and a non-recursive technique for the spectral analysis of deterministic signals. *J. Geophys. Res.* **84**, 679.

Swingler, D. N. (1980). Burg's maximum entropy algorithm versus the discrete Fourier transform as a frequency estimator for truncated real sinosoids. *J. Geophys. Res.* **85**, 1435.

Tanaka, H. and Tachibana, K. (1981). A geomagnetic reversal in the latest Brunhes Epoch discovered at Shibutarni, Japan. *J. Geomag. Geoelect.* **33**, 287.

Tanguy, J. C. and Wilson, R. L. (1973). Paleomagnetism of Mount Etna. *Phil. Trans. Roy. Soc. London* **A274**, 163.

Taylor, L. A. (1979). An effective sample preparation technique for paleointensity determinations at elevated temperatures. *Lunar and Planetary Science* **10**, 1209.

Taylor, S. (1975). *Lunar Science: A Post-Apollo View*. Pergamon Press, New York.

Taylor, S. R. (1982). *Planetary Science: A Lunar Perspective*. Lunar and Planetary Institute, Houston, Texas.

Thellier, E. (1937a). Aimantation des terres ciutes; application à la recherche de l'intensité du champ magnétique terrestre dans le passé. *C.R. Acad. Sci. Paris* **204**, 184.

Thellier, E. (1937b). Recherche de l'intensité du champ magnétique dans le passe: premier resultats. *Ann. de l'Institut de Phys. du Globe, Univ. Paris* **15**, 179.

Thellier, E. (1966). Le champ magnetique terrestre fossile. *Nucleus* **7**, 1.

Thellier, E. and Thellier, O. (1951). Sur la direction du champ magnétique terrestre, retrouvée sur des parois de fours des époques punique et romaine, á Carthage. *C.R. Acad. Sci. Paris* **233**, 1476.

Thellier, E. and Thellier, O. (1952). Sur la direction du champ magnétique terrestre, dans la region de Tréves, vers 380 aprés J.-C. *C.R. Acad. Sci. Paris* **234**, 1464.

Thellier, E. and Thellier, O. (1959a). Sur l'intensité du champ magnétique terrestre dans le passé historique et géologique. *Ann. Geophys.* **15**, 285.

Thellier, E. and Thellier, O. (1959b). The intensity of the geomagnetic field in the historical and geological past. *Akad. Nauk. SSR. Izv. Geophys. Ser.* 1296.

Thompson, R. (1975). Long period European secular variation confirmed. *Geophys. J. Roy. Astron. Soc.* **43**, 847.

Thompson, R. and Kelts, K. (1974). Holocene sediments and magnetic stratigraphy from Lakes Zug and Zurich, Switzerland. *Sedimentology* **21**, 577.

Thompson, R. and Turner, G. M. (1979). British geomagnetic master curve 10 000–0 yr B.P. for dating European sediments. *Geophys. Res. Lett.* **6**, 249.

Tolland, H. F. (1974). Thermal regime in the earth's core and lower mantle. *Phys. Earth Planet. Int.* **8**, 282.

Torbett, M. and Smoluchowski, R. (1980). The core and the magnetic field of Uranus. *Proc. 11th Lunar Sci. Conf.* 1907.

Tucker, P. and O'Reilly, W. (1980). Reversed thermoremanent magnetisation in synthetic titanomagnetites as a consequence of high temperature oxidation. *J. Geomag. Geoelect.* **32**, 341.

Turcotte, D. L. (1980). On the thermal evolution of the earth. *Earth Planet. Sci. Lett.* **48**, 53.

Turner, G. M. and Thompson, R. (1979). Behaviour of the earth's magnetic field as recorded in the sediment of Loch Lomond. *Earth Planet. Sci. Lett.* **42**, 412.

Ulrych, T. J. (1972). Maximum entropy power spectrum of long period geomagnetic reversals. *Nature* **235**, 218.

Ulrych, T. J. and Bishop, T. N. (1975). Maximum entropy spectral analysis and autoregressive decomposition. *Rev. Geophys. Space Phys.* **13**, 183.

Uyeda, S. (1958). Thermoremanent magnetism as a medium of paleomagnetism with special references to reverse thermoremanent magnetism. *Japan J. Geophys.* **2**, 1.

Valet, J. P. and Laj, C. (1981). Paleomagnetic record of two successive Miocene geomagnetic reversals in western Crete. *Earth Planet. Sci. Lett.* **54**, 53.

Van Allen, J. A., Ludwig, G. H., Ray, E. C. and McIlwain, C. E. (1959). Observations of high intensity radiation by satellites 1958 Alpha and Gamma. *Jet Propulsion* **28**, 588.

Van der Bos, A. (1971). Alternative interpretation of maximum entropy spectral analysis. *IEEE Trans. Inform. Theory* **17–17**, 493.

Van Zijl, J. S. V., Graham, K. W. T. and Hales, A. L. (1962). The paleomagnetism of the Stormberg lavas of South Africa, 1 and 2. *Geophys. J. Roy. Astron. Soc.* **7**, 23 and 169.

Veinberg, B. P. and Shibaev, V. P. (1970). *Catalogue. The results of magnetic determinations at equidistant points and epochs, 1500–1940* (Editor in Chief: A. N. Pushkov), IZMIRAN, Moscow, Translation: No. 0031 by Canadian Department of the Secretary of State, Translation Bureau.

Verhoogen, J. (1956). Ionic ordering and self-reversal of impure magnetites. *J. Geophys. Res.* **61**, 201.

Verhoogen, J. (1959). The origin of thermoremanent magnetisation. *J. Geophys. Res.* **64**, 2441.

Verhoogen, J. (1961). Heat balance of the earth's core. *Geophys. J. Roy. Astron. Soc.* **4**, 276.

Verhoogen, J. (1973). Thermal regime of the earth's core. *Phys. Earth Planet. Int.* **7**, 47.

Verhoogen, J. (1980). *Energetics of the Earth.* National Academy of Science, Washington, D.C.

Verosub, K. L. (1977). Depositional and postdepositional processes in the magnetisation of sediments. *Rev. Geophys. Space Phys.* **15**, 129.

Verosub, K. L. (1979). Paleomagnetism of varved sediments from western New England: secular variation. *Geophys. Res. Lett.* **6**, 245.

Verosub, K. L. and Banerjee, S. K. (1977). Geomagnetic excursions and their paleomagnetic record. *Rev. Geophys. Space Phys.* **15**, 145.

Vestine, E. H. (1954). The earth's core. *Trans. Amer. Geophys. Union* **35**, 63.

Vestine, E. H. (1967). Main geomagnetic field. In *Physics of Geomagnetic Phenomena* (S. Matsushita and W. H. Campbell, eds.), Vol. 1 (2), p. 181. Academic Press, London and New York.

Vestine, E. H., Lange, I., LaPorte, L. and Scott, W. E. (1947a). The geomagnetic field, its description and analysis. *Dept. Terres. Mag. Rept.* **580**, 390.

Vestine, E. H., LaPorte, L., Lange, I., Cooper, C. and Hendrix, W. C. (1947b). Description of the earth's main magnetic field, 1905–1945. *Dept. Terres. Mag. Rept.* **578**, 532.

Vine, F. J. and Matthews, D. H. (1963). Magnetic anomalies over oceanic ridges. *Nature* **199**, 947.

Vitorello, I. R. and Van der Voo, R. (1977). Magnetic stratigraphy of Lake Michigan sediments obtained from cores of lacustrine clays. *Quat. Res.* **1**, 398.

Wada, M. and Inoue, A. (1966). Relation between the carbon-14 production rate and the geomagnetic moment. *J. Geomag. Geoelect.* **18**, 484.

Wang, Chen-To. (1948). Discovery and application of magnetic phenomena in China, I. The lodestone spoon of the Han. *Chinese J. of Archaeology, Acad. Sinica* **3**, 119.

Wasilewski, P. J. (1972). Magnetic hysteresis classification for the lunar surface. *The Apollo 15 Lunar Samples.* The Lunar Science Inst., Houston, Texas.

Wasilewski, P. J. (1973). Magnetic hysteresis of natural materials. *Earth Planet. Sci. Lett.* **20**, 67.

Wasilewski, P. J. (1974). Magnetic remanence mechanisms in iron and iron-nickel alloys, metallographic recognition criteria and implications for lunar sample research. *The Moon* **9**, 335.

Wasson, J. T. (1974). *Meteorites: Classification and Properties.* Springer-Verlag, New York.

Watanabe, H. (1977). Bounds on the fluid velocity and the magnetic field in the earth's core imposed by hydromagnetic consideration of a dynamo. *J. Geomag. Geoelect.* **29**, 191.

Watanabe, H. (1981). Non-steady state of a hydromagnetic dynamo and its application to geomagnetic reversals. *J. Geomag. Geoelect.* **33**, 531.

Watkins, N. D. (1969). Non-dipole behaviour during an Upper Miocene geomagnetic polarity transition in Oregon. *Geophys. J. Roy. Astron. Soc.* **17**, 121.

Watkins, N. D. (1972). Review of the development of the geomagnetic polarity time scale and discussion of prospects for its finer definition. *Geol. Soc. Amer. Bull.* **83**, 551.

Watkins, N. D. (1973). Brunhes epoch geomagnetic secular variation on Reunion Island. *J. Geophys. Res.* **78**, 7763.

Watkins, N. D. and Walker, G. P. L. (1977). Magnetostratigraphy of eastern Iceland. *Amer. J. Sci.* **277**, 513.

Watkins, N. D., McDougall, I. and Krisjansson, L. (1977). Upper Miocene and Pliocene geomagnetic secular variation in the Borgarfjordur area of western Iceland. *Geophys. J. Roy. Astron. Soc.* **49**, 609.

Watson, D. E., Larson, E. E., Herndon, J. M. and Rowe, M. W. (1975). Thermomagnetic analysis of meteorites, 2. C2 chondrites. *Earth Planet. Sci. Lett.* **27**, 101.

Webb, E. N. and Chree, C. (1925). *Australasian Antarctic Exploration, 1911–14 Scientific Reports*, Series B, Vol. 1, Terrestrial Magnetism, p. 285. Govt. Printer, Sydney.

Wegener, A. (1924). *The Origin of Continents and Oceans* (English translation by J. G. A. Skerl). Methuen, London.

Wells, J. M. (1969). *Non-linear Spherical Harmonic Analysis of Paleomagnetic Data*. Ph.D. thesis, University of California, Berkley.

Wells, J. M. (1973). Non-linear spherical harmonic analysis of paleomagnetic data. In *Methods in Computational Physics*, Vol. 13, *Geophys.* (B. A. Bolt, ed.), p. 239. Academic Press, New York and London.

Wescott-Lewis, M. F. and Parry, L. G. (1971). Thermoremanence in synthetic rhombohedral iron–titanium oxides. *Aust. J. Phys.* **24**, 735.

Whang, Y. C. (1977). Magnetospheric magnetic field of Mercury. *J. Geophys. Res.* **82**, 1024.

Whitney, J., Johnson, H. P., Levi, S. and Evans, B. W. (1971). Investigations of some magnetic and mineralogical properties of the Laschamp and Olby flows, France. *Quat. Res.* **1**, 511.

Wilcox, J. M. and Ness, N. F. (1965). Quasi-stationary corotating structure in the interplanetary medium. *J. Geophys. Res.* **70**, 5793.

Williams, I. S. and Fuller, M. (1981a). Zonal harmonic models of reversal transition fields. *J. Geophys. Res.* **86**, 11657.

Williams, I. S. and Fuller, M. (1981b). A far sided R–N VGP path from a reversal recorded in the Agno batholith. *EOS* **62**, 853.

Wilson, R. L. (1962a). The paleomagnetic history of a double baked rock. *Geophys. J. Roy. Astron. Soc.* **6**, 397.

Wilson, R. L. (1962b). The paleomagnetism of baked contact rocks and reversals of the earth's magnetic field. *Geophys. J. Roy. Astron. Soc.* **7**, 194.

Wilson, R. L. (1970). Permanent aspects of the earth's non-dipole magnetic field over Upper Tertiary times. *Geophys. J. Roy. Astron. Soc.* **19**, 417.

Wilson, R. L. (1971). Dipole offset—the time-averaged paleomagnetic field over the past 25 million years. *Geophys. J. Roy. Astron. Soc.* **22**, 491.

Wilson, R. L. (1972). Paleomagnetic differences between normal and reversed field sources, and the problem of far-sided and right-handed pole positions. *Geophys. J. Roy. Astron. Soc.* **28**, 295.

Wilson, R. L. and Ade-Hall, J. M. (1970). Paleomagnetic indications of a permanent aspect of the non-dipole field. In *Palaeogeophysics* (S. K. Runcorn, ed.), p. 307, Academic Press, New York and London.

Wilson, R. L. and Haggerty, S. E. (1966). Reversals of the earth's magnetic field. *Endeavour* **25**, 104.

Wilson, R. L. and McElhinny, M. W. (1974). Investigation of the large scale paleomagnetic field over the past 25 million years; eastward shift of the Icelandic spreading ridge. *Geophys. J. Roy. Astron. Soc.* **39**, 571.

Wilson, R. L. and Watkins, N. D. (1967). Correlations of petrology and natural magnetic polarity in Columbia Plateau basalt. *Geophys. J. Roy. Astron. Soc.* **12**, 405.

Wilson, R. L., Dagley, P. and McCormack, A. G. (1972). Paleomagnetic evidence about the source of the geomagnetic field. *Geophys. J. Roy. Astron. Soc.* **28**, 213.

Wiskerchen, M. J. and Sonett, C. P. (1977). A lunar metallic core. *Proc. 8th Lunar Sci. Conf.* 515.

Wolfendale, A. W. (1963). *Cosmic Rays*. George Mewnes Ltd., London.

Woltjer, L. (1975). Astrophysical evidence for strong magnetic fields. *Ann. N.Y., Acad. Sci.* **257**, 76.

Wood, J. A. (1978). Pallasites and the growth of parent meteorite planets. *Lunar Planet. Sci.* **IX**, 1273.

Wood, J. A. (1979). *The Solar System*. Prentice-Hall, New Jersey.

Yashikawa, K., Nakajima, T., Kawai, N., Torii, M., Natsuhara, N. and Hore, S. (1973). Paleomagnetism of a core from Lake Biwa (1). *J. Geomag. Geoelect.* **25**, 447.

Yukutake, T. (1961). Archaeomagnetic study on volcanic rocks in Oshima Island, Japan. *Bull. Earthquake Res. Inst.* **39**, 467.

Yukutake, T. (1962). The westward drift of the magnetic field of the earth. *Bull. Earthquake Res. Inst.* **40**, 1.

Yukutake, T. (1965). The solar cycle contribution to the secular change in the geomagnetic field. *J. Geomag. Geoelect.* **17**, 287.

Yukutake, T. (1967) The westward drift of the earth's magnetic field in historic times. *J. Geomag. Geoelect.* **19**, 103.

Yukutake, T. (1970). Geomagnetic secular variation. *Comments on earth sciences. Geophys.* **1**, 55.

Yukutake, T. (1971). Spherical harmonic analysis of the earth's magnetic field for the 17th and 18th centuries. *J. Geomag. Geoelect.* **23**, 11.

Yukutake, T. (1977). Parallelism of secular change in the dipole and the quadupole components of the geomagnetic field (abstract). *EOS* **58**, 1129.

Yukutake, T. (1979). Geomagnetic secular variations on the historical time scale. *Phys. Earth Planet. Int.* **20**.

Yukutake, T. and Tachinaka, H. (1968). The non-dipole part of the earth's magnetic field. *Bull. Earthquake Res. Inst.* **46**, 1027.

Yukutake, T. and Tachinaka, H. (1969). Separation of the earth's magnetic field into drifting and standing parts. *Bull. Earthquake Res. Inst.* **47**, 65.

Zidarov, D. (1969). Geomagnetic field variations. *Geomag. Aeron.* (English trans.) **9**(6), 856.

Zidarov, D. (1970). Forecasting the geomagnetic field. *Geomag. Aeron.* (English trans.) **10**(4), 533.

Zidarov, D. (1974). Presentation of magnetic fields with multipolar fields and solution of the inverse magnetic problem. *Comptes rendus de l'Academie bulgare des Sciences* **27**(11), 1493.

Zidarov, D. and Bochev, A. (1965). Analysis of the geomagnetic field and its secular variation. *Geomag. Aeron.* (English trans.) **5**(5), 746.

Zidarov, D. and Bochev, A. (1969). Representation of secular geomagnetic field variation as variation of the field of optimum geomagnetic dipoles. *Geomag. Aeron.* (English trans.) **9**(2), 252.

Zidarov, D. P. and Petrova, T. (1974). Representation of the earth's magnetic field as a field of a circular loop. *Comptes rendus de l'Academie bulgare des Sciences* **27**, 203.

Index

International Geophysics Series

EDITED BY

Volume 1 BENO GUTENBERG. Physics of the Earth's Interior. 1959

Volume 2 JOSEPH W. CHAMBERLAIN. Physics of the Aurora and Airglow. 1961

Volume 3 S. K. RUNCORN (ed.). Continental Drift 1962

Volume 4 C. E. JUNGE. Air Chemistry and Radioactivity. 1963

Volume 5 ROBERT G. FLEAGLE AND JOOST A. BUSINGER. An Introduction to Atmospheric Physics. 1963

Volume 6 L. DUFOUR AND R. DEFAY. Thermodynamics of Clouds. 1963

Volume 7 H. U. ROLL. Physics of the Marine Atmosphere. 1965

Volume 8 RICHARD A. CRAIG. The Upper Atmosphere: Meteorology and Physics. 1965

Volume 9 WILLIS L. WEBB. Structure of the Stratosphere and Mesosphere. 1966

Volume 10 MICHELE CAPUTO. The Gravity Field of the Earth from Classical and Modern Methods. 1967

Volume 11 S. MATSUSHITA AND WALLACE H. CAMPBELL (eds). Physics of Geomagnetic Phenomena. (In two volumes.) 1967

Volume 12 K. YA. KONDRATYEV. Radiation in the Atmosphere. 1969

Volume 13 E. PALMEN AND C. W. NEWTON. Atmospheric Circulation Systems: Their Structure and Physical Interpretation. 1969

Volume 14 HENRY RISHBETH AND OWEN K. GARRIOTT. Introduction to Ionospheric Physics. 1969

Volume 15 C. S. RAMAGE. Monsoon Meteorology. 1971

Volume 16 JAMES R. HOLTON. An Introduction to Dynamic Meteorology. 1972

Volume 17 K. C. YEH AND C. H. LIU. Theory of Ionospheric Waves. 1972

Volume 18 M. I. BUDYKO. Climate and Life. 1974

Volume 19 MELVIN E. STERN. Ocean Circulation Physics. 1975

Volume 20 J. A. JACOBS. The Earth's Core. 1975

Volume 21 DAVID H. MILLER. Water at the Surface of the Earth: An Introduction to Ecosystem Hydrodynamics. 1977

Volume 22 JOSEPH W. CHAMBERLAIN. Theory of Planetary Atmospheres: An Introduction to Their Physics and Chemistry. 1978

Volume 23 James R. Holton. Introduction to Dynamic Meteorology. Second Edition. 1979

Volume 24 Arnett S. Dennis. Weather Modification by Cloud Seeding. 1980

Volume 25 Robert G. Fleagle and Joost A. Businger. An Introduction to Atmospheric Physics. Second Edition. 1980

Volume 26 Kuo-Nan Liou. An Introduction to Atmospheric Radiation. 1980

Volume 27 David H. Miller. Energy at the Surface of the Earth. 1981

Volume 28 Helmut E. Landsberg. The Urban Climate. 1981

Volume 29 M. I. Budyko. The Earth's Climate: Past and Future. 1982

Volume 30 Adrian E. Gill. Atmosphere–Ocean Dynamics. 1982

Volume 31 Paolo Lanzano. Deformations of an Elastic Earth. 1982

Volume 32 Ronald T. Merrill and Michael W. McElhinny. The Earth's Magnetic Field: Its History, Origin and Planetary Perspective. 1983

In preparation

Volume 33 John S. Lewis and Ronald G. Prinn. Planets and Their Atmospheres: Origin and Evolution. 1983.